High-Speed Countercurrent Chromatography

CHEMICAL ANALYSIS

A SERIES OF MONOGRAPHS ON
ANALYTICAL CHEMISTRY AND ITS APPLICATIONS

Editor
J. D. WINEFORDNER

VOLUME 132

A WILEY-INTERSCIENCE PUBLICATION

JOHN WILEY & SONS, INC.

New York / Chichester / Brisbane / Toronto / Singapore

High-Speed Countercurrent Chromatography

Edited by

YOICHIRO ITO

Laboratory of Biophysical Chemistry
National Institutes of Health
Bethesda, Maryland

WALTER D. CONWAY

Department of Pharmaceutics
SUNY Buffalo
Buffalo, New York

A WILEY-INTERSCIENCE PUBLICATION

JOHN WILEY & SONS

New York / Chichester / Brisbane / Toronto / Singapore

This text is printed on acid-free paper.

Copyright © 1996 by John Wiley & Sons, Inc.

All rights reserved. Published simultaneously in Canada.

Library of Congress Cataloging in Publication Data:

High speed countercurrent chromatography / edited by Yoichiro Ito and
    Walter D. Conway.
        p.   cm. — (Chemical analysis ; v. [132])
    Includes bibliographical references and index.
    ISBN 0-471-63749-1 (alk. paper)
    1. Countercurrent chromatography.   I. Ito, Yoichiro, 1928–
II. Conway, Walter D.   III. Series.
QP519.9.C68H54   1995   95-9529
543′.0894—dc20   CIP

Printed in the United States of America

10 9 8 7 6 5 4 3 2 1

CONTRIBUTORS

Hans J. Cahnmann, Genetics and Biochemistry Branch, National Institute of Diabetes, Digestive and Kidney Diseases, National Institutes of Health, Bethesda, Maryland 20892

Nancy L. Fregeau, Roger Adams Laboratory, University of Illinois, Urbana, Illinois 61801

Ken-Ichi Harada, Faculty of Pharmacy, Meijo University, Tempaku, Nagoya 468, Japan

Kurt Hostettmann, Institut de Pharmacognosie et Phytochimie, Ecole de Pharmacie, Université de Lausanne, CH-1015 Lausanne, Switzerland

Yoichiro Ito, Laboratory of Biophysical Chemistry, National Heart, Lung, and Blood Institute, National Institutes of Health, Bethesda, Maryland 20892

Eiichi Kitazume, Faculty of Humanities and Social Sciences, Iwate University, Ueda, Morioka, Iwate 020, Japan

Y. W. Lee, Research Triangle Institute, Chemistry and Life Sciences, Research Triangle Park, North Carolina 27709

Marc Maillard, Institut de Pharmacognosie et Phytochimie, Ecole de Pharmacie, Université de Lausanne, CH-1015 Lausanne, Switzerland

Andrew Marston, Institut de Pharmacognosie et Phytochimie, Ecole de Pharmacie, Université de Lausanne, CH-1015 Lausanne, Switzerland

Hisao Oka, Aichi Prefectural Institute of Public Health, Tsuji-machi, Kita-ku, Nagoya 462, Japan

Kenneth L. Rinehart, Roger Adams Laboratory, University of Illinois, Urbana, Illinois 61801

Daniel E. Schaufelberger, The R. W. Johnson Pharmaceutical Research Institute, Cilag Ltd., CH-8201 Schaffhausen, Switzerland

Yoichi Shibusawa, Tokyo College of Pharmacy, Hachioji, Tokyo, Japan

Adrian Weisz, Office of Cosmetics and Colors, Food and Drug Administration, Washington DC 20204

Tian-You Zhang, Beijing Institute of New Technology Application, Xizhimen, Beijing 100035, China

v

CONTENTS

CHAPTER 2   ANALYTICAL HIGH-SPEED COUNTERCURRENT CHROMATOGRAPHY

Daniel E. Schaufelberger

SPECIAL TECHNIQUES

CHAPTER 3   HIGH-SPEED COUNTERCURRENT CHROMATOGRAPHY/MASS SPECTROMETRY

Hisao Oka

APPLICATIONS

**CHAPTER 7 HIGH-SPEED COUNTERCURRENT
            CHROMATOGRAPHY OF NATURAL
            PRODUCTS   179**

Marc Maillard, Andrew Marston, and Kurt Hostettmann

PREFACE

The development of countercurrent chromatography (CCC), a support-free liquid–liquid partition chromatography, began in the mid-1960s. In its early stages of development, the method was hindered by a limited mobile-phase flow rate since the application of higher flow rates resulted in excessive loss of the stationary phase from the column. The first widely distributed commercial model of droplet CCC produced an unfavorable impression of the CCC technique as being time-consuming and inefficient. Despite the successive development of centrifugal CCC using a variety of coil planet centrifuge systems, this problem persisted until the late 1970s when a new CCC system was discovered. This new system utilized a combination of a particular type of planetary motion and coaxial orientation of the coiled column to generate a rapid movement of the mobile phase through a bulk of stationary phase under an efficient mixing of the two phases, thereby achieving both high-partition efficiency and an excellent retention of the stationary phase. Because of its rapid separation, this new CCC technique was named *high-speed CCC*. The method opened a rich domain of applications, triggering commercialization of the instruments in the United States. The present monograph is the first book entirely devoted to this efficient CCC technique.

During the past five years remarkable progress has been made in the high-speed CCC technique, which includes purification of recombinant proteins directly from a crude *E. coli* lysate, pH-peak-focusing CCC that led to the development of pH-zone-refining CCC that produces highly concentrated pure fractions comparable to displacement chromatography, and successful CCC/MS interfacing, among other achievements. None of these topics have been described in previous monographs. The present monograph also contains a number of recent applications such as separation and purification of natural products, marine products, medicinal products from Chinese herbs, antibiotics, dyes, and inorganic elements. The unique application of form CCC and liquid–liquid dual CCC are also introduced.

We hope that this monograph on high-speed CCC will stimulate the use of this chromatographic method as well as commercial production and improvement of high-speed CCC instruments worldwide.

YOICHIRO ITO
WALTER D. CONWAY

CHEMICAL ANALYSIS

A SERIES OF MONOGRAPHS ON
ANALYTICAL CHEMISTRY AND ITS APPLICATIONS

J. D. Winefordner, *Series Editor*

xix

INSTRUMENTATION

CHAPTER

1

PRINCIPLE, APPARATUS, AND METHODOLOGY OF HIGH-SPEED COUNTERCURRENT CHROMATOGRAPHY

YOICHIRO ITO

Laboratory of Biophysical Chemistry, National Heart, Lung, and Blood Institute, National Institutes of Health, Bethesda, Maryland 20892

1.1. INTRODUCTION

The partition of solutes between two immiscible solvent phases is an ideal method for separation and purification of natural products. By eliminating the various complications that arise from the interaction between solute molecules and the solid support matrix, this method can yield high purity fractions with high sample recovery rate and high reproducibility. In the 1950s, the counter-current distribution method was extensively used for separation of natural products (1). However, due to its inherent disadvantages, such as its time consuming operation and bulky fragile instrumentation, the method was quickly replaced by liquid chromatography during the 1960s.

In 1970, a new liquid–liquid partition method was introduced to perform countercurrent separation in a continuous mode, as in liquid chromatography: The method is named "Countercurrent chromatography (CCC)," since it combines the merits of the countercurrent distribution method and liquid chromatography (2). The successively introduced preparative partition method called droplet CCC uses unit gravity to move the droplets of the mobile phase through the column of the stationary phase in a tubular space (3). This simple CCC system produced a partition efficiency of 900 theoretical plates in the dinitrophenyl (DNP) amino acid separation, but it required a long separation time of 70 h. Continuous development of CCC technology soon introduced more efficient centrifugal CCC schemes that utilize a centri-fugal force field created by a planetary motion of the coil (4). This planetary motion fulfills two important tasks for solute partitioning: continuous mixing

High-Speed Countercurrent Chromatography, Edited by Yoichiro Ito and Walter D. Conway. Chemical Analysis Series, Vol. 132.
ISBN 0-471-63749-1 © 1996 John Wiley & Sons, Inc.

of the two phases and retention of the stationary phase in the coil. These centrifugal CCC systems require continuous elution of the mobile phase through a rotating coil. Although this may be accomplished by the use of a conventional rotary seal, the use of a low viscosity organic solvent under high back pressure would cause leakage and contamination through the seal. To solve this problem, a great effort has been made to develop a series of flow-through centrifuge systems that do not require the rotary seal device (5). A series of studies have been performed with the use of these centrifuge schemes to establish an ideal combination of the mode of planetary motion and the geometry of coiled columns on the holder. In the early 1980s, nearly a decade of steady research finally produced an efficient scheme called "high-speed CCC," which utilizes a particular mode of synchronous planetary motion (see Type J in Fig. 1.5) (6–8). As indicated by its name, the method produces high efficiency separations in a short period of time and has been widely used for separation and purification of natural and synthetic products. Further studies revealed that the method can be applied to dual CCC where two immiscible solvent phases can truly undergo countercurrent movement through the coiled column. This method has been applied not only to liquid–liquid dual CCC (9) (see Chapter 4 in this volume) but also to foam separation (see Chapter 5 in this volume) which utilizes the countercurrent movement of the foam and liquid phases through a long coiled column (10–14). It has recently been found that a new type of the coil planet centrifuge (CPC) system (the X–L type in Fig. 1.5) provides stable retention of the stationary phase for the viscous aqueous–aqueous polymer phase systems, which can be applied to separation of macromolecules and cell particles (15, 16). This cross-axis CPC has been successfully applied to purification of recombinant enzymes (17) (see Chapter 13 in this volume).

Most recently, a new technique called pH-zone-refining CCC has been developed, which may open a rich domain of applications in high-speed CCC (18–20). The method originated from an incidental finding that the presence of an organic acid in the sample solution produced an unusually sharp peak of acidic compounds, which were collected in a fraction of only a few milliliters (21). When the sample size was increased, each analyte formed a highly concentrated isotachic pH-zone in the column and eluted as a rectangular peak with minimum overlap, as observed in displacement chromatography. This pH-zone-refining CCC method has been successfully applied to a variety of organic acids and bases, including amino acids and peptides, dyes, alkaloids, structural, and geometrical and optical isomers. Because of its great impact on the CCC technology, Chapter 6 (in this volume) is entirely devoted to these new CCC techniques while Chapter 12 (in this volume) describes the application of pH-zone-refining CCC to the separation and purification of various hydroxyxanthene dyes.

1.2. PRINCIPLE OF HIGH-SPEED COUNTERCURRENT CHROMATOGRAPHY

1.2.1. Two-Phase Distribution in a Rotating Coil in Unit Gravity

Over 2000 years agò, a Greek mathematician named Archimedes used a rotating helical structure to lift river water to the bank. His device is called the "Archimedean screw." A similar principle can be applied to move two immiscible solvent phases one through the other by rotating the coiled tube in a gravitational field. Figure 1.1A–C illustrates this Archimedean screw effect using a coiled tube slowly rotating in a unit gravitational field. All coils are sealed at both ends before being rotated.

In Fig. 1.1A, the coil is first filled with water, and air bubbles and glass beads are introduced into the coil. By the effect of gravity, the air bubbles stay at the top of the coil and the glass beads stay at the bottom of the coil while the Archimedean screw force drives both objects toward one end (left) of the coil as the coil is rotated slowly in the gravitational field. The important fact is that all objects present in the coil, if they are either lighter or heavier than water, move through the coil in the same direction toward one end. This end of the coil is conventionally called the "head" and the other end is the "tail."

Figure 1.1B shows a similar experiment using two mutually equilibrated solvent phases. In the upper diagram, the coil is filled with the lighter phase and a small volume of the heavier phase is introduced at the tail. Then, the droplets of the heavier phase stay at the bottom of the coil and move toward the head of the coil as the glass beads in water. Similarly, in the lower diagram a small amount of the lighter phase suspended in the heavier phase in the coil moves toward the head as the air bubbles in the water.

In Fig. 1.1C, nearly equal volumes of the lighter and heavier phases are introduced in the coil. Then, the two phases are separated in each helical turn, the lighter phase in the upper portion and the heavier phase at the lower portion. The rotation of the coil forces both phases to competitively advance toward the head of the coil. Sooner or later, the two phases establish a hydrodynamic equilibrium from the head side of the coil where each phase occupies nearly equal space in each helical turn while any excess of either phase is pushed back toward the tail end of the coil. Once this hydrodynamic equilibrium is reached, further rotation of the coil results in mixing of the two solvent phases back and forth within each helical turn while the overall distribution of the two phases in the coil remains unaltered.

This hydrodynamic equilibrium condition may be used for performing CCC as follows: The coil is first completely filled with the stationary phase, either lighter or heavier phase, and the other phase is introduced from the head end of the coil while the coil is slowly rotated around its axis. Then, the two

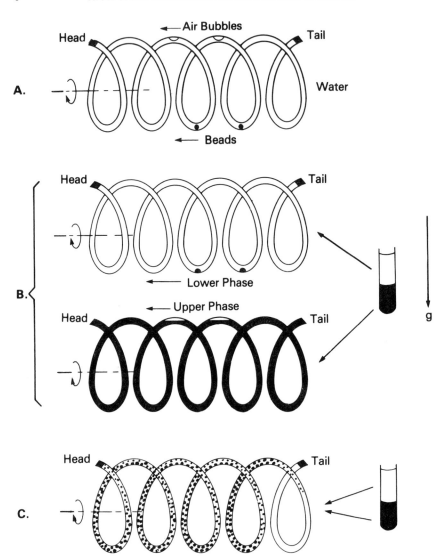

Figure 1.1. Motion of various objects in a slowly rotating coil.

solvent phases quickly establish the hydrodynamic equilibrium in each turn of the coil and the mobile phase finally emerges from the tail end of the coil, leaving some amount of the stationary phase permanently in the coil. Consequently, solutes locally introduced at the head of the coil are subjected to an efficient partition process between the two phases and eluted in the order of their partition coefficients.

This simple CCC system, however, has an inherent disadvantage in that the volume of the stationary phase retained in the coil is substantially less than 50% of the total capacity of the coil. This occurs because the system distributes two solvent phases evenly on the head side of the coil. In CCC, the amount of the stationary phase retained in the column is an important factor governing the resolution of solute peaks; in general, higher retention greatly improves peak resolution.

Further studies, however, revealed that the volume ratio of the two solvent phases in the hydrodynamic equilibrium is greatly altered by the rotational speed of the coil (22, 23). In Fig. 1.2, the volume percentage of the two phases occupying the head side of the coil is plotted against the applied rotational speed of the coil. The experiment was performed with a two-phase solvent system composed of chloroform, acetic acid, and water at a 2:2:1 volume ratio using a 20-mm i.d. glass coil with a 5-cm helical diameter.

At a slow rotation of 10–20 revolutions per minute (rpm), two solvent phases are rather evenly distributed in the coil. As the rotational speed increases, the heavier phase quickly occupies more space on the head side of the coil, and at the critical speed range of 60–100 rpm, the two solvent phases are completely separated along the length of the coil with the heavier phase on the head side and the lighter phase on the tail side. After this critical speed range, the amount of the heavier phase on the head side decreases sharply reaching substantially below the 50% level at 160 rpm. Further increase of the rotational speed again distributes the two solvent phases fairly evenly throughout the coil apparently due to the strong radial centrifugal force field produced by rotation of the coil. The phase distribution curve of chloroform/acetic acid/water (2:2:1) described above is also observed in many other solvent systems as illustrated in Fig. 1.3 (23).

The distribution of the two solvent phases in the coil produced at the critical rotation speed is called unilateral since one of the phases entirely occupies the head side. It may also be called bilateral indicating the distribution of the one phase on the head side and the other phase on the tail side. At any rate, this particular hydrodynamic equilibrium condition provides a great advantage in performing CCC, because the system permits retention of a large amount of stationary phase in the coil if the lighter phase is eluted in a normal mode (head-to-tail direction) or the heavier phase in a reversed mode (tail-to-head direction). The capability of this method has been successfully demonstrated in preparative separation of DNP amino acid samples using a multi-layer coil of 5 mm i.d., which was rotated at 80 rpm in the gravitational field (22).

The two-phase distribution in the rotating coil described above is a complex hydrodynamic phenomenon that has not been formulated mathematically. However, it may be worthwhile to describe some relationships

PHASE DISTRIBUTION DIAGRAMS FOR COAXIALLY ROTATED COIL

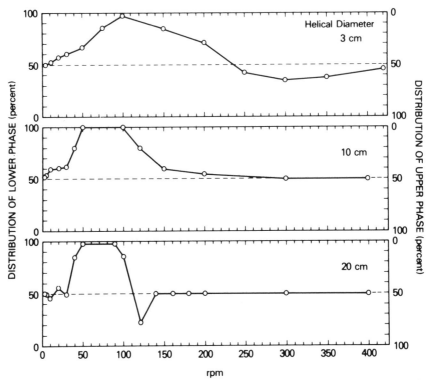

Figure 1.2. The effect of rotational speed on the two-phase distribution in a rotating coil. Distribution is expressed as percentage of head side of the coil occupied by heavier (left) or lighter (right) phase.

between the phase distribution pattern in the rotating coil and the acting force field at the various portions of the coil.

When the coil is slowly rotated (such as at 10 rpm), the centrifugal force generated by the rotation is quite negligible and every portion of the coil is subjected to the unit gravity. Under this condition, the Archimedean screw force acts evenly on the two solvent phases in the coil, that is, the upper phase moving toward the head in the upper portion of the coil and the heavier phase moving toward the head in the lower portion of the coil. Consequently, the two

→

Figure 1.3. Phase distribution diagrams for nine two-phase solvent systems obtained from glass coils with various dimensions. Solid lines indicate data obtained from plain glass coils; broken lines indicate data that were obtained from silicone-treated glass coils. A thin vertical line in each diagram indicates the rotation speed where the centrifugal force equals the unit gravity. Note that all solvent systems show a critical range of the rotation speed where one of the phases completely occupies the head side of the coil.

Two-Phase Distribution in Rotating Coils

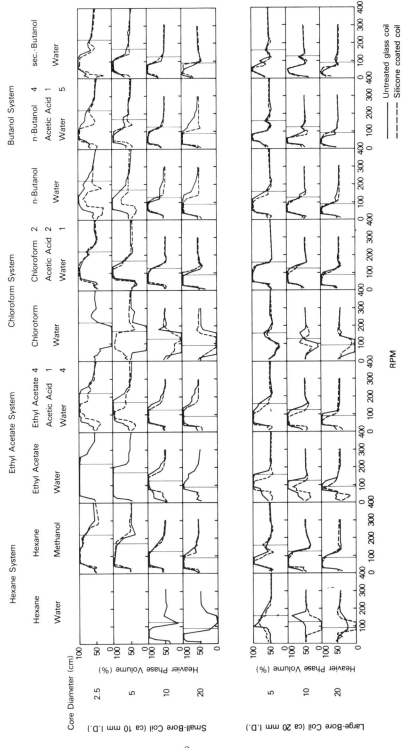

6

phases competitively establish a hydrodynamic equilibrium where each phase occupies about equal space on the head side of the rotating coil.

An increase of the rotational speed of the coil to the critical range of 80–100 rpm alters the balance of the hydrodynamic equilibrium by an enhanced radial centrifugal force field that increases the net force field acting at the bottom of the coil and decreases that acting at the top of the coil as illustrated in Fig. 1.4. Under this asymmetrical force distribution, the movement of the heavier phase toward the head is accelerated whereas the movement of the lighter phase toward the head is retarded. This results in a unilateral hydrodynamic phase distribution in the rotating coil. Further increase of the rotational speed of the coil above the critical range produces a stronger radial force field that finally redistributes the two solvent phases in such a way that the heavier phase occupies the outer portion and the lighter phase occupies the inner portion of the coil. This results in an even distribution of the two phases at each helical

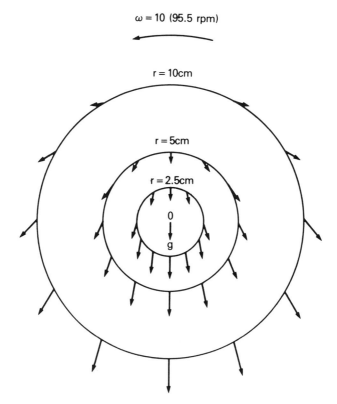

Figure 1.4. Distribution of force vectors acting on the coils rotating at ca 100 rpm.

turn throughout the length of the coil since equal volumes of the two phases were placed in the coil initially.

1.2.2.   Flow-Through Coil Planet Centrifuge Free of Rotary Seals

The simple CCC system based on the unilateral hydrodynamic equilibrium described above requires a relatively long separation time. This is apparently due to the fact that the retention of the stationary phase entirely relies on a relatively weak Archimedean screw force generated by unit gravity. In this situation, application of a high flow rate of the mobile phase would cause a depletion of the stationary phase from the coil resulting in a detrimental loss of peak resolution. This problem may be solved by the utilization of a centrifugal force field that would enhance the Archimedean screw force acting on the coil. Application of the centrifugal force, however, would increase the back pressure in the coil resulting in leakage of solvent through the rotary-seal joint of the flow-through centrifuge device at high mobile phase flow rates. To solve this problem, efforts have been made to develop various types of rotary-seal-free flow-through centrifuge systems, which permit continuous elution of the mobile phase through the rotating column without the use of conventional rotary seal devices.

The rotary-seal-free flow-through centrifuge systems provide the following advantages over the conventional centrifuge system:

1. The system is leak-free even under a high back pressure of several hundred pounds per square inch (psi).
2. It facilitates the use of multiple flow channels without a risk of cross contamination.
3. It has a negligible dead space.
4. It prevents mechanical damage to biological samples.
5. It permits the use of corrosive solvents.
6. There is no heat production.

These centrifuge systems are considered to be ideal for performing CCC because of these advantages.

Figure 1.5 illustrates a series of synchronous CPC systems free of rotary seals, which have been developed for performing CCC (8, 24–26). Each diagram shows orientation and motion of a cylindrical coil holder equipped with a bundle of flow tubes, one end of which is tightly supported on the central axis of the centrifuge. All these systems produce synchronous planetary motion: The holder revolves around the centrifuge axis and simultaneously rotates about its own axis at the same angular velocity ω, as illustrated by a pair of arrows.

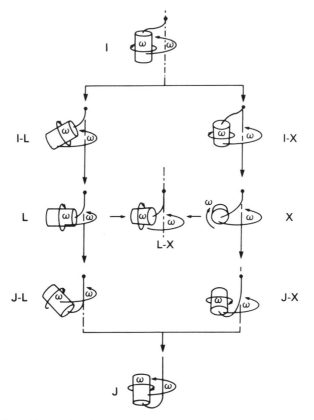

Figure 1.5. A series of the flow-through CPC systems free of rotary seals.

In the Type I planetary motion (top), the vertical holder revolves around the central axis of the centrifuge and synchronously counterrotates about its own axis. This counterrotation of the holder steadily unwinds the twist of the tube bundle caused by revolution, thus eliminating the need for the rotary seal.

This seal-free principle can be applied equally well to other types of synchronous planetary motions with tilted, horizontal, dipping, and inverted orientations of the holder. The I–L–J series (left column) is given by tilting the holder toward the central axis of the centrifuge, whereas the I–X–J series is obtained by tilting the holder sideways as illustrated. These two series also form various hybrid types at each level between Types I–L and I–X, Types L and X (shown in the diagram), and Types J–L and J–X.

1.2.3. Mechanism of High-Speed Countercurrent Chromatography

The effects of the centrifugal force fields on the hydrodynamic phase distribution have been studied in each type of planetary motion by varying orientations of the coil on the holder.

When the coil is mounted coaxially around the holder of Type I synchronous planetary motion, two solvent phases are distributed along the length of the coil in such a way that each phase occupies nearly equal space on the head side of the coil while any excess of either phase is accumulated at the tail of the coil. This phase distribution mode is quite similar to that observed in the slowly rotating coil in unit gravity described previously (22, 23).

On the other hand, the coil similarly mounted on the holder undergoing the Type J synchronous planetary motion produces a totally different phase distribution pattern. In the typical case, the two solvent phases are completely separated in the coil so that one phase entirely occupies the head end and the other phase occupies the tail end of the coil as observed in the rotating coil in unit gravity at the critical speed range of 80–100 rpm (Fig. 1.2) (22, 23). Similar unilateral phase distribution had been observed in the coaxially mounted coil in other types of planetary motions, such as Types X (24, 25), X–L (26) and J–L (27, 28) (see Fig. 1.5). Here, it is interesting to note that all these planetary motions producing unilateral phase distribution form an asymmetrical centrifugal force distribution that closely resembles that observed in the coil rotating at the critical speed in unit gravity (see Fig. 1.4).

This unilateral hydrodynamic equilibrium condition provides the basis for high-speed CCC (HSCCC). Figure 1.6 schematically illustrates the mechanism of HSCCC, where each coiled tube is drawn as a straight tube to indicate an overall distribution of the two solvent phases in the coil. In Fig. 1.6A, the two mutually immiscible solvent phases establish a unilateral distribution in an end-closed coil. The white phase is completely occupying the head side and the black phase is occupying the tail side, showing an interface at the middle portion of the coil. This hydrodynamic equilibrium condition suggests that the black phase, if introduced from the head, would move through the white phase toward the tail, and similarly that the white phase, if introduced from the tail of the coil, would move through the black phase toward the head. This hydrodynamic motion of the two phases can be effectively applied for performing CCC.

Figure 1.6B shows normal (top) and reversed (bottom) elution modes. In the normal elution mode, the coil is first completely filled with the white phase and the black phase is pumped into the head of the coil. In the reversed elution mode, the coil is first entirely filled with the black phase followed by pumping the white phase from the tail of the coil. In either case, the mobile phase rapidly flows through the stationary phase and emerges from the other end of the coil leaving a large volume of the stationary phase in the coil.

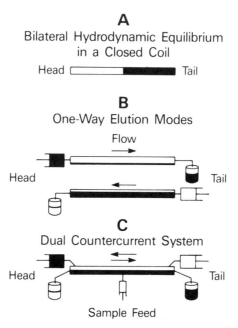

A

Bilateral Hydrodynamic Equilibrium
in a Closed Coil

Head ⬜⬛ Tail

B

One-Way Elution Modes

Flow
→

Head   Tail

←

C

Dual Countercurrent System

←→

Head   Tail

Sample Feed

Figure 1.6. Mechanism of HSCCC.

The present system also permits simultaneous feeding of two solvent phases through the respective terminals of the coil to produce the countercurrent movement of the two phases through the coil. As illustrated in Fig. 1.6C, this operation requires an additional flow tube at each terminal of the coil to collect the effluent and a sample feed tube at the middle portion of the coil. This dual countercurrent system has been applied to liquid–liquid dual CCC (9) and foam CCC (10–14).

Hydrodynamic motion and distribution of two immiscible solvent phases in the rotating column was first observed by Conway in our laboratory (29). The experiment was performed with a Type J CPC equipped with a spiral column and a transparent plastic cover using a two-phase solvent system composed of chloroform and water, each phase being stained with a dye to facilitate stroboscopic observation. The spiral column was first filled entirely with the upper aqueous phase and the lower chloroform phase was introduced through the internal head end while the apparatus was rotated at 750 rpm. After a steady-state hydrodynamic equilibrium was established, the strobo-scopic observation revealed two rather distinct zones in the spiral column: Approximately one fourth of the area (mixing zone) near the center of the centrifuge shows violent mixing of the two solvent phases. In the rest of the area (settling zone), the two solvent phases were separated in two layers: The heavier chloroform phase occupied the outer portion and the lighter aqueous phase occupied the inner portion of each segment, forming a linear interface of the two phases along the length of the spiral column.

Figure 1.7 schematically illustrates the motion and distribution of the two solvent phases in a rotating spiral column as observed under stroboscopic illumination. In the top diagram, the spiral column is drawn in successive positions as it revolves around the central axis of the centrifuge. As indicated by a pair of arrows, the spiral column revolves at angular velocity ω, and simultaneously rotates around its own axis at the same angular velocity in the same direction. The mixing zone is always fixed at the vicinity of the central axis of the centrifuge while the spiral column undergoes the planetary motion. In other words, the mixing zone is traveling through the spiral column toward

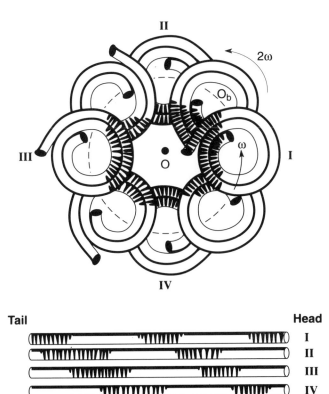

Figure 1.7. Mixing zone motion in the type J CPC. Upper diagram: Successive positions of the spiral column showing the mixing zone at the vicinity of the centrifuge axis. Lower diagram: Motion of the mixing zones through the spiral coil in one revolution cycle. The numbers I–IV correspond to the coil position on the upper diagram.

the head at a rate equal to the column rotation. In the bottom diagram, uncoiled columns numbered I–IV correspond to the column positions in the top diagram showing the movement of the mixing zone through the spiral column. Each mixing zone travels toward the head of the column (in analogy with the motion of waves over the sea) at a rate equal to the speed of revolution of the column. This indicates an extremely important fact that at any portion of the column the two solvent phases are subjected to a typical partition process of repetitive mixing and settling at an enormously high frequency of over 13 times per second at 800 rpm of column revolution, while the mobile phase steadily passes through the stationary phase. Consequently, this finding explains the high partition efficiency of HSCCC over a wide range of flow-rate of the mobile phase.

1.3. HIGH-SPEED COUNTERCURRENT CHROMATOGRAPH INSTRUMENTATION

1.3.1. Multilayer Coil Planet Centrifuge (Type J)

Figure 1.8 shows the design principle of the Type J synchronous CPC. A cylindrical column holder is equipped with a planetary gear that is inter-locked with an identical sun gear (shaded) affixed on the central axis of the centrifuge. This gear arrangement produces a particular mode of planetary motion to the holder: The holder revolves around the central axis of the centrifuge and simultaneously rotates about its own axis at the same angular velocity ω, in the same direction as indicated by a pair of arrows. The separation column is made simply by winding a single piece of polytetra-fluoroethylene (PTFE) tubing directly around the holder hub making multiple layers of coils. The flow tubes from both terminals of the coil are paired and exit the centrifuge system through the passage as indicated in the diagram. As mentioned earlier, these tubes maintain their integrity without twisting. Parameter β, computed by dividing the radius of the coil (r) by the orbital radius of revolution (R), is one of the important factors that determine the hydrodynamic distribution of the two immiscible solvent phases in the rotating coil.

Figure 1.9 schematically shows a cross-sectional view of the original HSCCC centrifuge based on the above design. The motor drives the rotary frame around the horizontal stationary pipe (shaded) of the centrifuge via a pair of toothed pulleys coupled with a toothed belt. The rotary frame of the centrifuge holds the column holder and the counterweight in symmetrical positions around the central axis of the centrifuge. A planetary gear on the column holder shaft is engaged with the identical stationary gear (shaded)

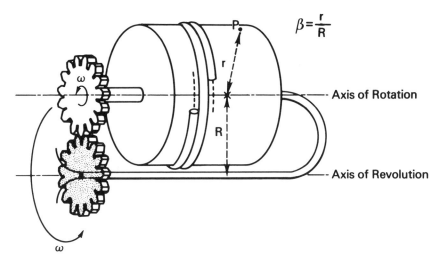

$$\beta = \frac{r}{R}$$

Figure 1.8. Design principle of the Type J CPC.

mounted around the central stationary pipe of the centrifuge to introduce the desired planetary motion of the holder. A multilayer coil is mounted coaxially around the holder. A pair of flow tubes from the coil first passes through an opening in the holder shaft, then forms an arch, enters the side hole of a short coupling pipe, and finally exits the centrifuge through the opening of the central stationary pipe.

Figure 1.10 is a photograph of the original model of the HSCCC centrifuge constructed in the NIH machine shop (7). The multilayer coil consists of a single piece of approximately 140-m long, 1.6-mm i.d. PTFE tubing with a total capacity of about 280 mL. The revolution speed is regulated up to 1000 rpm with a speed controller (Bodine Electric Co., Chicago, IL). Based on this prototype, several different models of the HSCCC centrifuge systems have been designed for particular applications: The analytical HSCCC centrifuge (Fig. 1.11) for analytical separation (30–34) and interfacing with a mass spectrometer (35–38), the foam CCC centrifuge for foam separation (10–14) (see Chapter 5), the eccentric CPC (Fig. 1.12) for separation of peptides and proteins (39, 40), and so on.

The original design of the HSCCC centrifuge uses a counterweight to balance the centrifuge system, which requires adjustment of the counter-weight mass according to the density of the two-phase solvent system used for the separation. Later, this problem was solved by a new design, as shown in Fig. 1.13 (41, 42). The top diagram (A) shows the conventional HSCCC centrifuge system, which utilizes a counterweight to balance the centrifuge

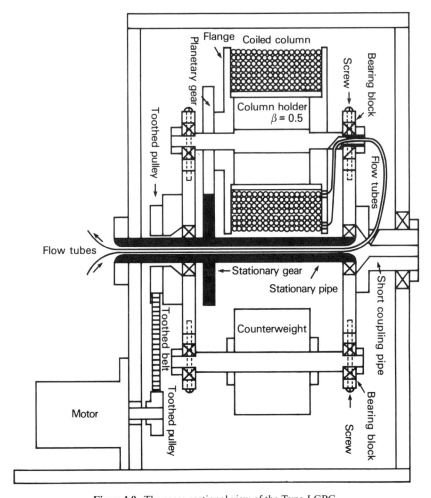

Figure 1.9. The cross-sectional view of the Type J CPC.

system. In the bottom diagrams (B and C), the counterweight was eliminated and, instead, the centrifuge system was balanced by placing multiple column holders symmetrically around the rotary frame. All columns can be serially connected on the rotary frame without risk of twisting. There are two options to arrange the flow tubes: In the first method (B) a pair of flow tubes returns through the same route (double passage) and in the second method (C) a single flow tube enters the centrifuge from one side and circles the rotary frame to exit the centrifuge from the other side (single passage).

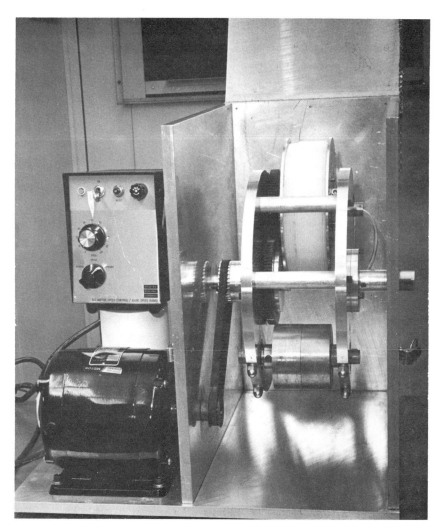

Figure 1.10. Photograph of the original model of the HSCCC centrifuge.

Figure 1.14 shows a photograph of the HSCCC centrifuge based on this improved design (42). This centrifuge is equipped with three column holders and each holder carries a set of two gears, one (left) engaging to the identical stationary sun gear (not seen) on the centrifuge axis, and the other (right) engaging to the identical gear mounted on the tube holder shaft. Thus, the tube holder shaft counterrotates against the coil holder to prevent the flow tubes

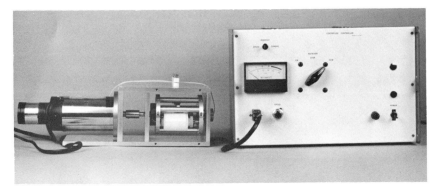

Figure 1.11. Photograph of the analytical HSCCC centrifuge.

Figure 1.12. Photograph of the eccentric multilayer CPC.

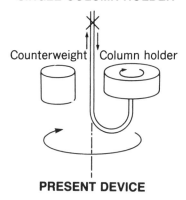

**CONVENTIONAL
SINGLE COLUMN HOLDER**

PRESENT DEVICE

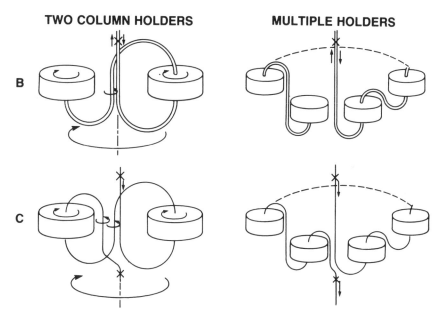

Figure 1.13. Design principle of the Type J multiholder CPC without rotary seals.

from twisting. Each multilayer coil on the holder consists of nine coiled layers of 1.6 mm i.d. PTFE tubing with a capacity of about 135 mL. All three columns are connected in series to provide a total capacity of about 400 mL. The revolution radius is 7.6 cm and the maximum revolution speed is 1200 rpm.

Figure 1.14. Photograph of the Type J multilayer CPC equipped with three column holders.

This unit has produced separations with partition efficiencies over 2000 theoretical plates (43).

1.3.2. Cross-Axis Synchronous Flow-Through Coil Planet Centrifuges

The cross-axis CPC is classified into Types X and L (see Fig. 1.5) and their hybrids, such as Types XL, XLL, and XLLL, according to the column position on the rotary shaft (see Fig. 13.2 in this volume) (15). All schemes have a common feature in that the axis of the column rotation is perpendicular to the axis of the revolution. Figure 1.15 shows a planetary motion of the column holder of the Type X cross-axis CPC (24, 25). The holder (shown as a disk) revolves around the central axis of the centrifuge and simultaneously rotates about its own axis at the same angular velocity ω as indicated by a pair of arrows. In doing so, the holder always maintains its axis in the horizontal position and at the same distance R from the centrifuge axis. As described earlier (Fig. 1.5), a bundle of flow tubes supported at the central axis of the centrifuge is free from twisting as the holder undergoes the planetary motion. Parameter β (r/R, where r is a distance from the holder axis to the coil and R, a distance from the holder axis to the centrifuge axis) is one of the major parameters that determine the hydrodynamic distribution of the two solvent phases in the rotating coil.

Figure 1.16 shows the original prototype of the Type X cross-axis CPC (24). The motor (a) drives the central shaft (b) and the rotary frame around the central axis of the centrifuge via a pair of toothed pulleys coupled with a toothed belt. The rotary frame consists of a pair of side-plates (c) rigidly bridged with links (d) to support a column holder (e) and a counterweight holder in symmetrical positions at a distance of 10 cm from the central axis of the centrifuge. The planetary motion of each holder is produced as follows: At the lower portion of the side-plates, a pair of countershafts is mounted radially and symmetrically on each side of the rotary frame through ball bearings. The

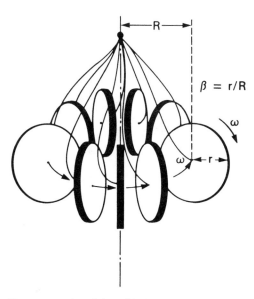

Figure 1.15. Planetary motion of the coil holder in the cross-axis CPC (Type X).

Figure 1.16. Photograph of the original cross-axis CPC (Type X).

24

stationary miter gear (45°) (f) coaxially mounted around the central shaft on the bottom plate of the centrifuge is coupled to the identical miter gear (g) on the proximal end of each countershaft. This gear coupling produces synchronous rotation of each countershaft on the revolving rotary frame. This motion is further conveyed to the column holder and the counterweight holder by coupling a pair of identical toothed pulleys, one (h) mounted on each holder shaft and the other (i) on the distal end of the respective countershaft. Consequently, both the column holder and counterweight holder undergo the desired synchronous planetary motion, that is, rotation about its own axis and revolution around the central axis of the centrifuge at the same angular velocity, as indicated in Fig. 1.15.

To facilitate column preparation and balancing of the centrifuge system, both the column holder and counterweight holder are designed so that they are easily removed from the rotary frame by loosening a pair of screws in each bearing block. The separation column was prepared by winding a long piece of PTFE tubing typically 2.6 mm i.d. directly onto the holder hub making multiple coiled layers. A pair of flow tubes from the separation column is first passed through the center hole of the holder shaft and then led through the side-hole of the central shaft. Finally, it exits the centrifuge system through the center of the top plate where it is tightly held by a pair of silicone-padded clamps. These flow tubes are thoroughly lubricated with grease and protected with a piece of Tygon tubing at each supported portion to prevent direct contact with metal parts. If this caution is followed, they maintain their integrity during many months of use.

This apparatus can be operated at revolution speeds ranging from 50 to 1000 rpm with a speed control unit (Bodine Electric Co., Chicago, IL). It was found that this prototype produced considerable noise at the site of the metal gears, thus limiting the use of a high revolution speed. This problem has been eliminated in later models by the use of plastic miter gears.

This original design of the prototype X-axis CPC has been improved by mounting a pair of identical column holders, one on each side of the rotary frame to balance the centrifuge system. A pair of multilayer coils mounted on these holders is connected in series with a flow tube to double the column capacity. In recent models, the position of the column holder is shifted laterally along the rotary shaft, as shown in Figs. 1.17 and 1.18. The degree of this holder shift is found to be critical for retention of viscous polar solvent systems that include aqueous–aqueous polymer phase systems useful for partition of macromolecules and cell particles (15, 16). The lateral shift of the holder on the rotary shaft is expressed by $\delta = L/R$, where L is the distance from the mid-portion of the rotary shaft to the column and R is the revolution radius or the distance from the central axis of the centrifuge to the rotary shaft. As illustrated elsewhere (Fig. 13.2 in this volume), various X-axis CPC models fabri-

Figure 1.17. Photograph of the Type XL cross-axis CPC.

cated at the NIH machine shop were named, according to parameter δ, Type X (A: $\delta = 0$) (the original model in Fig. 1.16); Type XL or LX (B: $\delta = 1$) (Fig. 1.17); Type XLL (C: $\delta = 2$); Type XLLL (D: $\delta = 3$) (Fig. 1.18); and Type L (E: $\delta \rightarrow \infty$).

As mentioned earlier, the X-axis CPC models with a large δ are suitable for application of the aqueous–aqueous polymer phase systems. Figure 1.18 shows a photograph of the Type XLLL cross-axis CPC equipped with a pair of multilayer coils. Because of the large δ value, the apparatus has been successfully used for the separation of proteins with polymer phase systems including both PEG–phosphate and PEG–dextran systems (44). The most recent model of Type L cross-axis CPC was found to retain a very viscous PEG–dextran polymer phase system over 50% of the total capacity of the multilayer coil at a relatively high flow rate of 1 mL/min (45).

1.4. PHASE DISTRIBUTION DIAGRAMS

In CCC, the volume of the stationary phase retained in the column is an important factor in determining the resolution of solute peaks. Generally

Figure 1.18. Photograph of the Type XLLL cross-axis CPC. In order to accommodate the rotary frame in a bench-top case, the revolution radius (distance from the centrifuge axis to the rotary shaft) of the apparatus is reduced to 3.8 cm.

27

speaking, the greater the volume of the stationary phase retained in the column, the higher the resolution obtained. A series of hydrodynamic studies has been performed to measure the retention of the stationary phase using both the Type J HSCCC centrifuge and X-axis CPC. Most of these studies were carried out with single-layer coils of either 1.6- or 2.6-mm i.d. PTFE tubing mounted on holders with various hub diameters.

In each measurement, the coil was first entirely filled with the stationary phase, either upper or lower phase of a mutually equilibrated two-phase solvent system, and the mobile phase was eluted through the coil at a flow rate of 2 mL/min (for 1.6-mm i.d. coil) or 4 mL/min (for 2.6-mm i.d. coil) while the apparatus was rotated at various revolution speeds ranging from 400 to 1000 rpm. The effluent through the outlet of the coil was collected in a graduated cylinder to measure the volume of the stationary phase eluted. The experiment was terminated after the total elution volume had reached the column capacity of the coil. Then, the percentage of the stationary phase volume retained in the coil relative to the total column capacity was computed using the conventional formula (46), $100(V_c + V_f - V_e)/V_c$, where V_c denotes the total column capacity; V_f is the dead space in the flow tube; and V_e is the volume of the stationary phase eluted from the coil. The results are conveniently summarized by a set of phase distribution diagrams constructed by plotting the percentage retention of the stationary phase as a function of the revolution speed for each mobile phase.

1.4.1. Retention of Stationary Phase in Type J Coil Planet Centrifuge

A set of phase distribution diagrams obtained from the standard Type J HSCCC centrifuge equipped with a 1.6-mm i.d. coil is illustrated in Fig. 1.19 (47).

These diagrams are arranged from left to right in the order of hydrophobicity of the major organic solvents employed as indicated at the top of each column. The upper panel shows the retention of the lower phase (the upper phase mobile) and the lower panel shows the retention of the upper phase (lower phase mobile). The first row in each panel was obtained from the coil with a 5-cm helical diameter ($\beta = 0.25$), the second row was obtained from the coil with a 10-cm helical diameter ($\beta = 0.5$), and the third row was obtained from the coil with a 15-cm helical diameter ($\beta = 0.75$) as indicated on the left margin. In each diagram, the solid curve indicates the retention obtained by the head-to-tail elution mode and the broken curve indicates the retention obtained by the tail-to-head elution mode. In general, the retention of over 50% is considered satisfactory.

The phase distribution curves display diverse patterns according to the hydrophobicity of the solvent system and within each solvent system the

PHASE DISTRIBUTION DIAGRAMS FOR VOLATILE SOLVENT SYSTEMS

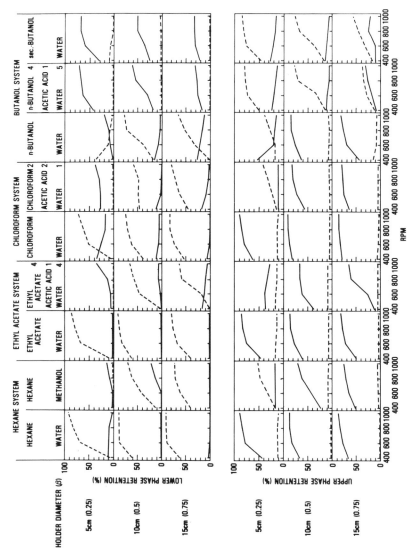

Figure 1.19. Phase distribution diagrams of nine volatile solvent systems obtained from the Type J CPC. Solid line: Head-to-tail elution; Broken line: Tail-to-head elution.

29

distribution pattern significantly changes with parameter β. From these results, the nine solvent systems may be classified into three groups according to their characteristic retention profile.

The hydrophobic binary solvent group including hexane/water, ethyl acetate/water, and chloroform/water, displays high retention if the lower phase is eluted in the normal mode from the head toward the tail (solid line), or the upper phase is eluted in the reversed mode from the tail toward the head (broken line). The hydrophilic solvent group including n-butanol/acetic acid/water (4:1:5) and sec-butanol/water displays a retention profile contrasting with that of the hydrophobic solvent group. These polar solvent systems yield high retention when eluting with the upper phase in the normal mode from the head toward the tail (solid line) or the lower phase in the reversed mode from the tail toward the head (broken line). The rest of the solvent systems belong to an intermediate solvent group characterized by moderate hydrophobicity of the organic phase. These intermediate solvent systems exhibit a very complex retention profile that is sensitively affected by a change of the β value. At a small β value of 0.25, better retention of the lower phase is obtained under conditions represented by the solid curve while the broken curve indicates better retention of the upper phases, the pattern being analogous to that observed in the hydrophilic solvent group. However, at $\beta > 0.5$ this retention profile is completely reversed, that is, the broken curve shows greater lower phase retention and the solid curve shows greater upper phase retention analogous to the hydrophobic solvent group. In brief, the hydrodynamic distribution of the intermediate solvent group approaches that of the hydrophilic solvent group at a small β value and becomes quite similar to that of the hydrophobic solvent group at a large β value.

In a practical separation, these phase distribution diagrams provide an important guideline for the choice of the solvent system and its eluting mode to ensure a satisfactory retention of the stationary phase. In most commercial HSCCC instruments, the multilayer coil has relatively large β values of over 0.5, which is suitable for the application of hydrophobic to intermediate solvent systems. In these solvent systems the lower phase should be eluted in the normal mode (head to tail) and the upper phase in the reversed mode (tail to head). Although the hydrophilic solvent systems can be used by modifying the elution mode, the retention of the stationary phase is usually much less than 50% of the column capacity. In addition, the sample solution, when introduced into the column, may change the two-phase distribution locally at the beginning of the column to adversely affect the retention of the stationary phase. For these polar solvent systems, the cross-axis CPCs will provide more reliable retention of the stationary phase as described in Section 1.4.2.

1.4.2. Retention of Stationary Phase in Cross-Axis
Coil Planet Centrifuges

As described earlier, the retention of the stationary phase in the X-axis CPC varies according to the position of the coil holder on the rotary shaft. By using various types of the X-axis CPC, a series of experiments has been performed to measure the retention of the stationary phase in a set of two-phase solvent systems at various β values.

A set of phase distribution diagrams obtained from the original X-axis CPC (Type X) is illustrated in Fig. 1.20 (24). According to the format described in Fig. 1.19, each diagram shows a pair of phase distribution curves, the solid line indicating the head-to-tail elution mode and the broken line, the tail-to-head elution mode (see Table 1.1). Overall results indicate that except for the hydrophilic solvent systems of n-butanol/acetic acid/water (4:1:5) and sec-butanol/water, all solvent systems give satisfactory levels of stationary phase retention at $\beta > 0.5$. In contrast with the phase distribution data obtained from the Type J CPC (Fig. 1.19), all intermediate solvent systems such as hexane/methanol, ethyl acetate/acetic acid/water (4:1:4), chloroform/acetic acid/water (2:2:1), and n-butanol/water, show a hydrodynamic trend similar to the hydrophilic solvent system, that is, a high phase retention is obtained from either the head-to-tail elution of the upper phase (solid lines in the upper panel) or the tail-to-head elution of the lower phase (broken lines in the lower panel).

As mentioned earlier, the retention of the hydrophilic solvent systems is dramatically improved by laterally shifting the column position along the rotary shaft. In the X-axis CPC with a lateral column position such as Types XL, XLL, and XLLL, the retention of the stationary phase is affected by three factors, that is, the direction of the planetary motion (P_I and P_{II}), head–tail elution mode, and inward–outward elution mode. Consequently, all possible experimental conditions are given by the combination of these three factors totaling eight composite elution modes, as listed in Table 1.2 (15, 48). The use of both right- and left-handed coils for examination of the hydrodynamic trend of all these combinations is required as indicated in the table.

Figure 1.21 similarly shows a set of phase distribution diagrams obtained from the Type XLL cross-axis CPC (15). In each diagram four phase distribution curves are drawn against the applied revolution speeds from 200 to 800 rpm. Among those four curves, two were obtained from the right-handed coil and the other two from the left-handed coil, while each group represents the highest retention among four possible combinations of the elution modes (see Table 1.2 for symbolic designs assigned for each elution mode).

All solvent systems show a satisfactory phase retention of over 50% at 800 rpm, regardless of the choice of the mobile phase or β values. However,

PHASE DISTRIBUTION DIAGRAMS FOR VOLATILE SOLVENT SYSTEMS
(2.6 mm i.d. Tube)

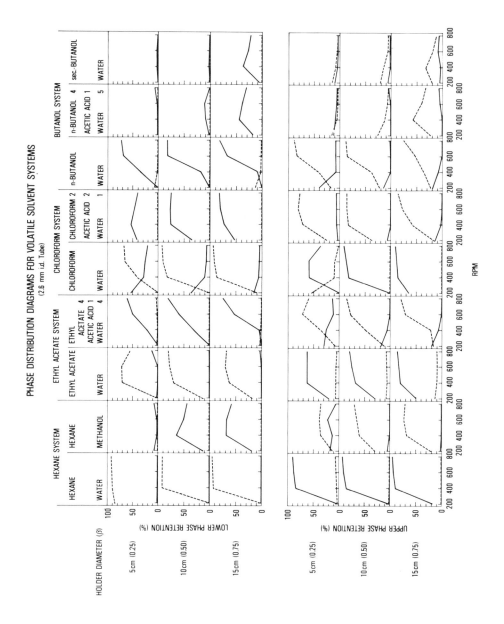

32

the most important finding is that the hydrophilic solvent systems, such as n-butanol/acetic acid/water (4:1:5) and sec-butanol/water show excellent phase retention ranging from 65 to 80% in the 7.6-cm helical diameter coil ($\beta = 0.5$). Here, note that the stationary phase retention is considerably affected by the handedness of the coil. In the majority of the retention diagrams, the left-handed coils (thin lines) yield the highest retention levels in the 7.6-cm helical diameter coils (49). In the 24-cm diameter coils, however, the difference in retention between the right- and left-handed coils becomes much less significant.

These phase distribution diagrams provide important guidance for construction of the multilayer coil according to the samples to be separated. For practical purposes, the multilayer coil separation column usually consists of right- and left-handed coils alternating in each layer. In this configuration, efficient separations are obtained only if each coiled layer retains a satisfactory volume of the stationary phase in a given elution mode. This may be the case for the use of solvent systems in the 24-cm helical diameter column except for the hydrophilic solvent systems. For the separation of polar samples, it becomes necessary to use the multilayer coil consisting entirely of left-handed coils with a small helical diameter. Such left-handed multilayer coils can be made by connecting each layer with a narrow transfer tube or using a continuous piece of tubing that is directly returned to the starting position after completing each coiled layer. Although the above modification causes some deformation of the multilayer coil by accommodating the connection tube between each layer, it provides a good retention of the stationary phase for viscous polar solvent systems, such as butanol systems, which tend to produce carryover problems in other types of the HSCCC apparatus.

Table 1.1. Two Elution Modes at Central Coil Position (Type X Cross-Axis CPC)

Planetary Motion	Head-Tail Elution Mode (Handedness of Coil[a])	Combined Elution Modes[b]	Design in PDD[c]
P_I 	HEAD → TAIL (R)	P_I–H	———
P_{II} 	TAIL → HEAD (R)	P_{II}–T	- - - - -

[a] Right handed.
[b] Head → tail = H; tail → head = T.
[c] Phase Distribution Diagram-PDD.

◄ ——————————————————————————————

Figure 1.20. Phase distribution diagrams of nine volatile solvent systems obtained from the original prototype of the Type X cross-axis CPC. For the applied elution modes, see Table 1.1.

Table 1.2. Eight Different Elution Modes in X-Axis CPC

Planetary Motion	Head-Tail Elution Mode	Inward–Outward Elution Mode (Handedness of Coil[a])	Combined Elution Mode[b]	Symbolic Signs in PDD[c]
P_I	Head → Tail	Inward (R)	P_I–H–I	○——○
	Head → Tail	Outward (L)	P_I–H–O	○——○
	Tail → Head	Inward (L)	P_I–T–I	●----●
	Tail → Head	Outward (R)	P_I–T–O	●----●
P_{II}	Head → Tail	Inward (L)	P_{II}–H–I	△——△
	Head → Tail	Outward (R)	P_{II}–H–O	△——△
	Tail → Head	Inward (R)	P_{II}–T–I	▲----▲
	Tail → Head	Outward (L)	P_{II}–T–O	▲----▲

[a]Right handed = R; Left handed = L.
[b]Head → tail = H; tail → head = T; inward = I; outward = O.
[c]Phase Distribution Diagram-PDD.

Recently, the Type XLL cross-axis CPC with small multilayer coils ($\beta = 0.25$–0.6) has been successfully applied to the separation of proteins with aqueous–aqueous polymer phase systems composed of polyethylene glycol (PEG) and potassium phosphate as described in Chapter 13 in this volume. The retention of the polymer phase systems can be further improved by increasing the δ value of the X-axis CPC. The preliminary tests on the Type XLLL and Type L CPC models showed satisfactory retention of most viscous PEG/dextran polymer phase systems, which can be applied to partition of cell particles and macromolecules (44, 45).

1.4.3. Settling Time

It has been reported that the hydrodynamic classification of the solvent system in the Type J CPC can be predicted by a simple settling time measurement (50). A 5-mL capacity graduated cylinder with a stopper (a 13-mm o.d., 100-mm long test tube with a polyethylene cap is also sufficient) containing 2-mL of each solvent phase is gently inverted five times and then placed on a flat table to measure the time required to form two clear layers. Our test results indicated that the settling times of the hydrophobic solvent group are no more than several seconds, those of the hydrophilic solvent group are over 30s, and those of the intermediate solvent group fall between the above two groups. This simple test can be used to classify various two-phase solvent systems, which are not included in Fig. 1.19.

The above method is also useful for preparation of sample solutions. In general, the sample solution is prepared by dissolving the solutes in about

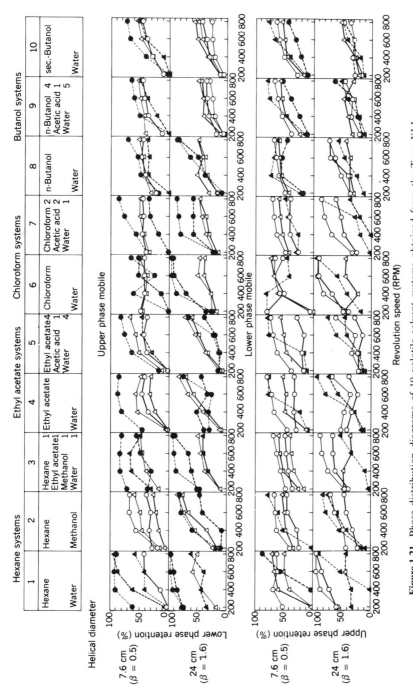

Figure 1.21. Phase distribution diagrams of 10 volatile solvent systems obtained from the Type XLL cross-axis CPC. The applied elution modes are described in Table 1.2. Note that the hydrophilic solvent systems, such as *n*-butanol/acetic acid/water (4:1:5) and *sec*-butanol/water show high retention of 60–80% in the 7.6-cm diameter coil.

35

equal volumes of each phase used for separation. Introduction of solutes in the solvent system tends to alter the physical properties, hence the settling time of the two phases. If the settling time of the sample solution becomes considerably longer than that of the original solvent system and exceeds 30 s, the hydrodynamic trend will be locally reversed in the sample compartment in the separation column. Such a reversed hydrodynamic trend in the sample solution will often cause steady carryover of the stationary phase from the column resulting in a detrimental loss of peak resolution. This problem can be alleviated by reducing the sample size and/or diluting the sample solution with the solvents used for separation as described later (see Section 1.5.2).

The above relationship between the settling time and the hydrodynamic trend of the solvent system does not apply to the X-axis CPCs, such as types XL and XLL, which hold multilayer coils in the lateral positions on the rotary shaft. These X-axis CPCs generate a strong centrifugal force field, which acts laterally across the diameter of the tube, thus preventing emulsification of the two solvent phases by violent phase mixing. Therefore, these X-axis CPCs are particularly suitable for preparative-scale separations of multigram quantities of samples.

1.5. GENERAL METHODOLOGY OF HIGH-SPEED COUNTERCURRENT CHROMATOGRAPHY

1.5.1. Partition Coefficient

Successful separation in CCC requires a proper choice of a two-phase solvent system that provides a suitable range of partition coefficients (K) for the desired compounds. In this chapter, the partition coefficient is expressed as $K = C_U/C_L$ or C_s/C_m, where C_U indicates solute concentration in the upper phase and C_L that in the lower phase. Similarly, C_s indicates the solute concentration in the stationary phase and C_m that in the mobile phase.

Generally speaking, the most suitable range of K values for HSCCC is 0.5–2 for C_U/C_L, where either upper or lower phase can be used as the mobile phase. The parameter $C_s/C_m \ll 0.5$ would result in a loss of peak resolution while $C_s/C_m \gg 1$ would require a long retention time with excessive sample band broadening.

One may conduct a literature search for the suitable solvent systems previously used for the similar compounds. If the nature of the compound is unknown or previous data for the similar compounds are not available, the search requires a time-consuming trial and error method. However, considerable time and effort may be saved by following a systematic solvent search (51, 52). In Table 1.3, various nonionic two-phase solvent systems are arranged according to the hydrophobicity of the nonaqueous phase. All these solvent systems provide nearly equal volumes of the upper and lower phases with reasonably short settling times, so that they can be applied to all centrifugal CCC systems including HSCCC.

The search for a solvent system may begin with the chloroform solvent system shown on the left. When the partition coefficient K (C_U/C_L), in $CHCl_3/MeOH/H_2O$ (2:1:1), where $Me = CH_3$, falls somewhere between 0.2 and 5, the desirable K value may be obtained by further modifying the volume ratio of each component, substituting acetic acid (HOAc) for methanol (MeOH), and/or partially replacing chloroform ($CHCl_3$) by carbon tetrachloride (CCl_4) or (methylene chloride (CH_2Cl_2). If the sample is more unevenly partitioned into one of the phases, the chloroform solvent system may be inadequate, and the search must be directed to other solvent systems that provide broader ranges of hydrophobicity and polarity, as illustrated on the right (Table 1.3).

When the sample is mostly distributed into the lower nonaqueous phase of the chloroform solvent system, a slightly more hydrophobic solvent system of n-hexane/EtOAc/MeOH/H$_2$O (where $Et = C_2H_5$) (1:1:1:1) should be tested, as indicated by an upward arrow in the table. Further search should be directed upward, if the sample is still largely distributed into the upper nonaqueous phase, or downward if the sample is more concentrated in the lower aqueous phase. If the continued search leads to the top solvent system of n-hexane/MeOH/H$_2$O (2:1:1) and still indicates the need for a more hydrophobic solvent system, the solvent composition can be further modified by reducing the volume of water and/or replacing methanol by ethanol.

On the other hand, if the sample is mostly distributed in the upper aqueous phase of the chloroform solvent system, the search should be directed in the opposite direction, toward the polar solvent systems as indicated by the downward arrow. If the most polar solvent system of n-BuOH/H$_2$O listed at the bottom of the table still partitions the sample largely into the lower aqueous phase, the n-butanol solvent system may be modified by adding a small amount of an acid and/or salt. The most commonly used among these modified solvent systems are n-butanol/acetic acid/water (4:1:5), n-butanol/trifluoroacetic acid/water (1:0.001–0.01:1), and n-butanol/0.2 M ammonium acetate (1:1), all of which have been extensively used for the separation of peptides (53). Among those, n-butanol/acetic acid/water (4:1:5) requires the application of a reversed elution mode for the multilayer coil in the Type J HSCCC centrifuge.

Partitioning of macromolecules and cell particles can be performed with a variety of aqueous–aqueous polymer phase systems (54). Among the various polymer phase systems that are available, the following two types are the most versatile for performing CCC: Both PEG–potassium phosphate and PEG–sodium phosphate systems produce satisfactory stationary phase retention at

Table 1.3. Search for the Suitable Two-Phase Solvent System

n-Hexane	:	EtOAc	:	MeOH	:	n-BuOH	:	H₂O
10	:	0	:	5	:	0	:	5
9	:	1	:	5	:	0	:	5
8	:	2	:	5	:	0	:	5
7	:	3	:	5	:	0	:	5
← 6	:	4	:	5	:	0	:	5
5	:	5	:	5	:	0	:	5
→ 4	:	5	:	4	:	0	:	5
3	:	5	:	3	:	0	:	5
2	:	5	:	2	:	0	:	5
← 1	:	5	:	1	:	0	:	5
0	:	5	:	0	:	0	:	5
→ 0	:	5	:	0	:	1	:	5
0	:	4	:	0	:	2	:	5
0	:	3	:	0	:	3	:	5
0	:	2	:	0	:	4	:	5
0	:	1	:	0	:	5	:	5
0	:	0	:	0	:	5	:	5

CHCl₃—MeOH—H₂O(2:1:1)

a relatively high flow rate of the mobile phase and also provide a convenient means of adjusting the partition coefficient of macromolecules by changing the molecular weight of PEG and/or pH of the phosphate buffer (see Chapter 14). The PEG 8000 (3.5–4%, w/w)/dextran 48 (5%, w/w) systems provide a physiological environment suitable for separating mammalian cells by optimizing osmolarity and pH with electrolytes.

For the separation of ionic compounds, the pH of the solvent may often become a critical factor in peak resolution. At the pH near pK_a, these molecules are present in two forms, ionized and unionized, each having its own partition coefficient: Since these two species constantly interchange to maintain their equilibrium state during separation, they tend to form a broad peak. This can be prevented by choosing a pH for the solvent remote from the pK_a of the compounds.

The partition coefficient of the compound can be determined by a simple test tube procedure: A known amount of the sample is thoroughly equilibrated with the two-phase solvent system, 1–2 mL of each phase, in a test tube. After the clear two layers are formed, an aliquot of each phase is mixed with a suitable solvent (such as water or methanol) and the absorbance is determined with a spectrophotometer (55). In addition to the ultraviolet (UV) and visible wavelengths, fluorescence, radioactivity, enzymatic activity, and various types of bioassay can also be used for determination of the partition coefficients.

When a pure compound is not available, the above method only gives an average K value of multiple components, which may not be useful for a practical separation. In this case, an aliquot of each phase preequilibrated with the sample is separately analyzed with high-performance liquid chromatography (HPLC) or thin-layer chromatography (TLC). From a pair of chromatograms obtained from each phase, the K value for each component is obtained by computing the ratio in height or area between the corresponding peaks (56).

1.5.2. Preparation of Sample Solution

The sample solution is usually prepared by dissolving the sample mixture in the solvent to be used for separation. When the sample size is small and the desired component has a low $K(C_s/C_m)$ value, the sample may be dissolved in the stationary phase. However, it is recommended to dissolve the sample in a solvent mixture consisting of equal volumes of both upper and lower phases for the followng reasons: When a large amount of sample is dissolved in the solvent, the physical properties of the two-phase solvent system is substantially altered and in the extreme case the solvent forms a single phase. Introduction of such sample solution into the separation column would result in a detrimental loss of the stationary phase. This can be prevented by dissolving the sample in a mixture of both phases, since one can observe single-phase formation in the sample bottle. Also, when the sample mixture contains multiple components with a wide range of polarity, the use of the two phases can minimize the volume of the sample solution, hence improving the peak resolution (57).

In contrast to HPLC, CCC permits introduction of a fair amount of undissolved material into the separation column. However, in some cases, the presence of solid particles in the sample solution may adversely affect the retention of the stationary phase. Therefore, when a large amount of suspended matter is present in the sample solution, the elimination of the particles by means of filtration or centrifugation will improve the results.

The amount of the sample that can be applied to the separation column without seriously affecting the peak resolution differs significantly in the two types of the apparatus: The Type J HSCCC centrifuge with the standard semipreparative column (1.6 mm i.d.) of about 300-mL capacity may take 200–300 mg of sample in a 10-mL sample volume, provided that the settling time of the sample solution is considerably below 30 s. In contrast, the Type XLL CPC with a pair of large preparative columns (2.6 mm i.d. and over 1000 mL in capacity) may separate multigram quantities of sample dissolved in over 50 mL of solvent, even if the settling time far exceeds 30 s (58–61).

1.5.3. Elution

Countercurrent chromatography separation is usually performed with the standard procedure as follows: The column is first entirely filled with the stationary phase followed by injection of the sample solution through the sample port. Then, the mobile phase is pumped into the column in a proper elution mode while the apparatus is rotated at the optimum revolution speed. Although the sample solution may be introduced after the column has been equilibrated with the two phases as in HPLC, the method usually fails to improve the results.

For both Type J and cross-axis CPCs, satisfactory retention of the stationary phase is obtained under a proper elution mode at an optimum flow-rate of the mobile phase. The optimum experimental conditions such as elution modes and revolution speeds to be applied for various two-phase solvent systems in these instruments are described in detail in Sections 1.4.1. and 1.4.2. and are summarized in the phase distribution diagrams (Figs. 1.19, 1.20 and 1.21). While the HSCCC permits a choice in two elution modes, the head-to-tail elution usually produces better peak resolution by retaining a larger volume of the stationary phase in the column. In addition, the tail-to-head elution may produce a reversed pressure gradient in the column where the outlet shows the highest pressure and the inlet shows the lowest. In some situations, the pressure at the inlet of the column drops below atmospheric, resulting in sucking an excess amount of the mobile phase from the reservoir through the one-way check valves of the metering pump. This tendency is enhanced at a high revolution speed especially when the applied solvent system has a large density difference between the two phases. These complications, however, can be easily eliminated by applying a piece of narrow-bore tubing, typically 0.3 mm i.d. × 1 m, at the outlet of the monitor (8, 62).

Analogous to HPLC, CCC fractionation permits the use of either stepwise or gradient elution provided that the following precautions are taken. In the CCC column, the mobile and stationary phases maintain a subtle hydrodynamic equilibrium that may be disturbed by introduction of a new mobile phase with a modified phase composition. This may in turn alter the volume ratio of the two solvent phases in the separation column resulting in steady carryover or depletion of the stationary phase. However, this complication can be minimized by a careful selection of an applied solvent that can maintain the volume of the stationary phase in the column unaltered. The most commonly used solvent systems for the gradient elution in HSCCC include n-butanol/aqueous solution for the gradient of dichloroacetic acid or trifluoroacetic acid and n-butanol/phosphate buffer for a pH gradient (7, 8).

1.5.4. Detection

In the past on-line detection in HSCCC has been almost entirely performed with a UV–VIS absorbance monitor. In our laboratory, an LKB Uvicord S, Shimadzu LC-SD, and ISCO model 1840 variable wavelength monitor have been satisfactorily employed. In all these monitors, the flow cells should be of a standard straight flow-through type with a vertical flow-path, since the U-shape flow-cell designed for the analytical HPLC has a tendency to trap droplets of the stationary phase causing extensive noise in recording. To avoid trapping the stationary phase in the flow-cell, the lighter (mobile) phase should be introduced from the top of the flow-cell downwards and the heavier (mobile) phase, from the bottom of the flow-cell upwards.

On many occasions, on-line monitoring in CCC is disturbed by carryover of the droplets of the stationary phase. In addition, the mobile phase is in a subtle equilibrium with the stationary phase and a slight change in temperature may cause cloudiness of the effluent. However, these complications can be largely eliminated by heating the effluent at the inlet of the monitor while applying a narrow-bore tubing at the outlet of the monitor to raise the back pressure sufficient to suppress gas bubble formation in the flow-cell (62).

Recently, various other detection methods have been reported that include postcolumn reaction (63), laser light scattering detection (64, 65), HSCCC/MS

(34), and HSCCC/FTIR (66, 67). Most of these detection methods are much less sensitive to the carryover of the stationary phase and, therefore, will improve the quality of tracing of the elution curves in HSCCC.

REFERENCES

1. L. C. Craig, *Comprehensive Biochemistry*, **Vol. 4**, Elsevier, Amsterdam, London, and New York, 1962, p. 1.

2. Y. Ito and R. L. Bowman, *Science*, **167**, 281 (1970).

3. T. Tanimura, J. J. Pisano, Y. Ito, and R. L. Bowman, *Science*, **169**, 54 (1970).

4. Y. Ito and R. L. Bowman, *Science*, **173**, 420 (1971).

5. Y. Ito, *J. Biochem. Biophys. Met.*, **5**, 105 (1981).

6. Y. Ito, *J. Chromatogr.*, **214**, 122 (1981).

7. Y. Ito, J. Sandlin, and W. G. Bowers, *J. Chromatogr.*, **244**, 247 (1982).

8. Y. Ito, *CRC Crit. Rev. Anal. Chem.*, **17**, 65 (1986).

9. Y.-W. Lee, C. E. Cook, and Y. Ito, *J. Liq. Chromatogr.*, **11**, 37 (1988).

10. Y. Ito, *J. Liq. Chromatogr.*, **8**, 2131 (1985).

11. M. Bhatnagar and Y. Ito, *J. Liq. Chromatogr.*, **11**, 21 (1988).

12. H. Oka, K.-I. Harada, M. Suzuki, H. Nakazawa, and Y. Ito, *Anal. Chem.*, **61**, 1998 (1989).

13. H. Oka, K.-I. Harada, M. Suzuki, H. Nakazawa, and Y. Ito, *J. Chromatogr.*, **482**, 197 (1989).

14. H. Oka, K.-I. Harada, M. Suzuki, H. Nakazawa, and Y. Ito, *J. Chromatogr.*, **538**, 213 (1991).

15. Y. Ito, E. Kitazume, M. Bhatnagar, and F. D. Trimble, *J. Chromatogr.*, **538**, 59 (1991).

16. Y. Shibusawa and Y. Ito, *J. Chromatogr.*, **550**, 695 (1991).

17. Y.-W. Lee, Y. Shibusawa, F. T. Chen, J. Myers, J. M. Schooler, and Y. Ito, *J. Liq. Chromatogr.*, **15**, 2831 (1992).

18. A. Weisz, A. L. Scher, K. Shinomiya, H. M. Fales, and Y. Ito, *J. Am. Chem. Soc.*, **116**, 704 (1994).

19. Y. Ito, K. Shinomiya, H. M. Fales, A. Weisz, and A. L. Scher, Abstract 54P in *the 1993 Pittsburgh Conference and Exposition on Analytical Chemistry and Applied Spectroscopy* in Atlanta GA on March 7, 1993.

20. Y. Ito, K. Shinomiya, H. M. Fales, A. Weisz, and A. L. Scher, Presented at the Symposium on Countercurrent Chromatography in *the 206th ACS National Meeting*, Chicago, IL, August 22–27, 1993.

21. Y. Ito, Y. Shibusawa, H. M. Fales, and H. J. Cahnmann, *J. Chromatogr.*, **625**, 177 (1992).

22. Y. Ito and R. Bhatnagar, *J. Liq. Chromatogr.*, **7**, 257 (1984).

23. Y. Ito, *J. Liq. Chromatogr.*, *11*, 1 (1988).

24. Y. Ito, *Sep. Sci. Tech.*, **22**, 1989 (1987).

25. Y. Ito, *Sep. Sci. Tech.*, **22**, 1971 (1987).

26. Y. Ito, H. Oka, and J. L. Slemp, *J. Chromatogr.*, **463**, 305 (1989).

27. Y. Ito, *J. Chromatogr.*, **358**, 313 (1986).

28. Y. Ito, *J. Chromatogr.*, **358**, 325 (1986).

29. W. D. Conway and Y. Ito, Presented at *the 1984 Pittsburgh Conference and Exposition on Analytical Chemistry and Applied Spectroscopy*, Abstract No. 475, 1984.

30. Y. Ito and Y.-W. Lee, *J. Chromatogr.*, **391**, 290 (1987).

31. Y.-W. Lee, Y. Ito, Q.-C. Fang, and C. E. Cook, *J. Liq. Chromatogr.*, **11**, 75 (1988).

32. H. Oka, Y. Ikai, N. Kawamura, M. Yamada, K.-I. Harada, M. Suzuki, F. E. Chou, Y.-W., Lee, and Y. Ito, *J. Liq. Chromatogr.*, **13**, 2309 (1990).

33. H. Oka, Y. Ikai, N. Kawamura, M. Yamada, J. Hayakawa, K.-I. Harada, K. Nagase, H. Murata, M. Suzuki, and Y. Ito, *J. High Resol. Chromatogr. Chromatogr. Commun.*, **14**, 306 (1991).

34. H. Oka, F. Oka, and Y. Ito, *J. Chromatogr.*, **479**, 53 (1989).

35. Y.-W. Lee, R. D. Voyksner, Q.-C. Fang, C. E. Cook, and Y. Ito, *J. Liq. Chromatogr.*, **11**, 153 (1988).

36. Y.-W. Lee, R. D. Voyksner, C. E. Cook, Q.-C. Fang, T.-W. Pack, and Y. Ito, *Anal. Chem.*, **62**, 244 (1990).

37. Y.-W. Lee, T.-W. Pack, R. D. Voyksner, Q.-C. Fang, and Y. Ito, *J. Liq. Chromatogr.*, **13**, 2389 (1990).

38. H. Oka, Y. Ikai, N. Kawamura, J. Hayakawa, K.-I. Harada, H. Murata, M. Suzuki, and Y. Ito, *Anal. Chem.*, **63**, 2861 (1991).

39. Y. Ito and H. Oka, *J. Chromatogr.*, **457**, 305 (1989).

40. M. Knight and Y. Ito, *J. Chromatogr.*, **484**, 319 (1989).

41. Y. Ito and F. E. Chou, *J. Chromatogr.*, **454**, 382 (1988).

42. Y. Ito, H. Oka, and J. L. Slemp, *J. Chromatogr.*, **474**, 219 (1989).

43. Y. Ito, H. Oka, and T.-W. Lee, *J. Chromatogr.*, **498**, 169 (1990).

44. Y. Shibusawa and Y. Ito, *J. Liq. Chromatogr.*, **15**, 2787 (1992).

45. Y. Shibusawa and Y. Ito, unpublished data.

46. Y. Ito, *J. Chromatogr.*, **301**, 377 (1984).

47. Y. Ito, *J. Chromatogr.*, **301**, 387 (1984).

48. Y. Ito and T.-Y. Zhang, *J. Chromatogr.*, **449**, 135 (1988).

49. Y. Ito, *J. Chromatogr.*, **538**, 67 (1991).

50. Y. Ito and W. D. Conway, *J. Chromatogr.*, **301**, 405 (1984).

51. F. Oka, H. Oka, and Y. Ito, *J. Chromatogr.*, **538**, 99 (1991).

52. Y. Ito, "Countercurrent chromatography," in *Journal of Chromatography Library*, **Vol. 51A**, E. Heftmann, Ed., Elsevier Scientific, Chapter 2, 1992, pp. 69–105.

53. M. Knight, "Countercurrent Chromatography for Peptides," in *Countercurrent Chromatography: Theory and Practice* N. Bhushan Mandava and Y. Ito, Eds., Marcel-Dekker, New York, Chapter 10, 1988, pp. 583–615.

54. P. Å. Albertsson, *Partition of Cell Particles and Macromolecules*, Wiley- Intersciences, New York, 1986.

55. W. D. Conway and Y. Ito, *J. Liq. Chromatogr.*, **7**, 275 (1984).

56. W. D. Conway and Y. Ito, *J. Liq. Chromatogr.*, **7**, 291 (1984).

57. J. L. Sandlin and Y. Ito, *J. Liq. Chromatogr.*, **7**, 323 (1984).

58. Y. Ito and T.-Y. Zhang, *J. Chromatogr.*, **449**, 153 (1988).

59. T.-Y. Zhang, Y.-W. Lee, Q.-C. Fang, C. E. Cook, R. Xiao, and Y. Ito, *J. Chromatogr.*, **454**, 185 (1988).

60. Y. Ito and T.-Y. Zhang, *J. Chromatogr.*, **455**, 151 (1988).

61. M. Bhatnagar, H. Oka, and Y. Ito, *J. Chromatogr.*, **463**, 317 (1989).

62. H. Oka and Y. Ito, *J. Chromatogr.*, **475**, 229 (1989).

63. E. Kitazume, M. Bhatnagar, and Y. Ito, *J. Chromatogr.*, **538**, 133 (1991).

64. D. E. Schaufelberger, T. G. McCloud, and J. A. Beutler, *J. Chromatogr.*, **538**, 87 (1991).

65. S. Drogue, M.-C. Rolet, D. Thiébaut, and R. Rosset, *J. Chromatogr.*, **538**, 91 (1991).

66. R. J. Romanach, J. A. de Haseth, and Y. Ito, *J. Liq. Chromatogr.*, **8**, 2209 (1985).

67. R. J. Romanach and J. A. de Haseth, *J. Liq. Chromatogr.*, **11**, 133 (1988).

CHAPTER

2

ANALYTICAL HIGH-SPEED COUNTERCURRENT CHROMATOGRAPHY

DANIEL E. SCHAUFELBERGER

The R. W. Johnson Pharmaceutical Research Institute, Cilag Ltd., CH-8201 Schaffhausen, Switzerland

2.1. INTRODUCTION

Countercurrent chromatography (CCC) plays an important role in separation science since recent improvements in instrumentation have increased speed and efficiency. Among various types of CCC, high-speed countercurrent chromatography (HSCCC) is the leading technique. Instrumentation for HSCCC is also known as the "multilayer coil planet centrifuge" (MLCPC) and was originally designed by Ito and co-workers (1, 2) at the National Institutes of Health (NIH). The HSCCC has since become a method of choice in natural products chemistry and has made possible the isolation of a number of biologically interesting natural products that were difficult, if not impossible, to isolate by other techniques. Theory and applications of HSCCC have been described in a number of review articles and are also the subject of other contributions to this volume.

Surprisingly, little attention has been given to analytical HSCCC. In this chapter, applications of analytical HSCCC are reviewed and the potential of this method is assessed.

2.2. DEVELOPMENT OF ANALYTICAL HIGH-SPEED COUNTERCURRENT CHROMATOGRAPHY

Analytical scale CCC was first carried out by Ito and Bowman (3) by means of the hydrostatic toroidal coil centrifuge (TCC). The instrument consisted of a helical coil of Teflon tubing (0.38 mm i.d.) with a length of 24 m and a total

High-Speed Countercurrent Chromatography, Edited by Yoichiro Ito and Walter D. Conway.
Chemical Analysis Series, Vol. 132.
ISBN 0-471-63749-1 © 1996 John Wiley & Sons, Inc.

capacity of 3 mL. The technique has recently been compared to analytical HSCCC (4). The TCC has also been used to measure partition coefficients (5). In another example, Kubo et al. (6) suggested the use of droplet countercurrent chromatography (DCCC) for the screening of phytoecdysteroids in medicinal plants. Generally, very little attention has been given to analytical CCC, whereas numerous applications of preparative CCC have been reported over the past decade.

Analytical HSCCC was first introduced in 1987 by Ito and Lee (7). The authors built a small-scale version of the MLCPC with a shorter revolutional radius, equipped with 0.85-mm i.d. tubing and suitable for operation at over 2000 rpm. At the same time, Romañach and deHaseth (8) studied the effect of the tubing diameter on the stationary-phase retention in HSCCC. They concluded that 1.2-mm i.d. tubing was best for the analytical HSCCC. So far, most applications have been carried out with coils made of 0.85-mm i.d. tubing and with user-built equipment. However, there are now commercially available instruments, the latest example being a coil that combines preparative and analytical columns. Further improvements in instrumentation may be expected for the near future.

2.3. THEORY

For the purpose of the following discussion, we define analytical HSCCC as a separation technique based on dynamic liquid–liquid partition chromatography using a MLCPC with tubing consisting of an internal diameter of $\leqslant 1.0$ mm and with a total coil volume of less than or equal to 50 mL.

This definition covers current applications but may have to be corrected towards smaller coil volumes, depending on future trends. By limiting the definition to technical parameters we can imagine two types of applications in analytical HSCCC: First, strictly analytical separations that provide information about the composition of a sample or about its chromatographic behavior and, second, "microscale" separations with recovery of fractions for further analysis. Basics of CCC have been studied extensively and are well documented (9–15). The theory of CCC also applies to analytical HSCCC, therefore we limit our discussion to the most important parameters.

The content of a coil (V_c) is composed of the volume of the stationary phase (V_s) and of the volume V_0 which the mobile phase takes up in the coil.

$$V_c = V_s + V_0 \tag{2.1}$$

The ratio of the stationary-phase volume over the coil volume is called the stationary-phase fraction (S_F).

$$S_F = \frac{V_s}{V_c} \tag{2.2}$$

The stationary-phase fraction depends on the solvent system and on experimental parameters, such as the flow-rate, the speed of revolution, and the β value of the coil (see Section 2.4.1). Typical values for S_F range between 0.5 and 0.9. Often, the stationary-phase fraction is expressed as "percent retention" $[=(V_s/V_c) \times 100\%]$. Conway and Ito (16) pointed out the importance of achieving a high S_F since resolution increases with higher S_F. Small bore tubing in analytical instrumentation leads to higher solvent–wall interactions, which in turn result in a lower stationary-phase retention.

In the coil planet centrifuge (CPC) stationary and mobile phases form zones of fine dispersion alternating with zones of a more side-by-side arrangement of the two phases. There is a dynamic process of mixing and settling, mimicking the shaking and the settling of two phases in a separatory funnel. This effect depends on differences in the centrifugal forces along the coiled column. The centrifugal force distribution depends on the coil geometry (number of turns), as well as on the speed and on the type of rotation (13).

The chromatographic process in CCC is based on the partition of a solute between the two liquids that are used as the mobile and stationary phase, respectively. The partition coefficient (K) is therefore the most important parameter in HSCCC. The parameter K is defined as the ratio of the concentration of a solute in the stationary phase (C_s) to the concentration in the mobile phase (C_m).

$$K = \frac{C_s}{C_m} \tag{2.3}$$

Partition coefficients of around 1 are desirable in HSCCC. Higher K values provide higher resolution, but practical considerations, such as increased run time and peak broadening due to diffusion speak against solvent systems with $K > 3$. Partition coefficients less than 1 result in faster separations and are indicated if less resolution is needed. As the retention of a solute in CCC primarily depends on its partition coefficient, retention times and retention volumes are predictable. It is also possible to calculate the K value of a solute from its CCC chromatogram.

For a better understanding of the theory of HSCCC, we use the help of a hypothetical separation of five compounds (1–5), as illustrated by the chromatogram shown in Fig. 2.1. For simplicity, we assume a total coil volume (V_c) of 40 mL, a flow rate of 1 mL/min, and a stationary-phase fraction of 0.75. Under such conditions the nonretained solute 1 $(K = 0)$ elutes with the "solvent front" at t_0 with the corresponding elution volume $V_0 = 10$ mL. The volume V_0 is equal to the volume of the mobile phase in the coil. Compound 2 with a partition coefficient of 1 elutes at $V_R = 40$ mL. This volume corre-

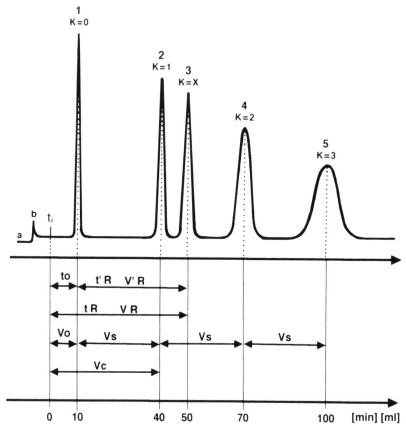

Figure 2.1. Hypothetical separation of "compounds" **1–5** by analytical HSCCC; the following parameters are assumed: a coil volume of 40 mL, a flow-rate of 1 mL/min, and a stationary phase retention of 75%. Partition coefficients of compounds **1–5** are noted in the chromatogram; a = displacement of stationary phase; b = break through of mobile phase; t_i = sample injection.

sponds to the total coil volume (V_c) according to the general rule that a solute with $K = 1$ elutes once the equivalent of one coil volume (V_c) of mobile phase has passed through the coil.

$$V_R = V_c \quad \text{if} \quad K = 1 \tag{2.4}$$

The elution at V_c is independent of the stationary phase fraction and of the flow-rate. Conway (12) considers V_c as the "hub of the CC chromatogram" and emphasizes the centrality of $K = 1$. The retention time of a solute is expressed in Eqs. (2.5) and (2.6).

$$t_R = \frac{V_c}{f} \quad \text{if} \quad K = 1 \tag{2.5}$$

$$t_R = \frac{V_c}{f}[1 + S_F(K - 1)] \quad \text{if} \quad K \neq 1 \tag{2.6}$$

In the example of Fig. 2.1, Compound 4 ($K = 2$) elutes at $V_R = 70\,\text{mL}$. This volume corresponds to $V_0 + 2V_s$, as calculated by Eq. (2.6). Under the same conditions, a compound with $K = 4$ would elute at $V_R = 130\,\text{mL}$.

The partition coefficient K of a solute can be calculated from chromatographic parameters using Eq. (2.7). Consequently, HSCCC represents a convenient method to determine partition coefficients (see Section 2.5.2).

$$K = \frac{t_R - t_0}{t_0}\left(\frac{1 - S_F}{S_F}\right) = \frac{V_R - V_0}{V_0}\left(\frac{1 - S_F}{S_F}\right) \tag{2.7}$$

In the example of Fig. 2.1, Compound 3 has a retention volume of 50 mL. Using Eq. (2.7), a partition coefficient of approximately 1.3 is calculated for Compound 3. For more precise calculations, especially if small analytical coils are used, contributions of the feeding lines and of the sample loop to the retention volumes have to be taken into account.

The capacity factor k' is related to K and to the stationary-phase fraction as expressed in Eq. (2.8).

$$k' = \frac{V_R - V_0}{V_0} = K\frac{V_s}{V_0} = K\frac{S_F}{1 - S_F} \tag{2.8}$$

The separation factor or selectivity (α) is defined (16) as

$$\alpha = \frac{K_1}{K_2} \quad \text{where} \quad K_2 > K_1 \tag{2.9}$$

The efficiency (number of theoretical plates) can be calculated by Eq. (2.10), where t_R is the retention time of the solute and W is its 4σ base width.

$$N = 16\left(\frac{t_R}{W}\right)^2 \tag{2.10}$$

Efficiencies reported for preparative HSCCC separations are in the order of $N = 300$–1000. So far, analytical instrumentation yielded disappointingly low

plate numbers, often not exceeding 1000. High plate numbers in the order of
5000–10000 have been reported by Ito et al. (17) and by Oka et al. (18).
Calculations of theoretical plate numbers using Eq. (2.10) are misleading
in separations with a low stationary phase retention. In the case of a low
S_F, values for t_0 drastically increase because a nonretained solute has to
pass through a large section of the coil filled with mobile phase. Consequently, retention times (t_R) increase by the amount of t_0 and lead to artifi-
cially high plate numbers pretending high efficiencies. The term N_{ccc}
(theoretical plates corrected for CCC) compensates for this effect and is
presented here as an alternative to calculate theoretical plate numbers in
HSCCC.

$$N_{ccc} = 16\left(\frac{t'_R}{W}\right)^2 \qquad \text{where} \quad t'_R = t_R - t_0 \qquad (2.11)$$

Berthod (19) estimates that the maximum theoretical efficiency in CCC is
equal to one plate per turn. Under this aspect it is very unlikely that HSCCC
could become an efficient method comparable to analytical high-performance
liquid chromatography (HPLC).

The term for resolution (R), known from HPLC, also applies to
HSCCC(10).

$$R = 2\left(\frac{t_2 - t_1}{W_2 + W_1}\right) \qquad (2.12)$$

The influence of selectivity (α), efficiency (N), partition coefficient (K), and
stationary phase retention (S_F) has been summarized by Conway (12) in Eq.
(2.13):

$$R = \tfrac{1}{4}(\alpha - 1)N\left[K_1/K_1\left(\frac{\alpha+1}{2}\right) + \left(\frac{1-S_F}{S_F}\right)\right] \qquad (2.13)$$

where K_1 is the partition coefficient of the first eluting solute of a pair of two. It
follows from Eq. (2.13) that a high stationary phase fraction is required to
obtain high resolution. Beside this parameter, inherent to CCC, Eq. (2.13) is
similar to the "Knox" equation used in HPLC. In both methods, selectivity
has a very pronounced impact on resolution. This is also the reason why
HSCCC has an important advantage over other techniques: The almost unlimited choice of solvent systems and the option to add modifiers to one or to
both phases provides an exellent control over selectivity. For a more detailed
discussion of the theory of HSCCC the reader is referred to the articles of

Conway and co-worker (11, 12, 16), Ito et al. (14), and Foucault and co-workers (20).

2.4. PRACTICAL CONSIDERATIONS

2.4.1. Instrumentation

Instrumentation for analytical HSCCC consists of a multilayer CPC and of auxillary equipment, such as injectors, pumps, and detectors. Single-layer coils have been used, but primarily for stationary-phase retention studies. In HSCCC, the "column"—more commonly referred to as the "coil"—consists of polytetrafluoroethylene (PTFE) tubing (or of other suitable material),

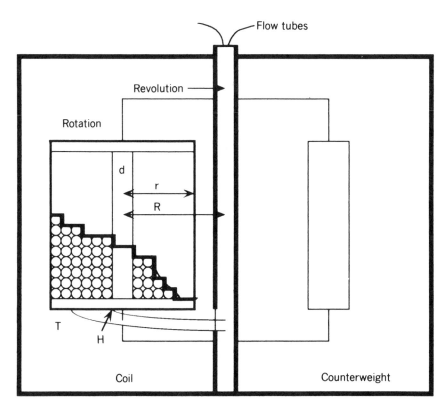

Figure 2.2. Schematic cross section of a MLCPC for analytical HSCCC; β = ratio r/R; d = holder diameter; H = "head" inlet/outlet; r = helical radius; R = revolutional radius; T = "tail" inlet/outlet.

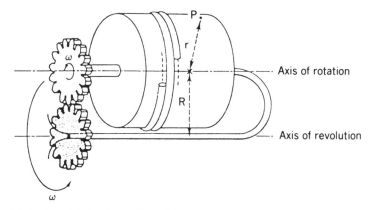

Figure 2.3. Design of the HSCCC-4000 coil. [Reproduced with permission from Oka et al. (4)].

which is wound in numerous turns onto a column holder. The coil, when operated, is rotating in a planetary motion around its own axis of rotation, as well as around the central axis of revolution (see Figs. 2.2 and 2.3). The geometry of a coil is best expressed by its β value, which is defined as the ratio of the revolutional radius (R) to the helical radius (r). Compared to preparative HSCCC, helical radii are smaller in analytical coils in order to accommodate the higher rotational speeds. The interaction between solvents and the tubing wall is increased in narrow-bore tubing, and therefore higher centrifugal forces (speeds) are required to obtain sufficient stationary-phase retention. Rotational speeds are in the order of 1000–2000 rpm, or even higher as it is the case for the HSCCC-4000 coil (21). A schematic drawing of the HSCCC-4000 coil is reproduced in Fig. 2.3 and the performance of this coil is illustrated by the separation of indole auxins (Fig. 2.4). In general, column lengths for analytical HSCCC range from 1 to 100 m depending on how many turns and how much coil volume is required. Different sizes of small-bore tubing (0.38–1.2 mm i.d.) have been evaluated by Romañach and de Haseth (8). Currently, tubing of 0.85 mm i.d. is a standard for analytical coils. Each "coil" has a "head" inlet/outlet and a "tail" inlet/outlet. By convention, the "tail" of a coil is the end of the tubing that experiences the higher centrifugal forces (=end towards which the heavy phase has a tendency to flow). This "tail" inlet is selected if the upper phase is used as the mobile phase. Table 2.1 summarizes typical parameters for an analytical multilayer coil centrifuge of 30–40 mL of coil volume. Parameters like stationary-phase retention, rotation, and flow-rate depend on each other, while other parameters depend rather on the dimensions of the coil. For a better understanding of parameters, the classification by Berthod (22) is recommended (see Table 2.2).

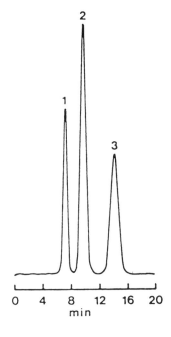

Figure 2.4. Separation of indole auxines by analytical HSCCC. Instrument: HSCCC-4000 with 0.55-mm i.d. tubing and 5.0-mL coil volume; solvent: n-hexane/ethyl acetate/MeOH/H_2O (1:1:1:1), lower phase = mobile phase; flow-rate 0.3 mL/min; detection: UV 260 nm; sample: **1** = indole-3-acetamide, **2** = indole-3-acetic acid, **3** = indole-3-butyric acid, 10–100 μg, each compound. [Reproduced with permission from Oka et al. (4)].

Table 2.1. Typical Parameters in Analytical HSCCC

Column (= coil)	Multilayer coil planet centrifuge
Coil volume	30–40 mL
Tubing	50–70 m × 0.85 mm i.d. PTFE
Flow-rate	0.5–1.0 mL/min
Back pressure	< 300 psi
Retention	50–90% (S_F:0.5–0.9)
Rotation	1500–2000 rpm
Sample	1 μg–10 mg
Run time	20–60 min

Table 2.2. Classification of Parameters in HSCCC[a]

Configurational Parameters	Operating Parameters	
	Active	Passive
Tube diameter (internal)	Flow rate	Driving pressure
Coil radius	Rotation	Stationary Phase retention
Number of turns		
Total internal volume		

[a]After Berthod (22).

Table 2.3. Applications of Analytical HSCCC

Configurational Parameters	Operational Parameters	Sample/Solvent	References
1. Single-Layer Coils			
1.1. 0.10–0.55 mm i.d.[a] 0.05–1.2 mL	0.025–0.5 mL/min; 3500 rpm	Stationary-phase retention study	Oka et al. (18)
1.2. 0.85 mm i.d. × 3.5–4 m ∼2 mL; β = 0.28–0.77	1 mL/min; 2000–4000 rpm	Stationary-phase retention study	Oka et al. (21)
1.3. 0.85 mm i.d. × 5 m β = 0.5–0.8	0.2–2.0 mL/min; 500–2000 rpm	Stationary-phase retention study	Ito and Lee (7)
2. Multilayer Coils			
2.1. 0.4 mm i.d.; 1.5 mL, or 0.55 mm i.d.; 5 mL, or 0.85 mm i.d.; 8 mL	0.1–0.5 mL/min; 3500 rpm	Indole auxins *n*-hexane/EtOAc/MeOH/H_2O (1:1:1:1)	Oka et al. (4)
2.2. 0.30 mm i.d.; 6 mL[a] β = 0.51–0.77	0.1–0.3 mL/min; 3500 rpm	Indole auxins *n*-hexane/EtOAc/MeOH/H_2O (1:1:1:1)	Oka et al. (18)
2.3. 0.85 mm i.d. × ∼15 m 8 mL; β = 0.51–0.77; R = 2.5 cm 4 coiled layers	2 mL/min; 3500 rpm	Flavonoids $CHCl_3$/MeOH/H_2O (4:3:2)	Oka et al. (21)
2.4. 1.07 mm i.d. × 13.7 m[b] 14.7 mL; β = 0.85	1.9 mL/min; 1200 rpm	*p*-Nitrophenols 1-BuOH–0.2 *M* $KHCO_3$ pH = 8.6 (1:1)	Conway et al. (23)
	1.2 mL/min; 1200 rpm	Phenylamines heptane/1-BuOH/H_2O	
2.5. 0.85 mm i.d. × ∼50 m[c] ∼30 mL	0.4 mL/min; 1200 rpm	Nucleosides 1-BuOH/H_2O	Conway et al. (39)
	0.5 mL/min; 1800 rpm	Podophyllotoxin hexane/EtOAc/MeOH/H_2O (1:1:1:1)	Schaufelberger et al. (59)

54

No.	Coil	Flow; rotation	Application / solvent system	Reference
2.6.	0.85 mm i.d.; 38 mL β=0.5–0.7	1 mL/min; 2000 rpm	Bryostatins (macrolides) hexane/isoPrOH/MeOH/H_2O (4:1:1.6:0.4)	Schaufelberger (62)
			Indole auxins n-hexane/EtOAc/MeOH/H_2O (3:7:5:5)	Ito and Lee (7)
2.7.	0.85 mm i.d. × ~70 m[d] ~40 mL; β=0.4–0.75 R = 6.4 cm	1 mL/min; 1800 rpm	Anthraquinones n-hexane/EtOAc/MeOH/H_2O (9:1:5:5)	Zhang et al. (55)
			Alkaloids n-hexane/EtOAc/MeOH/H_2O (1:1:1:1), (3:7:5:5)	Zhang et al. (54)
2.8.	0.85 mm i.d.; ~38 mL	1–5 mL/min; 1800 rpm 2000 rpm	Flavonoids $CHCl_3$/MeOH/H_2O	Zhang et al. (37)
			s-Triazines n-hexane/EtOAc/MeOH/H_2O (8:2:5:5)	Lee et al. (56)
			Vinca alkaloids hexane/EtOH/H_2O (6:5:5)	
			Lignans hexane/EtOH/H_2O (6:5:5)	Lee et al. (57)
2.9.	0.80 mm i.d. × 93.5 m[e] 47 mL; β=0.32–0.7	0.5 mL/min; 1850 rpm	(Nitro-)phenols $CHCl_3$/MeOH/H_2O (3:1:3)	Drogue et al. (34)
2.10.	0.85 mm i.d. × 140 m 90 mL; β=0.5	0.5–1.0 mL/min; 600–1000 rpm	Aromatic hydrocarbons hexane/MeOH/H_2O	Romañach and de Haseth (8)
2.11.	1.6 mm i.d.; 107 mL[f]	3.8 mL/min; 1000–1200 rpm	Partition coefficient measurements 1-octanol/H_2O	Vallat et al. (51)

[a] HSCCC-4000 (21).
[b] "Tripple"™ Coil (contains 14, 75, and 215-mL columns in one coil), P.C. Inc., Potomac, MD.
[c] "Micro" high-speed countercurrent chromatograph, P.C. Inc., Potomac, MD.
[d] CCC-2000 instrument, Pharma-tech, Baltimore, MD.
[e] CPHV 2000 HSCCC system (only one coil used of three connected in series), SFCCC, Neuilly-Plaisance, France.
[f] Ito multilayer coil separator–extractor, coil No. 14 shortened, P.C. Inc., Potomac, MD.

Fast rotation results in an increased wear and tear effect on the tubing material, especially in the section between the coil and the central shaft. Ito and Lee (7) solved this problem by using a pair of PTFE flow tubes of 0.5 mm i.d. and of 0.5-mm wall thickness. These flow-tubes were connected to the coil terminals by means of two connector pieces of 1.7-mm i.d. tubing. Additionally, the flow-tubes were lubricated with silicone grease and protected with a sheath of Tygon tubing. Oka et al. (18) used a twisted piece of PTFE tubing with 0.30 mm i.d. and 0.22-mm wall thickness and reinforced with a single piece of alkene heat-shrinkable tubing for the same purpose.

A new approach to analytical HSCCC has been proposed by Conway et al. (23) who used a coil that combines three coils within the same column holder: two preparative and one analytical unit. Capacities of these coils are 215, 75, and 15 mL, respectively. The analytical coil consists of 13.7-m × 1-mm i.d. tubing and has a β value of 0.85.

Table 2.3 lists, although not exhaustively, the different types of instruments used for analytical HSCCC. Some of the coils are commercially available.

2.4.2. Solvent Systems

Selecting a solvent system for HSCCC means simultaneously choosing the column and the eluent. This underlines the fact that the selection of a solvent system is the most important step in performing HSCCC. Despite a large number of procedures, this step is still somehow empirical, if not to say arbitrary. Analytical instrumentation with its speed and minimized solvent consumption offers a very promising way to carry out methods development in HSCCC. Analytical instrumentation may also provide an opportunity to study gradient elution, a very promising technique that had been introduced in CCC by Foucault and Nakanishi (24, 25).

In general, solvent systems known from preparative HSCCC can also be used for analytical scale separations. Stationary-phase retentions may be lower in analytical instruments due to higher solvent–wall interactions. Good results have been obtained with $CHCl_3/MeOH/H_2O$ and hexane/H_2O solvent systems. High viscosity as observed with $BuOH/H_2O$ systems may cause problems of high back pressure. To our knowledge analytical scale separations with aqueous–aqueous solvent systems have not yet been reported.

2.4.3. Handling

This section discusses some practical aspects of analytical HSCCC. Solvent mixtures are equilibrated at least 1 h prior to use. Filtration of the solvents is not necessary. In the case of hexane or other volatile solvents it is recommended to add to the reservoir containing one phase a few milliliters of the corresponding other phase. The lines are first flushed with stationary phase and the coil is filled with stationary phase (check at the coil outlet that the effluent forms two layers when mixed with mobile phase; residual solvents used for rinsing of the coil may disturb the phase equilibrium and may result in a failure to retain the stationary phase). The coil is rotated while stationary phase is still being pumped into it. Adjusting the counterweight may not be necessary because of the small coil volumes. Next, mobile phase is pumped into the coil and the effluent is monitored. The volume of stationary phase, which is displaced before the mobile phase breaks through (solvent front), is an indicator for the stationary-phase retention. The flow-rate is adjusted and once the detector response is stable, the sample is injected (see Fig. 2.1). It is recommended to dissolve the sample in a mixture of stationary and mobile phase. Sample loops and injector-valves known from HPLC may also be convenient for analytical HSCCC. Filtration of the sample is not necessary, however, it is important that the sample solution still shows phase separation. After the run, the stationary-phase fraction is determined by displacing the coil contents with a flow of nitrogen gas and by measuring the volumes of stationary and mobile phases. Instead of using nitrogen, the column contents may also be displaced by pumping mobile phase into the coil (rotation stopped). In this case, the volume of the displaced stationary phase is measured and the stationary-phase fraction is calculated. This additional rinsing of the tubing (instead of using air) is advantageous if crude samples were separated. The stationary-phase fraction represents an important parameter to describe the quality of a chromatographic system and should always be reported in the experimental section. It is also recommended to add a nonretained compound to the sample as a marker for t_0.

2.4.4. Detection

Several types of detectors have been used for analytical HSCCC (26). Ultraviolet (UV) detectors are commonly used, but carryover of droplets of non-retained stationary phase may cause a high noise level. Schaufelberger (27) suggested pumping an adjuvant solvent (e.g., methanol or isopropanol) to the coil effluent prior to the detector by means of a mixing tee. Thus, droplets of nonretained stationary phase are dissolved and detector noise is reduced. This is illustrated by the separation of naphthalene, benzophenone, o-nitrophenol, and acetophenone in hexane/methanol/water with isopropanol as the adjuvant solvent (Fig. 2.5). The UV spectrum of each compound was recorded on-line by means of a photodiode array detector. Spectra were similar to those recorded with single compounds in pure methanol. In another effort to reduce detector noise, a piece of small-bore PTFE tubing (3 m × 0.5 mm i.d.) was inserted between the coil and the UV detector and heated to 30° C (28).

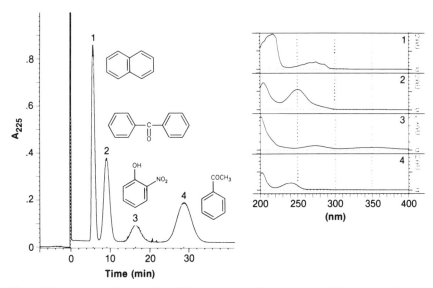

Figure 2.5. Separation of naphthalene (**1**), benzophenone (**2**), *o*-nitrophenol (**3**), and acetophenone (**4**); UV spectra were recorded on-line by means of a photodiode array detector. "Micro" HSCCC (0.85-mm i.d. tubing; 30-mL coil volume); solvent: hexane/MeOH/H$_2$O (3:3:2), upper phase = mobile phase; flow-rate: 1.0 mL/min; sample: 10–25 μg each compound); postcoil reactor with additional flow of isopropanol at 0.5 mL/min; detection: 225 nm. [Reprinted with permission from Schaufelberger (27)].

Analytical instruments have also been connected to a mass spectrometer (MS) (29, 30) or to a Fourier transform infrared (FT–IR) detector (31). These two detectors provide important structural information about solutes, but complex instrumentation (MS) or large sample sizes (IR) may restrict their use. Diallo et al. (32) combined HSCCC and thin-layer chromatography (TLC) for the automated analysis of fractions by means of a flow-splitter and a spray-on apparatus.

Recently, Schaufelberger et al. (33) and Drogue et al. (34) recommended evaporative laser-light scattering detection for HSCCC. Evaporative laser-light scattering (ELSD) represents a universal detection method that is commonly used in the HPLC of lipids and carbohydrates. The mobile phase is evaporated in the nebulizer compartment of the detector leaving behind fine particles of the solute that will scatter the light of a laser beam. The ELSD is a very promising and easy to use alternative to UV detection. Figure 2.6 shows the separation of microgram quantities of four natural products by HSCCC equipped with an ELSD. The nebulizer compartment of the detector was heated at 110° C to evaporate nonretained droplets of the aqueous stationary phase.

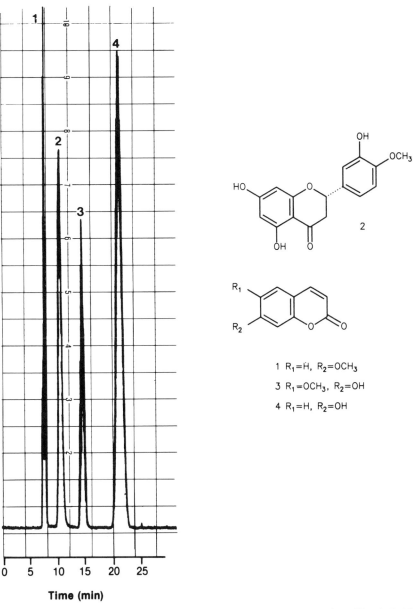

Figure. 2.6. Analytical HSCCC with laser-light scattering detection: separation of herniarin (**1**), hesperetin (**2**), scopoletin (**3**), and umbelliferone (**4**). "Micro" HSCCC (0.85-mm i.d. tubing; 30 mL); solvent; $CHCl_3/MeOH/H_2O$ (13:7:8), lower phase = mobile phase; flow-rate: 0.8 mL/min; sample: 300 μg (**1**), 10 μg (**2**), 35 μg (**3**), and 20 μg (**4**); detector: Varex ELSD detector; nebulizer and exhaust temperatures set at 110 and 80°C, respectively; carrier gas: nitrogen. [Reprinted with permission from Schaufelberger et al. (33)].

2.5. APPLICATIONS

2.5.1. Methods Development

Analytical and preparative instrumentation is available for most separation methods. This has not been the case in CCC, where until very recently, only apparatus for preparative separations have been known. Consequently methods development had to be carried out by alternative methods. The best approach, although time consuming, was to partition the sample between the two phases of a solvent system, followed by (semi-)quantitative analysis of the two layers. Under certain conditions even TLC provided useful information for the solvent selection process (35, 36). This disadvantage of CCC has now been put aside by the introduction of analytical HSCCC. Several contributions have already shown that analytical HSCCC is very useful for methods development. Zhang et al. (37, 38) studied the separation of flavonoids from Hippophaë rhamnoides by analytical and preparative HSCCC using 43- and 280-mL coils, respectively. Chromatograms are reproduced in Fig. 2.7a and b. The arrows indicate where a compound with a partition coefficient of $K = 1$

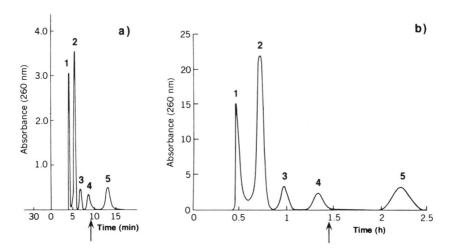

Figure 2.7. Analytical (a) and preparative (b) HSCCC of flavonoids in *Hippophae rhamnoides* extract: (a) CCC-2000 analytical apparatus with 0.85-mm i.d. tubing; coil volume: 43 mL; solvent: $CHCl_3/MeOH/H_2O$ (4:3:2), lower phase = mobile phase; flow-rate: 5 mL/min; 1800 rpm; stationary-phase retention: 68%; sample: 3 mg of crude extract. (b) Ito multilayer coil separator extractor (for preparative HSCCC) with 1.6-mm i.d. tubing; coil volume: 280 mL; solvent: same as above; flow-rate: 3.3 mL/min; 800 rpm; stationary-phase retention: 66%; sample: 100 mg of crude extract. Peak 2: isorhamnetin, peak 4: quercetin, peaks 1,3,5: unknown. Arrows indicate in both chromatograms the retention time that a compound with $K = 1$ would have. [After Zhang et al. (37) with modifications).

would elute. The two separations reveal an excellent correlation as in both chromatograms the mark for V_R ($K = 1$) is on the downslope of the quercetin peak. Sample loads were approximately 0.1 mg/mL of the retained stationary phase for the analytical instrument and 0.5 mg/mL of the retained stationary phase for the preparative instrument. The analytical separation was carried out with 3 mg of sample and was achieved in 15 min. This example clearly demonstrates the advantage of analytical instrumentation for methods development in preparative HSCCC. In other applications, Conway et al. (23, 39) used the Tripple™ coil to evaluate 1-BuOH/H₂O and heptane/1-BuOH/H₂O solvent systems for the separation of nucleosides and of drug metabolites. Figure 2.8 illustrates the separation of *p*-nitrophenol (PNP), of PNP-sulfate ester, and of PNP-glucoronide using this type of coil.

Various approaches to a systematic search for solvent systems have been proposed, among others by Abbot and Kleiman (40), Conway (41), Foucault and Nakanishi (24, 25), Ito (14, 42), Kantoci et al. (43), and Oka et al. (44). Besides the classical solvents in CCC, attention is also focusing on BuOH/H₂O solvent systems and on biphasic aqueous–aqueous mixtures. These solvents are particularly interesting for the separation of peptides (45, 46) and of proteins (47). Further progress in the development of solvent systems can be expected from the use of fast and solvent-saving analytical instrumentation.

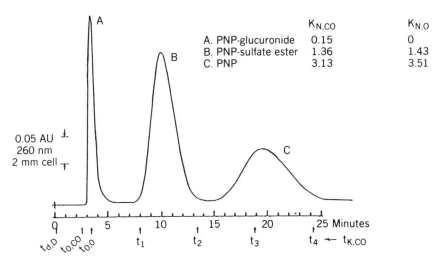

Figure 2.8. Separation of PNP-glucuronide (**A**), PNP-sulfate ester (**B**), and of PNP (= p-nitrophenol) (**C**) by HSCCC with a Tripple ™ analytical coil: 1.07 mm i.d. × 13.7 m, 14.7 mL; solvent: 1-BuOH–0.2 M KHCO₃ pH 8.6 (1:1); aqueous phase = mobile phase; 1.89 mL/min; retention: 70%. [Reproduced with permission from Conway et al. (23)].

2.5.2.   Measurement of Partition Coefficients

The octanol–water partition coefficient (P_{ow}) is generally accepted in medicinal chemistry as a way to characterize the lipophilic–hydrophilic nature of a compound. Since the partition coefficient of a solute can be calculated from its chromatogram [see Eq. (2.7)], CCC has been suggested as an alternative to the shake flask method to measure partition coefficients (48, 49). Refined methods like backflushing (50) or filling the coil with a defined ratio of stationary and mobile phase (51) accommodate a broad range of P values for direct measurements. Recently, El Tayar et al. (52) reviewed the various techniques of CCC for the measurements of partition coefficients. Best results were obtained by HSCCC (log P range, precision, effectiveness) and excellent correlations with published log P values (shake flask method) were found (see Fig. 2.9). Berthod and Bully (53) measured heptane–aqueous methanol partition coefficients of alkylbenzenes using an HSCCC apparatus with three coils connected in series and under controlled temperature.

So far, coils with $V_c > 100\,\text{mL}$ have been used for partition coefficient measurements. It is expected that further improvements can be achieved using smaller, analytical devices.

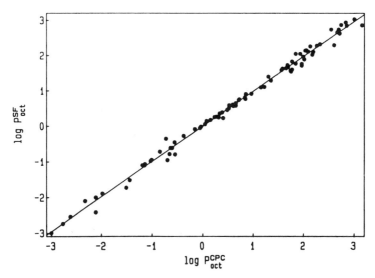

Figure 2.9. Linear relationship between 1-octanol–water partition coefficients (P_{oct}) determined by HSCCC (P^{cpc}) and by the shake flask method (P^{SF}). [Reprinted with permission from El Tayar et al. (52)].

2.5.3. Natural Products Isolation

Preparative HSCCC is now a standard method in natural products isolation. It is therefore not surprising that most of the analytical applications of HSCCC are encountered in natural products research (Table 2.4).

Analytical instrumentation has been used for the microscale separation of isoquinoline alkaloids from Stephania tetrandra S. Moore (54) and of anthraquinones from Rheum palmatum L. (55). In both cases, the elution mode was reversed after 50–70 min of run time. The elution was continued with the phase that was initially selected as the stationary phase. This procedure allows us to separate compounds of a wide range of polarity. Lee et al. (56) separated the Vinca alkaloids vincamine and vincine in hexane/ EtOH/H_2O (6:5:5) and they compared the separation with that achieved by reversed-phase HPLC (RP-HPLC). The selectivity of CCC is nicely illustrated by the separation of ligans from Schisandra rubriflora Rhed et Wils (57), a traditional Chinese herbal medicine for the treatment of chronic hepatitis. The plant contains a number of bioactive lignans, among these are schisanhenol and the corresponding acetate. The two compounds coelute in RP-HPLC, whereas they are base line separated in analytical HSCCC with hexane/EtOH/H_2O (6:5:5).

Separations of polyphenolic natural products are difficult because these compounds tend to show "peak tailing" in RP-HPLC, as well as to irreversibly adsorb on silica gel. These difficulties do not exist in CCC and are the reason why CCC has been recognized as a most valuable technique for the isolation of polyphenolics (58). Zhang et al. (37, 38) separated flavonoids by analytical HSCCC and demonstrated that preparative scale separations can be predicted by analytical HSCCC. In another example of separating polyphenolics, microgram quantities of coumarins were base line separated by analytical HSCCC using a "micro" HSCCC with $CHCl_3$/MeOH/H_2O (13:7:8) as solvent (27).

Crude extracts of plants or of other organisms are often too complex for the direct analysis by HPLC. Certain material may irreversibly bind to the packing material or may plug the column inlet filters, and hence reduce column life. These restrictions do not apply to analytical HSCCC, which represents an interesting method for sample purification and sample enrichment. This finding is illustrated by the separation of the antitumor macrolides bryostatin 1 and 2 (59). Figure 2.10a shows the separation by HPLC of a crude extract of Bugula neritina containing bryostatin 1 and 2. The two compounds are present in small quantities and their elution peaks are partially masked by other constituents of the extract (60). A 10-mg sample of the extract was separated by analytical HSCCC using hexane/isoPrOH/MeOH/H_2O with the lower phase as the mobile phase. Fractions eluting between 30–34 min

Table 2.4. Analytical HSCCC of Natural Products

Class of Compound	Source	Solvent System	References
Alkaloids	*Stephania tetrandra* (Menispermaceae)	Hexane/EtOAc/MeOH/H_2O (3:7:5:5)	Zhang et al. (54)
Anthraquinones	*Vinca minor* (Apocynaceae)	Hexane/EtOH/H_2O (6:5:5)	Lee et al. (56)
	Rheum palmatum (Polygonaceae)	*n*-Hexane/EtOAc/MeOH/H_2O (9:1:5:5)	Zhang et al. (55)
Coumarins		$CHCl_3$/MeOH/H_2O (13:7:8)	Schaufelberger (27)
Flavonoids	*Hippophaë rhamnoides* (Elaeagnaceae)	$CHCl_3$/MeOH/H_2O (4:3:2)	Oka et al. (21)
Lignans	*Podophyllum peltatum* (Podophyllaceae)	Hexane/EtOAc/MeOH/H_2O (1:1:1:1)	Schaufelberger et al. (59)
	Schisandra rubriflora (Schisandraceae)	Hexane/EtOH/H_2O (6:5:5)	Lee et al. (30, 57)
Macrolides	*Bugula neritina* (Bugulidae)	Hexane/*iso*PrOH/MeOH/H_2O (4:1:1.6:0.4)	Schaufelberger (26, 62)
Triterpenoids	*Boswellia carterii* (Burseraceae)	Hexane/EtOH/H_2O (6:5:1)	Lee et al. (63)

64

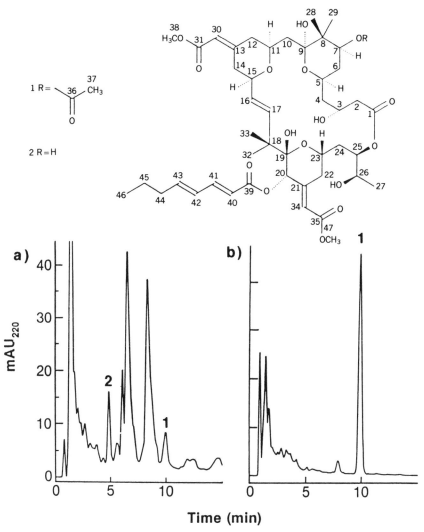

Figure 2.10. The HPLC of *Bugula neritina* crude extracts before (a) and after (b) purification with analytical HSCCC. HPLC: RP-18, 3 μm, 100 × 4.6 mm i.d.; acetonitrile/H$_2$O (78:22), 1 mL/min; detection: 220 nm; **1** = bryostatin 1, **2** = bryostatin 2. Analytical HSCCC: same as Fig. 2.5; solvent: hexane/isoPrOH/20% aqueous MeOH (4:1:2), lower phase = mobile phase, 0.6 mL/min; sample: 10 mg of crude extract; fractions eluting between 30 and 34 min were pooled and analyzed by HPLC (= chromatogram b).

were pooled and analyzed by HPLC (Fig. 2.10b); they were enriched in bryostatin 1 and were free of bryostatin 2.

Analytical HSCCC may also represent a very useful technique for the screening of new bioactive compounds in crude extracts and in other complex samples. For a more detailed discussion of analytical HSCCC and of natural products chemistry see Schaufelberger (26).

2.6. DISCUSSION

By now HSCCC has attracted more attention than any other form of liquid–liquid partition chromatography. The HSCCC is well established in natural products isolation and the technique is also being used for the separation of biochemicals and pharmaceuticals. Applications of HSCCC have also been reported from areas such as the isolation of radiolabeled compounds (61) and the separation of rare earth metal ions (17). In the past, very little attention has been given to analytical applications of CCC and only recently HSCCC has emerged as an analytical method. In this chapter instrumentation and applications of analytical HSCCC are reviewed.

For the purpose of this discussion, the term "analytical HSCCC" has been restricted to separations carried out with coils of less than 50-mL volume and with tubing of internal diameters of 1 mm or less. Most separations have been carried out with coils made of 0.85-mm i.d. tubing. Presently, there are two trends in the development of analytical instrumentation: (1) building instruments with small-bore tubing and designed for high rotational speeds, and (2) building instruments that can be used for analytical and preparative separations. The second goal can be achieved by connecting several coils in series or by building combined coils that contain analytical and preparative coils on the same holder. Flow tubes still represent a technical problem as they can get worn out by frequent and fast rotation. Further improvements in the design and in the instrumentation, for example, automation, is expected.

Efficiencies reported in analytical HSCCC clearly indicate that this technique will not compete with highly efficient methods, such as analytical HPLC. As of this point, analytical HSCCC may become a useful method for the following applications: (1) methods development, (2) microscale separations, and (3) measurements of partition coefficients.

Analytical instrumentation ideally supports methods development in preparative HSCCC. A compound with a partition coefficient of one elutes in both analytical and in preparative coils at a retention volume equal to the coil volume. Retention volumes and retention times are also predictable if $K \neq 1$: In this case the stationary phase fraction has to be taken into account [see

Eqs. (2.5) and (2.6)]. An excellent correlation between analytical and preparative HSCCC is presented by the separation of flavonoids illustrated in Fig. 2.7. Analytical instrumentation with its speed and low solvent consumption represents a practical way to evaluate solvent systems for HSCCC. Further progress in gradient elution and in the use of aqueous–aqueous solvent systems may be expected.

Microscale isolations represent another application where the advantages of analytical HSCCC are recognized. Sample sizes range between micrograms and a few milligrams. Crude extracts and other complex samples can be separated without filtration prior to injection. The HSCCC carries no risks of loosing compounds through irreversible adsorption onto stationary-phase material. This finding makes the method very attractive for the screening of bioactive molecules and for sample clean-up procedures. Figure 2.10 shows the enrichment of the antitumor compound bryostatin 1 from a crude extract using analytical HSCCC.

Measurement of partition coefficients is the third field where analytical HSCCC is expected to become a very useful method. So far, only preparative coils have been used. Partition coefficients, particularly between octanol and water, are important to describe the lipophilic behavior of pharmaceuticals and of chemicals. The partition coefficient of a solute is calculated from its retention time using Eq. (2.7). Exellent correlations have been reported between partition coefficients measured by the traditional shake flask method and those determined by HSCCC.

Because of relatively low efficiencies HSCCC will not become an analytical method as such. However, the application of analytical HSCCC in methods development in preparative HSCCC, in microisolations, and in partition coefficient measurements is very promising. Analytical HSCCC certainly has an important role to play in the future development of HSCCC.

Symbols and Abbreviations

C_m	Concentration of a solute in the mobile phase
C_s	Concentration of a solute in the stationary phase
CCC	Countercurrent chromatography
DCCC	Droplet countercurrent chromatography
f	Flow rate
K	Partition coefficient
k'	Capacity factor
N	Number of theoretical plates
N_{ccc}	Number of theoretical plates, corrected for CCC
P_{oct}	Octanol–water partition coefficient
S_F	Stationary-phase fraction

R	Revolutional radius of the coil
r	Rotational radius of the coil
t_0	Retention time of a solute with $K = 0$
t_i	Time of sample injection
t_R	Retention time of a solute (elapsed time since t_i)
t_R'	Corrected retention time $(t_R - t_0)$
V_0	Retention volume of a solute with $K = 0$; volume of the mobile phase fraction
V_c	Total volume of the coil
V_R	Retention volume of a solute (eluted volume since t_i)
V_R'	Corrected retention volume $(V_R - V_0)$
V_s	Volume of the stationary phase in the coil
α	Separation factor
β	Ratio of rotational radius to revolutional radius

REFERENCES

1. Y. Ito, *J. Chromatogr.*, **214**, 122 (1981).

2. Y. Ito, J. Sandlin and W. G. Bowers, *J. Chromatogr.*, **244**, 247 (1982).

3. Y. Ito and R. L. Bowman, *Anal. Biochem.*, **85**, 614 (1978).

4. H. Oka, Y. Ikai, N. Kawamura, M. Yamada, K.-I. Harada, M. Suzuki, F. E. Chou, Y.-W. Lee, and Y. Ito, *J. Liq. Chromatogr.*, **13**, 2309 (1990).

5. R.-S. Tsai, N. El Tayar, B. Testa, and Y. Ito, *J. Chromatogr.*, **538**, 119 (1991).

6. I. Kubo, A. Matsumoto, F. J. Hanke, and J. F. Ayafor, *J. Chromatogr.*, **321**, 246 (1985).

7. Y. Ito and Y.-W. Lee, *J. Chromatogr.*, **391**, 290 (1987).

8. R. J. Romañach and J. A. de Haseth, *J. Liq. Chromatogr.*, **11**, 91 (1988).

9. D. W. Armstrong, *J. Liq. Chromatogr.*, **11**, 2433 (1988).

10. W. D. Conway, in N. Mandava and Y. Ito, Eds., *Countercurrent Chromatography, Theory and Practice; Chromatographic Science Series*, Vol. 44, Marcel-Dekker, New York, 1988, p. 443.

11. W. D. Conway, *Countercurrent Chromatography Apparatus, Theory and Applications*. VCH, New York, 1990.

12. W. D. Conway, *J. Chromatogr.*, **538**, 27 (1991).

13. Y. Ito, *CRC Crit. Rev. Anal. Chem.*, **17**, 65 (1986).

14. Y. Ito, in N. Mandava and Y. Ito, Eds., *Countercurrent Chromatography, Theory and Practice; Chromatographic Science Series*, Vol. 44, Marcel-Dekker, New York, 1988, p. 79.

15. A. Foucault, *Anal. Chem.*, **63**, 569A (1991).

16. W. D. Conway and Y. Ito, *J. Liq. Chromatogr.*, **8**, 2195 (1985).

17. Y. Ito, H. Oka, E. Kitazume, M. Bhatnagar, and Y.-W. Lee, *J. Liq. Chromatogr.*, **13**, 2329 (1990).

18. H. Oka, Y. Ikai, N. Kawamura, M. Yamada, J. Hayakawa, K.-I. Harada, K. Nagase, H. Murata, M. Suzuki, and Y. Ito, *J. High Res. Chromatogr.*, **14**, 306 (1991).

19. A. Berthod, *Eur. Chromatogr. Anal.*, February, 13 (1991).

20. O. Bousquet, A. P. Foucault, and F. Le Goffic, *J. Liq. Chromatogr.*, **14**, 3343 (1991).

21. H. Oka, F. Oka, and Y. Ito. *J. Chromatogr.*, **479**, 53 (1989).

22. A. Berthod, *J. Chromatogr.*, **550**, 677 (1991).

23. W. D. Conway, H. C. Yeh, and N. J. Tedesche, Pittsburgh Conference and Exposition on Analytical Chemistry and Applied Spectroscopy, New Orleans, LA, March 9–13, 1992, Abstract No. 387.

24. A. Foucault and K. Nakanishi, *J. Liq. Chromatogr.*, **12**, 2587 (1989).

25. A. Foucault and K. Nakanishi, *J. Liq. Chromatogr.*, **13**. 3583 (1990).

26. D. E. Schaufelberger, *J. Chromatogr.*, **538**, 45 (1991).

27. D. E. Schaufelberger, *J Liq. Chromatogr.*, **12**, 2263 (1989).

28. H. Oka and Y. Ito, *J. Chromatogr.*, **475**, 229 (1989).

29. Y.-W. Lee, R. D. Voyksner, Q.-C. Fang, C. E. Cook, and Y. Ito, *J. Liq. Chromatogr.*, **11**, 153 (1988).

30. Y. W. Lee, R. D. Voyksner, T. W. Pack, C. E. Cook, Q. C. Fang and Y. Ito, *Anal. Chem.*, **62**, 244 (1990).

31. R. J. Romañach and J. A. de Haseth, *J. Liq. Chromatogr.*, **11**, 133 (1988).

32. B. Diallo, R. Vanhaelen-Fastré, and M. Vanhaelen, *J. Chromatogr.*, **558**, 446 (1991).

33. D. E. Schaufelberger, T. G. McCloud, and J. A. Beutlèr, *J. Chromatogr.*, **538**, 87 (1991).

34. S. Drogue, M. C. Rolet, D. Thiébaut, and R. Rosset, *J. Chromatogr.*, **538**, 91 (1991).

35. B. Domon, M. Hostettmann, and K. Hostettmann, *J. Chromatogr.*, **246**, 133 (1982).

36. K. Hostettmann, *Planta Med.*, **39**, 1 (1980).

37. T.-Y Zhang, R. Xiao, Z.-Y. Xiao, L. K. Pannell, and Y. Ito, *J. Chromatogr.*, **445**, 199 (1988).

38. T.-Y. Zhang, X. Hua, R. Xiao, and S. Knog, *J. Liq. Chromatogr.*, **11**, 233 (1988).

39. W. D. Conway, G. R. Watkins, T. I. Kalman, X. J. Jiang, and S. G. Kerr, Pittsburgh Conference and Exposition on Analytical Chemistry and Applied Spectroscopy, Chicago, IL, March 4–8, 1991, Abstract No. 1059.

40. T. P. Abbot and R. Kleiman, *J. Chromatogr.*, **538**, 109 (1991).

41. W. D. Conway, *J. Liq. Chromatogr.*, **13**, 2409 (1990).

42. Y. Ito, in N. Mandava and Y. Ito Eds., *Countercurrent Chromatography, Theory and Practice; Chromatographic Science Series*, Vol. 44, Marcel-Dekker, New York, 1988, p. 79.

43. D. Kantoci, G. R. Pettit, and Z. Cichacz, *J. Liq. Chromatogr.*, **14**, 1149 (1991).

44. E. Oka, H. Oka, and Y. Ito, *J. Chromatogr.*, **538**, 99 (1991).

45. M. Knight, in N. Mandava and Y. Ito, Eds., *Countercurrent Chromatography, Theory and Practice; Chromatographic Science Series*, Vol. 44, Marcel-Dekker, New York, 1988, p. 583.

46. Y. Shibusawa and Y. Ito, *J. Liq. Chromatogr.*, **14**, 1575 (1991).

47. Y. Shibusawa and Y. Ito, *J. Chromatogr.*, **550**, 695 (1991).

48. H. Terada, Y. Kosuge, W. Murayama, N. Nakaya, Y. Nunogaki, and K.-I. Nunogaki, *J. Chromatogr.*, **400**, 343 (1987).

49. A. Berthod and D. W. Armstrong, *J. Liq. Chromatogr.*, **11**, 547 (1988).

50. R. A. Menges, G. L. Bertrand, and D. W. Armstrong, *J. Liq. Chromatogr.*, **13**, 3061 (1990).

51. P. Vallat, N. El Tayar, B. Testa, I. Slacanin, A. Marston, and K. Hostettmann, *J. Chromatogr.*, **504**, 411 (1990).

52. N. El Tayar, R.-S. Tsai, P. Vallat, C. Altomare, and B. Testa, *J. Chromatogr.*, **556**, 181 (1991).

53. A. Berthod and M. Bully, *Anal. Chem.*, **63**, 2508 (1991).

54. T.-Y. Zhang, L. K. Pannell, D.-G. Cai, and Y. Ito, *J. Liq. Chromatogr.*, **11**, 1661 (1988).

55. T.-Y. Zhang, L. K. Pannell, Q.-L. Pu, D.-G. Cai, and Y. Ito, *J. Chromatogr.*, **442**, 455 (1988).

56. Y.-W. Lee, Y. Ito, Q.-C. Fang, and C. E. Cook, *J. Liq. Chromatogr.*, **11**, 75 (1988).

57. Y.-W. Lee, C. E. Cook, Q.-C. Fang, and Y. Ito, *J. Chromatogr.*, **477**, 434 (1989).

58. A. Marston, I. Slacanin, and K. Hostettmann, *Phytochem. Anal.*, **1**, 3 (1990) and references cited therein.

59. D. E. Schaufelberger, M. P. Koleck, and G. M. Muschik, Pittsburgh Conference and Exposition on Analytical Chemistry and Applied Spectroscopy, New York, March 5–9, 1990, Abstract No. 297.

60. D. E. Schaufelberger, A. B. Alvarado, P. Andrews, and J. A. Beutler, *J. Liq. Chromatogr.*, **13**, 583 (1990).

61. C. Baker, C. Bowlen, D. Koharski, and P. McNamara, *J. Chromatogr.*, **484**, 347 (1989).

62. D. E. Schaufelberger, 30th Annual Meeting of the American Society of Pharmacognosy, San Juan, PR, Aug. 6–10, 1989, Abstract Nr. O:26.

63. Y.-W. Lee, T. W. Pack, R. D. Voyksner, Q.-C. Fang, and Y. Ito, *J. Liq. Chromatogr.*, **13**, 2389 (1990).

SPECIAL TECHNIQUES

CHAPTER

3

HIGH-SPEED COUNTERCURRENT CHROMATOGRAPHY/MASS SPECTROMETRY

HISAO OKA

Aichi Prefectural Institute of Public Health, Tsuji-machi, Kita-ku, Nagoya 462, Japan

3.1. INTRODUCTION

Over the past 20 years, countercurrent chromatography (CCC), which totally eliminates the use of a solid support, has been developing. Countercurrent chromatography technology has advanced in various directions including preparative and trace analysis, dual CCC, foam CCC and, more recently, partition of macromolecules with polymer phase systems (1, 2). However, during the preceding decade CCC has been used almost exclusively for preparative separation of natural products, due to the relatively long separation time required. In order to fully explore the potential of the CCC method, considerable effort has recently been made to develop analytical models of high-speed CCC (HSCCC). Furthermore, interfacing HSCCC to the mass spectrometer (MS) has also been tried, because it integrates the advantages of HSCCC with the low detection limit and identification capability of MS. Table 3.1 summarizes analytical HSCCC apparatus. Among them, HSCCC-2000 with a 5-cm revolution radius and 0.85-mm i.d. column has been interfaced to thermospray (TSP) MS. Although it has been successfully applied for analyses of alkaloids, triterpenoic acids, and lignans from plant natural products (3–5), several improvements must be made with respect to chromatographic resolution and the interfacing system as described in Section 3.2.

Recently, we developed an analytical HSCCC-4000 with a 2.5-cm revolutional radius and a 0.3-mm i.d. column, which is capable of operating at a maximum speed of 4000 rpm (6, 7). When indole auxin mixtures were analyzed by this HSCCC, they were completely separated in 28 min with excellent theoretical plate numbers ranging from 10500 to 5500 (8). Also, the

High-Speed Countercurrent Chromatography, Edited by Yoichiro Ito and Walter D. Conway.
Chemical Analysis Series, Vol. 132.
ISBN 0-471-63749-1 © 1996 John Wiley & Sons, Inc.

Table 3.1. HSCCC Systems for Analytical Works

Apparatus	Revolutional Radius (cm)	Revolution (rpm)	Column i.d. (mm)	Column Capacity (mL)	Subject	Theoretical Plate Number	Published Year
HSCCC-2000	5.0	2000	0.85	38	Indole mixture	1000–1300	1987[a]
CCC-2000	6.35	2000	0.85	40	Flavonoids	1000–1200	1988[b]
CCC-2000 (Twin columns)	6.35	1600	0.85	45	Indole mixture	1600–2700	1988[c]
HSCCC-4000	2.5	3500	0.85	8	Flavonoids	280–380	1989[d]
HSCCC-4000	2.5	3500	0.55	5	Indole mixture	800–1300	1990[e]
HSCCC-4000	2.5	3500	0.40	1.5	Indole mixture	600–1000	1990[e]
HSCCC-4000	5	3500	0.30	6	Indole mixture	5500–10500	1990[f]

[a] Y. Ito and Y. W. Lee, *J. Chromatogr.*, **391**, 290 (1987).
[b] T.-Y. Zhang et al., *J. Chromatogr.*, **445**, 199 (1988).
[c] Y. Ito and F. E. Chou, *J. Chromatogr.*, **454**, 382 (1988).
[d] H. Oka et al., *J. Chromatogr.*, **479**, 53 (1989).
[e] H. Oka et al., *J. Liq. Chromatogr.*, **13**, 2309 (1990).
[f] H. Oka et al., *J. High Resolut. Chromatogr.*, **14**, 306 (1991).

feasibility of interfacing HSCCC to MS was suggested. Therefore, this paper is focused on interfacing of the HSCCC-4000 to a MS.

3.2. PREVIOUS INTERFACING OF HSCCC WITH THERMOSPRAY MASS SPECTROMETRY

Lee et al. (3–5) tried using analytical HSCCC-2000 with a 5-cm revolutional radius and a 0.85-mm i.d. column for interfacing to TSP MS. Figure 3.1 shows their interfacing system. When HSCCC is directly coupled with TSPMS, the column often breaks due to high back pressure generated by the thermospray vaporizer. To overcome this problem, a high-performance liquid chromato-graphic (HPLC) pump (HPLC pump I) was inserted at the interface junction to protect the column from the high back pressure. The effluent from the CCC (0.8 mL/min) was introduced into the HPLC I pump through a zero dead volume tee fitted with a reservoir. The HPLC pump I was operated at 0.7 mL/min with the reservoir providing extra solvent or venting excess solvent from the CCC system. The effluent from the HPLC pump I was mixed with 0.3 M ammonium acetate added at 0.3 mL/min and was introduced into the thermospray interface. Although this system has been successfully applied to the analyses of alkaloids, triterpenoic acids, and lignans from plant natural product, and it gave much useful structural information, such as HSCCC-UV and HSCCC/TSPMS total ion current (TIC) chromatograms of plant alkaloids (see in Fig. 3.2), a large dead space in the pump adversely affected the resulting chromatogram. This was seen by the loss of a minor peak that represents an isomer of vincine. Therefore, we considered interfacing HSCCC to MS without an additional pump.

3.3. DIRECT INTERFACING OF HSCCC WITH FRIT ELECTRON IONIZATION, CHEMICAL IONIZATION, AND FAST ATOM BOMBARDMENT MASS SPECTROMETRY

Desired performance of the HSCCC/MS interface is summarized as follows:

1. High enrichment of sample in ion source.
2. High yield of sample reaching MS.
3. No peak broadening.
4. High applicability to nonvolatile sample.

In view of these points, various interfaces are listed and compared in Table 3.2. Namely, Frit fast atom bombardment (FAB) including continuous flow (CF)

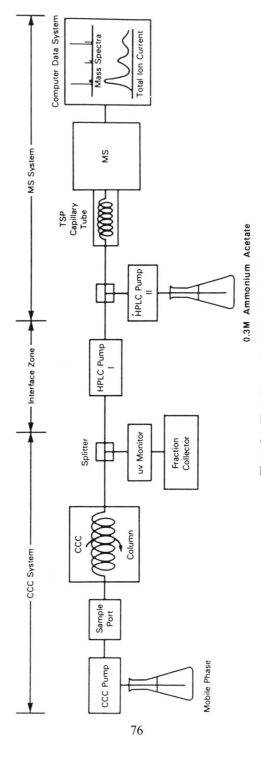

Figure 3.1. Flow diagram for HSCCC/TSPMS.

76

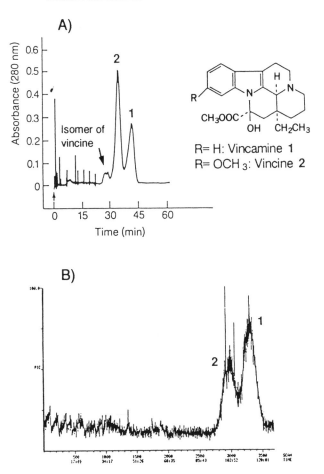

Figure 3.2. HSCCC/UV chromatogram (A) and total ion current chromatogram (B) of plant alkaloids under HSCCC/TSPMS conditions.

FAB, Frit electron ionization (EI), Frit chemical ionization (CI), TSP, atmospheric pressure chemical ionization (APCI), and electrospray ionization (ESI). Each interface has its own characteristics, so it is desired to use them to complement each other. However, among them we selected Frit MS to directly interface to HSCCC, because it does not generate high back pressure, only $2\,kg/cm^2$, which is one tenth of TSP, and Frit MS can be applied to analytes with broad polarity. We tried to interface the HSCCC-4000, which has excellent separation efficiency, and its performance was evaluated in terms of chromatography and MS. Figure 3.3 is the flow diagram for HSCCC/Frit MS. The HSCCC-4000 produces excellent separation efficiency at flow-rates be-

Table 3.2. Comparison of Interfaces for HSCCC/MS

	Frit			TSP	APCI	ESI
	FAB (CFF)	EI	CI			
Heating (°C)	30–70	100–300	100–300	200–350	200–300	60
Splitter	+	+	+	−	−	+
Back pressure (kg/cm²)	2	2	2	20	20	−

Applicable range of these interfaces to analytes with various polarity

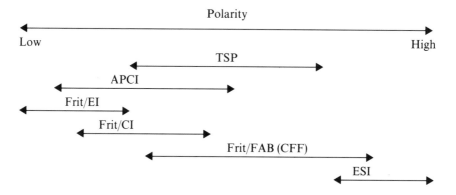

tween 0.1 and 0.2 mL/min, however, it is suitable to introduce effluent from HSCCC into Frit MS only at a flow-rate between 1 and 5 µL/min. Therefore, the effluent from HSCCC was introduced into Frit MS through a splitting tee, which was adjusted to a split ratio of 1:40. Illustration of the split tee is shown in Fig. 3.4 (9.10). A 0.06-mm i.d. fused silica tube is led to the MS while a 0.5-mm i.d. stainless steel tube is connected to the HSCCC column. The other side of the fused silica tube extends deeply into the stainless steel tube to receive a small portion of the effluent from the HSCCC column while the rest of the effluent is discarded through the 0.1-mm i.d. polytetrafluoroethylene (PTFE) tube. The split ratio of the effluent depends on the flow-rate of the effluent and the length of the 0.1-mm i.d. PTFE tube. A 2 m length of the 0.1-mm i.d. tube is needed to adjust the split ratio at 1:40.

3.4. HSCCC/FRIT EIMS OF INDOLE AUXINS

First, we analyzed indole auxins under Frit EIMS conditions (9). Their structures and the analytical conditions are shown in Fig. 3.5 and Table 3.3,

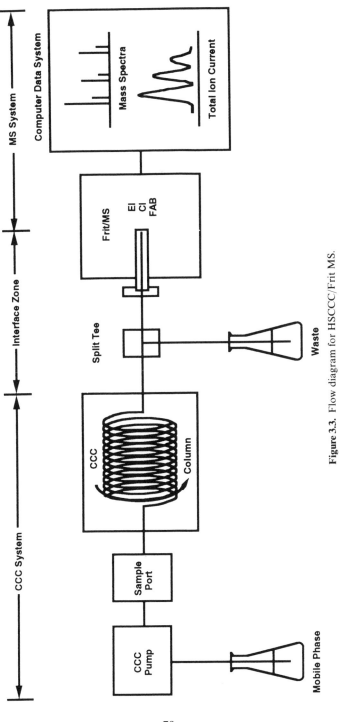

Figure 3.3. Flow diagram for HSCCC/Frit MS.

79

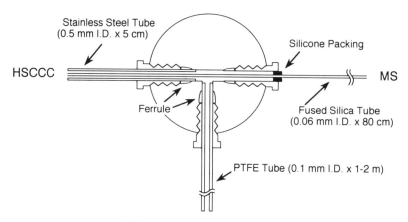

Figure 3.4. Illustration of split tee.

respectively. The column i.d. is 0.3 mm, the column capacity is 6 mL, the solvent system used is *n*-hexane/ethyl acetate/methanol/water (1:1:1:1), and the lower aqueous phase is used as a mobile phase. For MS conditions, the Frit EI–CI ion source is used under EI mode at a temperature of 220°C. Figure 3.6 shows the TIC, the mass chromatogram, and the UV chromatogram. Comparing the TIC with the UV chromatogram, both showed the same chromatographic resolution with excellent theoretical plate numbers ranging from 12000 to 5500. Therefore, the interfacing to MS does not adversely affect chromatographic resolution. Figure 3.6B shows mass chromatograms for individual molecular ions. The ions at m/z 203, 175, and 174 are molecular ions of IBA, IAA, and IA as labeled. Peak IA appears in the middle and bottom chromatograms at the same retention time. Peak IA in the middle chromatogram at m/z 175 indicates the protonated IA molecule, which was formed when the sample was introduced into the MS with the mobile phase. In this case the mobile phase behaves like a reagent gas in chemical ionization mass spectrometry (CIMS). Figure 3.7 shows mass spectra of indole auxins at the tops of each peak on TIC. Both molecular ions and protonated molecules appear in all mass spectra and these facts are very useful for estimating molecular weight. Common ions are found at m/z 116 and 130, which originated from indole nuclei. Therefore, these mass spectra provide some useful structural information.

3.5. HSCCC/FRIT CIMS OF MYCINAMICINS

Mycinamicins were analyzed under HSCCC/Frit CIMS conditions (9). Mycinamicins are 16-member macrolide antibiotics and are used mainly for

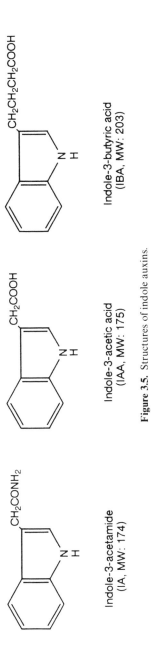

Indole-3-acetamide
(IA, MW: 174)

Indole-3-acetic acid
(IAA, MW: 175)

Indole-3-butyric acid
(IBA, MW: 203)

Figure 3.5. Structures of indole auxins.

Table 3.3. Summary of Analytical Conditions for HSCCC/Frit MS

	Indole Auxins	Mycinamicins	Colistins
HSCCC			
Apparatus	HSCCC-4000	HSCCC-4000	HSCCC-4000
Column	0.3-mm i.d. PTFE tube, 640 turns, 10 layers	0.3-mm i.d. PTFE tube, 640 turns, 10 layers	0.55-mm i.d. PTFE tube, 258 turns, 6 layers
Column capacity (mL)	7	7	6
Solvent system	n-Hexane/ethyl acetate/methanol/water (1:1:1:1)	n-Hexane/ethyl acetate/methanol/8% aq. ammonia (1:1:1:1)	n-Butanol/0.04 M TFA (1:1)
Mobile phase	Lower phase	Lower phase	Lower phase
Flow rate (mL/min)	0.2	0.1	0.16
Retention of stationary phase (%)	27.2	40.4	34.3
MS			
Apparatus	JEOL JMS-AX505	JEOL JMS-AX505	JEOL JMS-AX505
Ion source	Frit EI-CI (EI mode)	Frit EI-CI (CI mode)	Frit FAB
Ion source temperature (°C)	220	220	70
Ionization voltage (eV)	70	—	—
Acceleration voltage (kV)	3	3	5 (primary) 3 (secondary)
Reagent gas	—	Mobile phase	—
Emission current (μA)	—	300	—
Primary beam	—	—	Xe^0

82

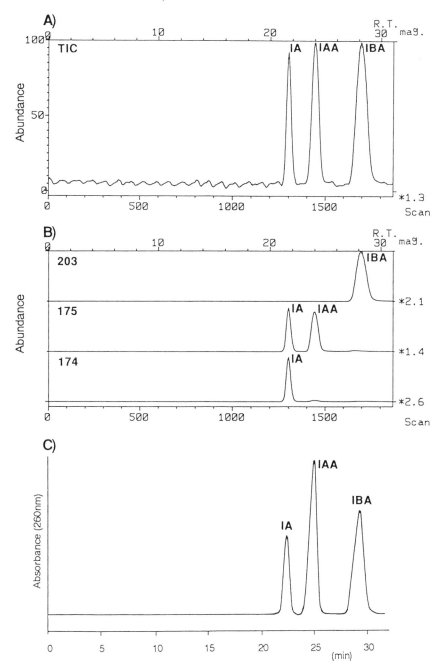

Figure 3.6. Total ion current (A) and mass chromatograms (B) under HSCCC/Frit EIMS conditions and HSCCC/UV chromatogram (C) of indole auxins. Indole-3-acetamide = 1A (MW 174), indole-3-acetic acid = IAA (MW 175), indole-3-butyric acid = IBA (MW 203).

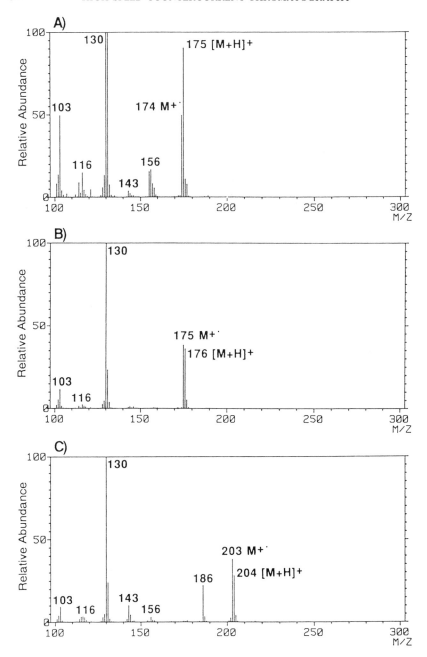

Figure 3.7. HSCCC/Frit EI mass spectra of indole auxins. (A) IA, (B) IAA, (C) IBA.

animals. Mycinamicins consist of six components, mycinamicins I–VI, but in the present study we used isolated mycinamicins IV and V. The structural difference comes from the site indicated by R (Fig. 3.8). The analytical conditions shown in Table 3.3 are identical to those applied to the indole auxins. For MS conditions, the ion source was used under CI mode and the mobile phase was used as a reagent gas. Figure 3.9 shows TIC (A) and mass chromatograms (B) at m/z 696 and 712 corresponding to protonated molecules of mycinamicins IV and V, respectively. The theoretical plate numbers of mycinamicins calculated from them are 3200. We consider that the lower theoretical plate numbers than the results observed in the case of indole auxins, are caused by the difference in stationary-phase retention, because UV chromatograms also showed the same separation efficiency and it is known that stationary-phase retention percent affects the theoretical plate number (11). Therefore, the interfacing to MS does not affect peak resolution in the analysis of mycinamicins. Because protonated ethyl acetate (m/z 89) and its dimer (m/z 177) occurred very intensely, mass spectra of mycinamicins above m/z 200 were obtained at the top of each peak on TIC as shown in Fig. 3.10. Probably the solvent ions operated as reactant ions for CI reactions. In both spectra, protonated molecules clearly appear as base peaks and they are useful in determining molecular weights.

3.6. HSCCC/FRIT FABMS OF COLISTIN COMPLEX

To apply the present system to less volatile compounds, we examined the potential capability under Frit FABMS conditions (9). Colistin complex, a mixture of peptide antibiotics, was analyzed under Frit FABMS conditions

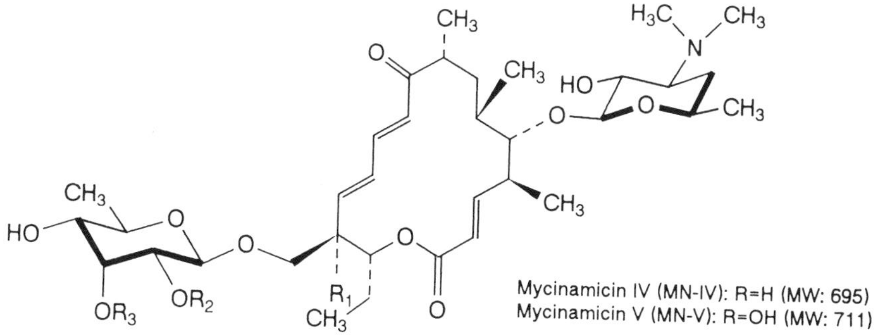

Figure 3.8. Structures of mycinamicine.

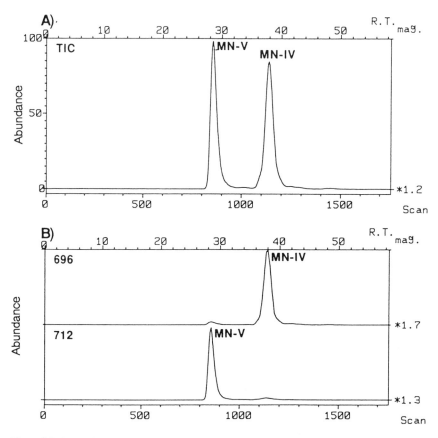

Figure 3.9. Total ion current (A) and mass chromatograms (B) of mycinamicins under HSCCC/Frit CIMS conditions. Mycinamicin IV = MN-IV and mycinamicin V = MN-V.

(Table 3.3). The colistin complex consists of major component colistins A (CL-A, MW 1168) and B (CL-B, MW 1154), and a number of minor components of unknown nature. The difference between CLs-A and B is due to their acyl moiety fatty acid, as indicated in Fig. 3.11. For HSCCC analysis, a 0.55-mm i.d. column was used, because the 0.3-mm i.d. column did not give satisfactory stationary-phase retention for the *n*-butanol/trifluoroacetic acid solvent system used in the present study. To obtain the FAB mass spectrum it is necessary to introduce a sample with an appropriate matrix such as glycerol, thioglycerol, and *m*-nitrobenzyl alcohol into the FABMS ion source. We added glycerol as a matrix to the mobile phase at a concentration of 1%. The use of a two-phase solvent system containing glycerol was the first experiment for

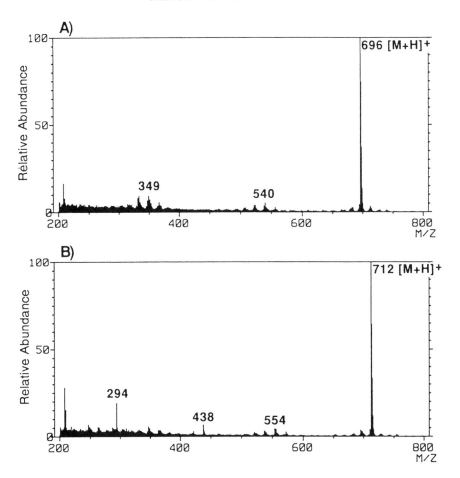

Figure 3.10. HSCCC/Frit CI mass spectra of mycinamicins. (A) MN-IV and (B) MN-V.

HSCCC study, so we carefully observed stationary-phase retention and separation efficiency. The results showed the same retention and separation efficiency as in the case without glycerol. Figure 3.12 shows TIC (A) and mass chromatograms (B). Because we used a 0.55-mm i.d. column and *n*-butanol/trifluoroacetic acid (TFA) as a solvent system, it shows a less efficient separation than the results obtained with indoles and mycinamicins. However, peaks corresponding to CLs-A and B clearly appear. Furthermore, unknown peak X can be also separated from CL-B. Mass chromatograms of individual protonated molecules show symmetrical peaks and a significant drop of peak resolution was not produced by interfacing with MS. Figure 3.13 shows mass

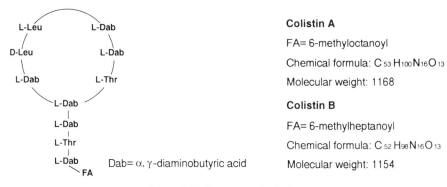

Colistin A

FA= 6-methyloctanoyl

Chemical formula: $C_{53}H_{100}N_{16}O_{13}$

Molecular weight: 1168

Colistin B

FA= 6-methylheptanoyl

Chemical formula: $C_{52}H_{98}N_{16}O_{13}$

Molecular weight: 1154

Dab= α, γ-diaminobutyric acid

Figure 3.11. Structures of colistins.

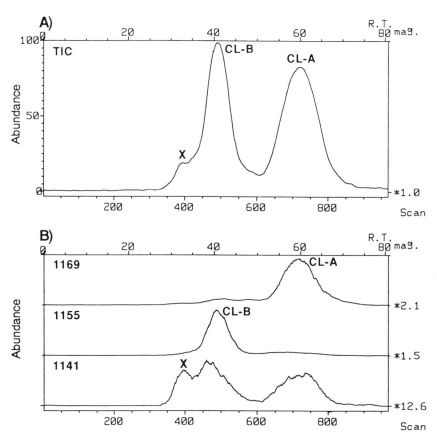

Figure 3.12. Total ion current (A) and mass chromatograms (B) of colistin complex under HSCCC/Frit FABMS conditions. Colistin A = CL-A, colistin B = CL-B, unknown component = X.

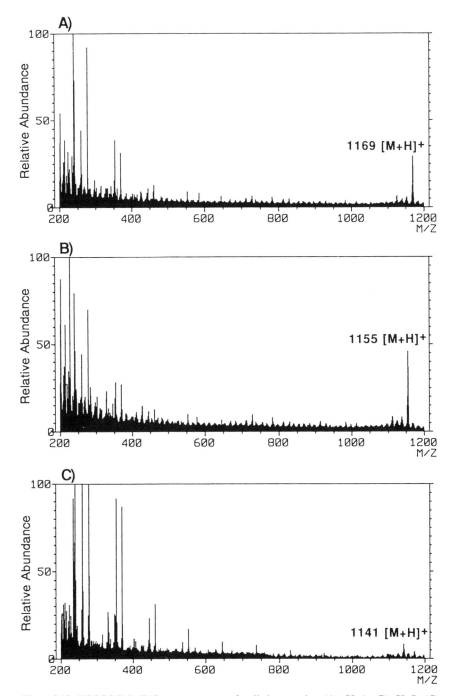

Figure 3.13. HSCCC/Frit FAB mass spectra of colistin complex. (A) CL-A, (B) CL-B, (C) unknown component X.

spectra at peak tops for each peak on the TIC. In all spectra, protonated molecules clearly appear and we consider that they are good spectra for determining molecular weight. The molecular weight corresponding to peak X is estimated to be 1140 from this mass spectrum. This component is the same as that found by Elverden et al. in 1981 (12).

3.7. CONCLUSION

An analytical HSCCC-4000 can be directly interfaced to Frit/EI, CI, and FABMS without an additional HPLC pump. The data obtained indicate that the interfacing with Frit/MS does not adversely affect the chromatographic resolution and the mass spectra obtained give useful structural information. The HSCCC-Frit/MS may become a uniquely useful tool for the investigation of natural products. Although this chapter has mainly described the interfacing of HSCCC-4000 to Frit MS, HSCCC-2000/TSP MS has also been successfully applied to structural characterization of natural products. Therefore, interfacing analytical HSCCC with MS including TSP MS and Frit MS will offer a new dimension in the separation of biologically important substances.

REFERENCES

1. Y. Ito, *CRC Crit. Rev. Analyt. Chem.*, **17**, 65 (1986).

2. N. B. Mandava and Y. Ito, Eds., *Countercurrent Chromatography, Theory and Practice* (*Chromatographic Science Series*, Vol. 44), Marcel-Dekker, New York, 1988.

3. Y.-W. Lee, R. D. Voyksner, Q.-C. Fang, C. E. Cook, and Y. Ito, *J. Liq. Chromatogr.*, **11**, 153 (1988).

4. Y.-W. Lee, T. W. Pack, R. D. Voyksner, Q.-C. Fang, and Y. Ito, *J. Liq. Chromatogr.*, **13**, 2389 (1990).

5. Y.-W. Lee, R. D. Voyksner, T. W. Pack, C. E. Cook, Q.-C. Fang, and Y. Ito, *Anal. Chem.*, **62**, 244 (1990).

6. H. Oka, F. Oka, and Y. Ito, *J. Chromatogr.*, **479**, 53 (1989).

7. H. Oka, Y. Ikai, N. Kawamura, M. Yamada, K.-I. Harada, M. Suzuki, F. E. Chou, Y.-W. Lee, and Y. Ito, *J. Liq. Chromatogr.*, **13**, 2309 (1990).

8. H. Oka, Y. Ikai, N. Kawamura, M. Yamada, J. Hayakawa, K.-I. Harada, K. Nagase, H. Murata, M. Suzuki, and Y. Ito, *J. High Resol. Chromatogr.*, **14**, 306 (1991).

9. H. Oka, Y. Ikai, N. Kawamura, J. Hayakawa, K.-I. Harada, H. Murata, M. Suzuki, and Y. Ito, *Anal. Chem.*, **63**, 2861 (1991).

10. Y. Ikai, H. Oka, J. Hayakawa, K.-I. Harada, and M. Suzuki, *Mass Spectrosc.*, **39**, 199 (1991).

11. H. Oka, Y. Ikai, J. Hayakawa, K.-I. Harada, K. Nagase, H. Murata, M. Suzuki, H. Nakazawa, and Y. Ito, *J. Liq. Chromatogr.*, **15**, 2707 (1992).

12. I. Elverdam, P. Larsen, and E. Lund, *J. Chromatogr.*, **218**, 653 (1981).

CHAPTER

4

DUAL COUNTERCURRENT CHROMATOGRAPHY

Y. W. LEE

Research Triangle Institute, Chemistry and Life Sciences,
Research Triangle Park, North Carolina 27709

4.1. INTRODUCTION

The development in the 1980s of modern high-speed countercurrent chromatography (HSCCC), which was based upon the fundamental principles of liquid–liquid partition, has caused a resurgence of interest in the separation sciences. The advantages of applying continuous liquid–liquid extraction, a process for separation of a multicomponent mixture according to the differential solubility of each component in two immiscible solvents have long been recognized. For instance, the countercurrent distribution method, which prevailed in the 1950s and 1960s, was applied successfully to fractionate commercial insulin into two subfractions that differed only by one amide group with a molecular weight of 6000 (1).

In recent years, significant improvements have been made to enhance the performance and efficiency of liquid–liquid partitioning (2–8). The high-speed coil planet centrifuge (CPC) technique utilizes a particular combination of coil orientation and planetary motion to produce a unique hydrodynamic, unilateral phase distribution of two immiscible solvents in a coiled column. The hydrodynamic properties can effectively be applied to perform a variety of liquid–liquid partition chromatographies including HSCCC (2), foam CCC (8, 9) and the dual countercurrent chromatography (DuCCC) (10, 11).

In most cases, where the two-phase solvent system is selected for HSCCC, one liquid phase serves as a stationary phase and the second phase is used as a mobile phase. An efficient separation can be achieved by continuous partitioning of a mixture between the stationary phase and the mobile phase. By definition, this mode of separation ought to be called high-speed liquid–liquid partition chromatography, because only one solvent phase is mobile (2).

High-Speed Countercurrent Chromatography, Edited by Yoichiro Ito and Walter D. Conway.
Chemical Analysis Series, Vol. 132.
ISBN 0-471-63749-1   © 1996 John Wiley & Sons, Inc.

In the case of DuCCC, where the two-phase solvents are countercrossing each other inside the coiled column from opposite directions, both phases are mobile and there is no stationary phase involved. The name of DuCCC is redundant, however, it is useful to distinguish it from ordinary HSCCC.

DuCCC shares several common advantages with other types of liquid–liquid partition chromatography. For instance, there is an unlimited number of two-phase solvent systems that can be employed and there are no sample losses from irreversible adsorption or decomposition on the solid support. In addition, DuCCC is extremely powerful in the separation of crude natural products that usually consist of multicomponents with an extremely wide range of polarities. In a standard operation, the crude sample is fed through the middle portion of the column. The extreme polar and nonpolar components are readily eluted from the opposite ends of the column followed by components with decreasing orders of polarity in one phase and increasing order of polarity in the other phase. A component with a partition coefficient equal to 1 will remain inside the coiled column. Basically, the DuCCC resembles a highly efficient performance of classic countercurrent distribution. They differ in that DuCCC is a dynamic process, whereas CCD is an equilibrium process.

The principles, of DuCCC and its capabilities in natural products isolation, are illustrated in Section 4.2.

4.2. PRINCIPLES AND MECHANISM

The fundamental principle of separation for modern DuCCC is identical to the classic countercurrent distribution. It is based on the differential partitions of a multicomponent mixture between two countercrossing and immiscible solvents. The separation of a particular component within a complex mixture is based on the selection of a two-phase solvent system that provides an optimized partition coefficient difference between the desired component and the impurities. In other words, DuCCC and HSCCC cannot be expected to resolve all the components with one particular two-phase solvent system. Nevertheless, it is always possible to select a two-phase solvent system that will separate the desired component.

In general, the crude sample is applied to the middle of the coiled column and the extreme polar and nonpolar components are readily eluted by two immiscible solvents to opposite outlets of the column. Contrary to the classic countercurrent distribution method, modern DuCCC allows the entire operation to be carried out in a continuous and highly efficient manner. DuCCC is based on the ingenious design of Ito (8) and is illustrated in Figs 4.1 and 4.2. In Fig. 4.1, a cylindrical coil holder is equipped with a planetary gear that is coupled to an identical stationary sun gear (shaded) placed around the central axis of the centrifuge. This gear arrangement produces an epicyclic motion; the

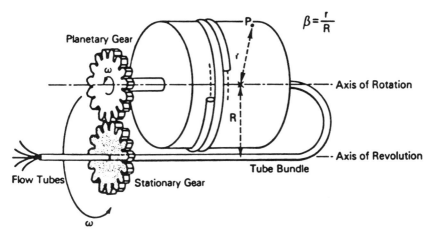

Figure 4.1. Epicyclic rotation of DuCCC column holder.

holder rotates about its own axis relative to the rotating frame and simultaneously revolves around the central axis of the centrifuge at the same angular velocity indicated by the pair of arrows. The epicyclic rotation of the holder is necessary to unwind the twist of the five flow tubes caused by the revolution, eliminating the use of rotary seals to connect each flow tube. As shown in Figs. 4.1 and 4.2, this unique design enables the performance of DuCCC using five flow channels connected directly to the column without using a rotary seal. When a column with a particular coil orientation is subjected to an epicyclic rotation, it produces a unique hydrodynamic phenomenon in the coiled column in which one phase entirely occupies the head side and the other phase occupies the tail side of the coil column. This unilateral phase distribution enables the performance of DuCCC in an efficient manner. A theoretical calculation of the hydrodynamic forces resulting from such an epicyclic rotation is very complicated and has not been elucidated.

4.3.   METHODS AND APPARATUS

The DuCCC experiments are performed with a table top model high-speed planet centrifuge equipped with a multilayer coiled column connected to five flow channels. The multilayer coiled column is prepared from 2.6-mm i.d. PTFE tubing by winding it coaxially onto the holder to a total volume capacity of 400 mL. The multilayer coiled column is subjected to an epicyclic rotation at 500–800 rpm. The fractions are collected simultaneously from both

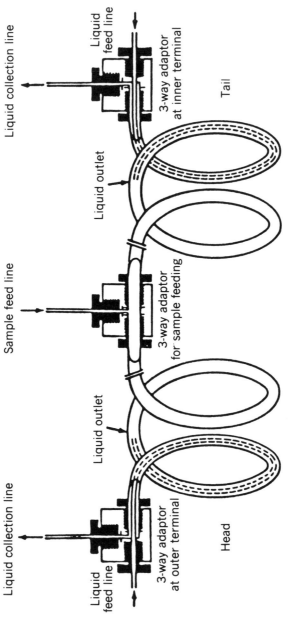

Liquid collection line

Liquid feed line

3-way adaptor at inner terminal

Tail

Liquid outlet

Sample feed line

3-way adaptor for sample feeding

Liquid outlet

Liquid collection line

Liquid feed line

3-way adaptor at outer terminal

Head

Figure 4.2. Column design for DuCCC.

96

ends of the column and analyzed by thin-layer chromatography (TLC) or high-performance liquid chromatography (HPLC) (8, 12).

4.4. APPLICATIONS

In the past decade, the rapid development of sophisticated spectroscopic techniques, including various two-dimensional nuclear magnetic resonance (2D-NMR) methods, automated instrumentation, and the routine availability of X-ray crystallography has greatly simplified structural elucidation in natural product investigations. Consequently, the challenge to today's chemists has shifted to one's capability of isolating the bioactive components from crude

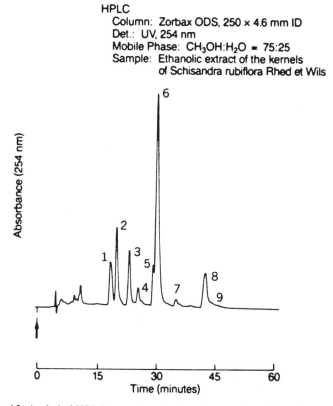

HPLC
Column: Zorbax ODS, 250 × 4.6 mm ID
Det.: UV, 254 nm
Mobile Phase: $CH_3OH:H_2O$ = 75:25
Sample: Ethanolic extract of the kernels
        of Schisandra rubiflora Rhed et Wils

Figure 4.3. Analytical HPLC trace of crude ethanol extract from *Schisandra rubriflora.*

extracts of either plants or animals. The extract of crude natural products usually is comprised of hundreds of components over a wide range of polarities. In isolating these natural products, it is essential to preserve the biological activity while performing chromatographic purifications. DuCCC represents one of the most efficient methods for isolation of the desired compound from a complex mixture.

DuCCC has several advantages over conventional HSCCC (13) in dealing with crude natural products. One distinct feature of DuCCC is the capability of performing normal phase and reverse-phase elutions simultaneously. This provides a highly efficient and unique method for separation of crude natural products. In many instances, fractions eluted from DuCCC are pure enough for recrystallization or structural study. For example, Fig. 4.3 shows an analytical HPLC trace of the crude ethanol extract of *Schisandra rubriflora*. Because the major bioactive lignan schisanhenol 6 is closely eluted with its acetate 5, it has been a major problem to obtain the pure lignan 6. The fractions collected from DuCCC after injection of a crude ethanol extract of *S. rubriflora* (125 mg) were analyzed by TLC (Fig. 4.4). Figure 4.5 shows the reverse-phase HPLC analyses of the DuCCC fractions eluted from the lower

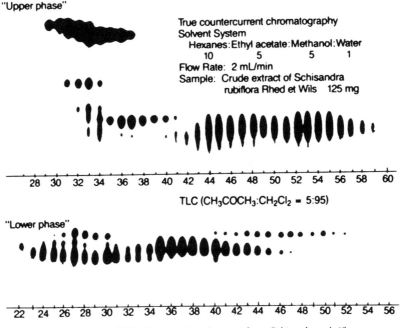

Figure 4.4. Dual CCC of crude ethanol extract from *Schisandra rubriflora*.

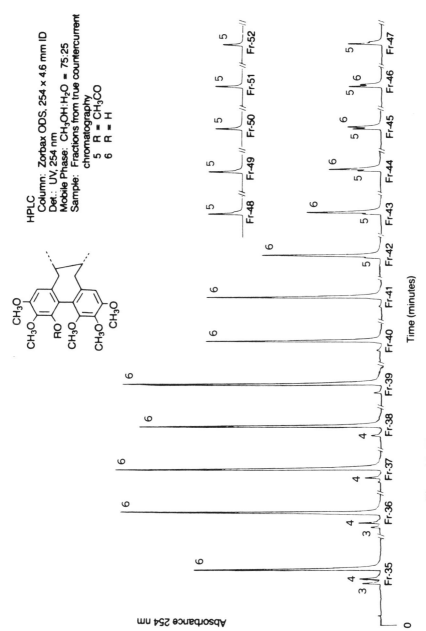

Figure 4.5. The HPLC traces of the fractions from DuCCC.

100

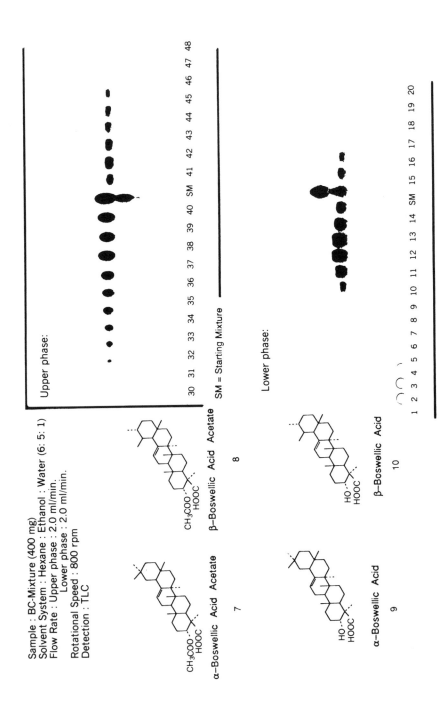

Figure 4.6. DualCCC of triterpenoic acids.

phase. The solvent system employed for DuCCC was hexane/ethyl acetate/ methanol/water (10:5:5:1). The upper phase, being less polar than the lower phase, results in a sequence of elutions similar to normal phase chromatography, while the lower phase provides a sequence of elution resembling reverse-phase chromatography. The bioactive components, schisanhenol acetate 5 and schisanhenol 6, were eluted in the lower phase. Reverse-phase HPLC analyses of fractions 36–40 accounted for 32 mg of almost pure schisanhenol 6. A total of 4 mg of schisanhenol acetate 5 was also obtained from fractions 50–57. As evidenced by this experiment, DuCCC offers an excellent method for semipreparative isolation of bioactive components from very crude natural products (11).

The isolation of the topoisomerase inhibitor boswellic acid acetate from its triterpenoic acid mixture has also been accomplished by DuCCC (12). As shown in Fig. 4.6, when an isomeric mixture of triterpenoic acids (400 mg) was subjected to DuCCC, using hexane/ethanol/water (6:5:1) as the solvent system, 215 mg of the boswellic acid acetates ($7 + 8, \alpha + \beta$ isomer) and 135 mg of the corresponding boswellic acid ($9 + 10, \alpha + \beta$ isomer) were obtained. Some highly polar impurities were eluted immediately in the solvent front, from fraction 1 to 4. The isomeric boswellic acids ($7 + 8$) were eluted in the lower phase solvent and the less polar acetates ($9 + 10$) were eluted simultaneously in the upper phase solvent. Although the α and β isomers were only partially

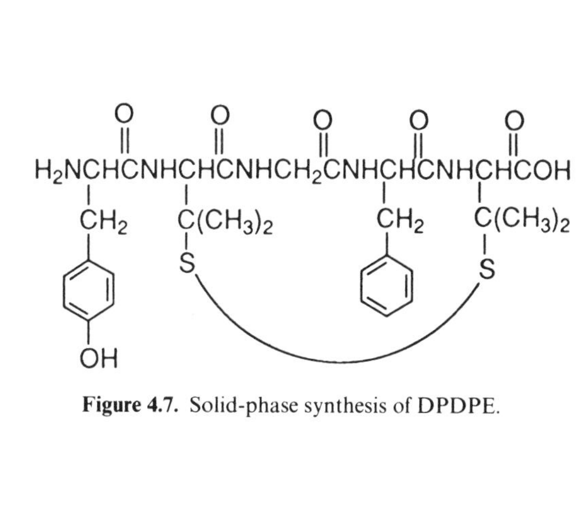

Figure 4.7. Solid-phase synthesis of DPDPE.

HPLC:

Sample: Crude DPDPE
Column: Zorbax-ODS (4.6mm x25 cm)
Solvent System: 60% aqueous CH_3CN (0.1% TFA)
Flow Rate: 1.0 mL/min
UV: 220 nm   Authentic Sample of DPDPE

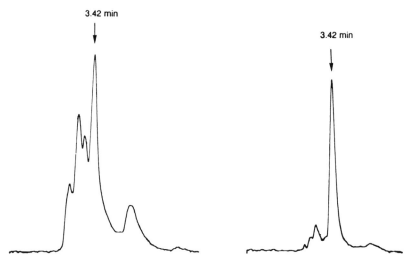

Figure 4.8. Analytical HPLC trace of crude DPDPE.

resolved by DuCCC, this experiment demonstrates that DuCCC is a highly
efficient system for preparative purification.

The conformationally restricted cyclic disulfide-containing enkephalin
analog [D-Pen, D-Pen] enkephalin (DPDPE), shown in Fig. 4.7, was synthesized by solid-phase methods. Its purification was accomplished previously
by partition on Sephadex G-25, using the solvent system [1-butanol/acetic
acid/water (4:1:5)], followed by gel filtration on Sephadex G-15 with 30%
acetic acid as the eluent (14). DuCCC demonstrated a highly efficient one step
method for the purification of DPDPE. As shown in Figs. 4.8 and 4.9, the
crude DPDPE (500 mg), which contained impurities and salts, was purified by
DuCCC with a two-phase solvent system consisting of 1-butanol containing
0.1% trifluoroacetic acid (TFA) and water also containing 0.1% TFA in a 1:1
(v/v) ratio. The desired DPDPE was eluted from the upper phase in fractions
15–19. The purity of each fraction collected was monitored by HPLC. A total
of 24 mg of pure DPDPE was obtained within 2 h. DuCCC can be a highly
cost effective procedure for purification of polypeptides.

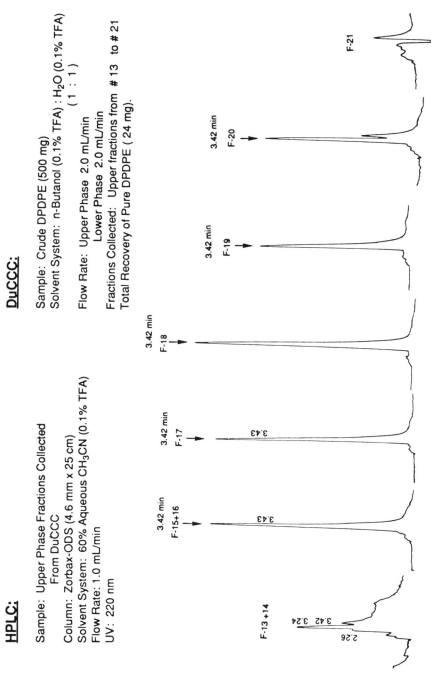

HPLC:

Sample: Upper Phase Fractions Collected
From DuCCC

Column: Zorbax-ODS (4.6 mm x 25 cm)

Solvent System: 60% Aqueous CH_3CN (0.1% TFA)

Flow Rate: 1.0 mL/min

UV: 220 nm

DuCCC:

Sample: Crude DPDPE (500 mg)

Solvent System: n-Butanol (0.1% TFA) : H_2O (0.1% TFA)
(1 : 1)

Flow Rate: Upper Phase 2.0 mL/min
Lower Phase 2.0 mL/min

Fractions Collected: Upper fractions from # 13 to # 21

Total Recovery of Pure DPDPE (24 mg).

Figure 4.9. The HPLC analysis of fractions collected from DuCCC.

103

4.5. CONCLUSION

Dual countercurrent cromatography (DuCCC) is a unique separation method that allows the performance of classic countercurrent distribution in a highly efficient manner. The system consists of a multilayer coiled column integrated with two inlet and two outlet flow tubes for solvent phases and a sample feed line. Subjecting the system to a particular combination of centrifugal and planetary motions produces a unique hydrodynamic effect that allows two immiscible liquids to flow countercurrently through the coiled column. The sample solution is fed at the middle portion of the column and is eluted simultaneously through the column in opposite directions by the two solvents. This distinct feature of maintaining a constant fresh two phases within the coiled column permits a rich domain of applications.

The capability and efficiency of DuCCC in performing classic countercurrent distribution has been demonstrated in the isolation of bioactive lignans and triterpenoic acids from crude natural products and in the purification of synthetic polypeptides. Besides the resolution and sample loading capacity, DuCCC offers a unique feature for elution of the nonpolar components in the upper phase solvent (assuming upper phase is less polar than the lower phase) and concomitant elution of the polar components in the lower phase. This capability results in an efficient and convenient preparative method for purification of the crude complex mixture.

The capability of DuCCC has not yet been explored. For instance, a particular solvent system can be selected to give the desired bioactive component a partition coefficient of 1. This will allow the "stripping" of the crude extract with DuCCC to remove the impurities or inactive components. Consequently, the bioactive component will be concentrated inside the column for subsequent collection. This strategy can also be applied to extract and concentrate certain metabolites in biological fluids such as urine or plasma. A large amount of sample can also be processed by DuCCC because there is no saturation of the stationary phase and the system can be easily automated with computer-assisted sample injection and fractionation.

REFERENCES

1. L. C. Craig, W. hausmann, P. Ahrens, and E. J. Harfenist, *Anal. Chem.*, **23**, 1326 (1951).
2. Y. Ito, *CRC Crit. Rev., Anal. Chem.*, **17**, 65 (1986).
3. Y. W. Lee, Y. Ito, Q. C. Fang, and C. E. Cook, *J. Liq. Chromatogr.*, **11**(1), 75 (1988).
4. T. Y. Zhang, X. Hua, R. Xiao, and S. Kong, *J. Liq Chromatogr.*, **11**(1), 233 (1988).

5. Y. W. Lee, C. E. Cook, Q. C. Fang, and Y. Ito, *J. Chromatogr.*, **477**, 434 (1989).

6. G. M. Brill, J. B. McAlpine, and E. J. Hochlowski, *J. Liq. Chromatogr.*, **8**, 2259 (1985).

7. D. G. Martin, R. E. Peltonen, and J. W. Nielsen, *J. Antibiot.*, **39**, 721 (1986).

8. Y. Ito, *J. Liq. Chromatogr.*, **8**(12), 2131 (1985).

9. H. Oka, K.-I. Harada, M. Suzuki, H. Nakazawa, and Y. Ito, *J. Chromatogr.*, **482**, 197 (1989).

10. Y. W. Lee, C. E. Cook, and Y. Ito. *J. Liq. Chromatogr.*, **11**(1), 37 (1988).

11. Y. W. Lee, Q. C. Fang, Y. Ito, and C. E. Cook, *J. Nat. Prod.*, **52**(1), 706 (1989).

12. Y. W. Lee, *True countercurrent chromatography of triterpenoic acids*, in preparation (1995).

13. W. Murayama, Y. Kosuge, N. Nakaya, Y. Nunogaki, N. Nunogaki, J. Cazes, and H. Nunogaki, *J. Liq. Chromatogr.*, **11**(1) 283 (1988).

14. H. J. Mosberg, R. Hurst, V. J. Hruby, K. Gee, H. I. Yamamura, J. J. Galligan, and T. F. Burks, *Proc. Natl. Acad. Sci. USA*, **80**, 5871 (1983).

CHAPTER

5

FOAM COUNTERCURRENT CHROMATOGRAPHY OF BACITRACIN COMPLEX

HISAO OKA

Aichi Prefectural Institute of Public Health, Tsuji-machi, Kita-ku, Nagoya 462, Japan

5.1. INTRODUCTION

Foam separation methods have been used for the separation of various samples ranging from metal ions to mineral particles (1). The separation is based on a unique parameter of foaming capacity or foam affinity for samples in aqueous solution. This technique has a great potential for an application to biological samples. However, utility of this method has been extremely limited, mainly due to a lack of efficient instruments. The conventional instruments for foam separation are equipped with a simple short separation column and the separation is carried out under unit gravity. Therefore, the separation is very inefficient. Recently, a remarkable improvement of foam separation technology has been achieved by development of foam countercurrent chromatography (CCC), which uses a long coiled column in a centrifugal force field (2).

The experimental results with foam CCC have shown better separation than the conventional foam separation method (2–4). However, complicated procedures are required to remove the surfactants and additives after fractionation. Many natural products have foaming capacity, and therefore foam CCC may be possible without surfactant or other additives for isolation and enrichment of natural products having foam producing capacity.

Bacitracin complex (BC) was selected as a test sample, because it has a strong foaming capacity. Foam CCC for separation and enrichment of BC has been conducted using nitrogen gas and distilled water entirely free of surfactant or other additives. This paper describes chromatographic fractionation of BC with batch sample loading, as well as concentration enrichment of foam active compounds from a bulk liquid on continuous sample feeding.

High-Speed Countercurrent Chromatography, Edited by Yoichiro Ito and Walter D. Conway.
Chemical Analysis Series, Vol. 132.
ISBN 0-471-63749-1 © 1996 John Wiley & Sons, Inc.

5.2. APPARATUS OF FOAM COUNTERCURRENT CHROMATOGRAPHY

The foam countercurrent chromatograph used is illustrated in Fig. 5.1 (2, 3). The motor drives the rotary frame around the central axis of the centrifuge. The rotary frame holds a coiled separation column and a counterweight symmetrically at a distance of 20 cm from the central axis of the centrifuge. A set of gears and pulleys produces synchronous planetary motion of the

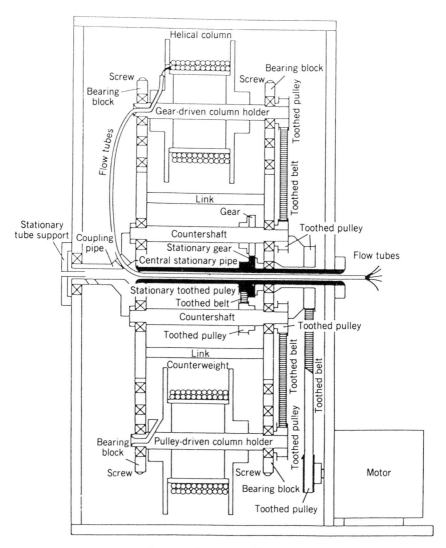

Figure 5.1. Illustration of foam CCC.

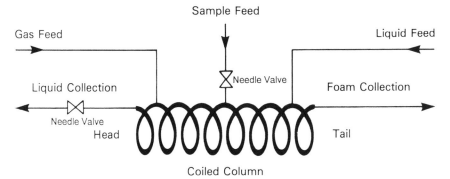

Figure 5.2. Column design for foam CCC.

coiled column. The planetary motion induces a true countercurrent movement between foam and its mother liquid through a long, narrow coiled tube. Introduction of the sample mixture into the coil results in separation of the sample components. Foam active components quickly move with the foaming stream and are collected from one end of the coil while the rest are carried with the liquid stream in the opposite direction and collected from the other end of the coil.

Figure 5.2 shows the column design for foam CCC. The coil consists of a 10-m long, 2.6-mm i.d. Teflon tube with a 50-ml capacity. The column is equipped with five flow channels. The liquid is fed from the liquid feed line at the tail and collected from the liquid collection line at the head. Nitrogen gas is fed from the gas feed line at the head and discharged through the foam collection line at the tail while the sample solution is introduced through the sample feed line at the middle portion of the coil. The head-tail relationship of the rotating coil is conventionally defined by an Archimedean screw force, where all objects of different density are driven toward the head. Both liquid and sample feed rates are each separately regulated with a needle valve while the foam collection line is left open to the air.

5.3.   PREVIOUS FOAM COUNTERCURRENT CHROMATOGRAPHY WORK

The previous applications of foam CCC are summarized in Table 5.1 (2–4). Ionic compounds were collected with suitable surfactants (2–4) and surface active proteins were separated in a phosphate buffer to prevent denaturation (2). Namely, in the cases of separations of rhodamine B and evans blue, DNP-leu and methylene blue, and methylene blue and rhodamine B, sodium

Table 5.1. Previous Application Works on Foam Countercurrent Chromatography

Subjects[a]	Additives
Rhodamine B (F) and evans blue (L)	SDS[b]
BSA (F) and sheep hemoglobin (L)	Na$_2$HPO$_4$[b]
DNP-leu (F) and methylene blue (L)	SDS[c]
Methylene blue (F) and DNP-leu (L)	CPC[c]
Methylene blue (F) and rhodamine B (F)	SDS and Na$_2$ HPO$_4$[d]

[a]Effluent from foam line = F; effluent from liquid line = L.
[b]Y. Ito, *J. Liq. Chromatogr.*, **8**, 2131 (1985).
[c]M. Bhatnagar and Y. Ito, *J. Liq. Chromatogr.*, **11**, 21 (1988).
[d]Y. Ito, *J. Chromatogr.*, **403**, 77 (1987).

dodecyl sulfate (SDS), which is an anion surfactant, was used as a collector and positively charged compounds were collected in the foam fraction. In the case of separation of methylene blue and DPN-leu, cetylpyridinium chloride, which is a cation surfactant, was used as a collector and negatively charged compounds were collected in the foam fraction. Several kinds of proteins including bovine serum albumin (BSA) (2), human and sheep hemoglobins, and ovalbumin were subjected to foam CCC. Only BSA showed an inherent foam-producing ability and was collected through foam collection line, whereas other proteins were mostly eluted through the liquid collection line without any denaturation. However, the BSA eluted through the foam line was denatured by exposure to gas–liquid interfaces. The use of disodium hydrogen phosphate for the separation of BSA and sheep hemoglobin prevented the denaturation of BSA. In order to explore the possibility of separation based on the subtle differences in foam affinity, addition of salt to surfactant solution was tried for the separation of two basic dyes, rhodamine B and methylene blue, by foam CCC (4). The foam affinity of these two dyes was reduced by adding a high concentration (0.5 M) of salt such as disodium hydrogen phosphate to the surfactant solution, yielding complete peak resolution between these two dyes.

5.4. SEPARATION OF BACITRACIN COMPONENTS WITH NITROGEN AND ADDITIVE-FREE WATER

Bacitracin is a basic cyclic peptide antibiotic consisting of more than 20 components. This antibiotic is commonly used worldwide as a feed additive for livestock. Chemical structures of these components are still unknown except for BCs-A and F. Figure 5.3 shows a high-performance liquid chromatogram (HPLC) of commercial BC (5). In this study, special attention

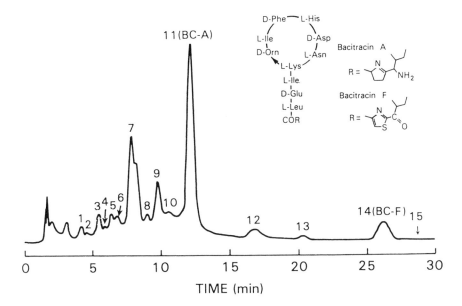

Figure 5.3. High-performance liquid chromatographic separation of commercial bacitracin components.

Chromatographic conditions

Column: Capcell Pak C18 (5 μm, 150 × 4.6 mm, i.d.)

Mobile phase: methanol-0.04M aqueous disodium phosphate solution 68-32

Flow-rate: 1.0ml/min; Detection: 234nm

was paid to four peaks, 3, 7, 11, and 14, to evaluate their separation and enrichment in foam CCC (6, 7).

First, operational conditions for foam fractionation of BC components were optimized. With the use of a set of fixed conditions for nitrogen gas flow-rate, sample size, fractionation rate, and the column rotation at 500 rpm, we investigated the effects of the liquid rate, opening of the needle valve at the liquid outlet, and standing time after sample injection, on the separation efficiency.

Separation was initiated by simultaneous introduction of distilled water from the tail and nitrogen gas from the head into the rotating column while the needle valve at the liquid collection line was fully open. After a steady-state hydrodynamic equilibrium was reached, the pump was stopped and the sample solution was injected through the sample port. After the desired

Table 5.2. Optimization of Operational Conditions for Foam Fractionation of Bacitracin

Conditions[a]		Results
Liquid flow-rate	$<3.2\,\text{mL/min}$	Failing to elute foam
	>3.2	Less efficiency of separation (peaks 11 and 14)
Needle valve at liquid outlet		
	<0.5 turn open	Less efficiency of separation (all peaks)
	0.5–0.8	Peak 7 elute from foam outlet
	0.8–1.2	Peak 7 elute from liquid outlet
	>1.2	No foam
Standing time after sample injection		
	<5 min	Less efficiency of separation (peaks 11 and 14)
	>5 min	Intermittent foam elution

[a]Sample size, 0.5 mg in 0.5 mL of distilled water; nitrogen gas pressure, 80 psi; revolution speed, 500 rpm; fractionation-rate, 15 s tube.

standing time, the needle valve opening was adjusted to the desired level and pumping was started again. Effluents from both outlets were collected at 15-s intervals. As shown in the optimization summary in Table 5.2, liquid flow rates lower than 3.2 mL/min failed to elute foam while those higher than 3.2 mL/min gave less efficient separation between peaks 11 and 14. Therefore, 3.2 mL/min was selected as the liquid flow-rate.

When the needle valve at the liquid outlet was opened less than 0.5 turn, all peaks showed less efficient separation. Opening the valve between 0.5 and 0.8 turn, eluted peak 7 from the foam outlet, while the same peak eluted from the liquid outlet with the valve opening between 0.8 and 1.2 turns. Valve opening over 1.2 turns gave no foam fraction. Therefore, we adjusted the needle valve opening at 0.8 turn in the subsequent experiments.

With respect to the standing time after sample injection, a standing time less than 5 min gave inefficient separation between peaks 11 and 14, while no foam eluted for a standing time over 5 min. Therefore, 5 min was selected for the standing time after sample injection. The separation of bacitracin components was carried out using these conditions. Elution curves of BC components from the foam and liquid outlets were drawn by spectrophotometric analysis of fractions, which were also analyzed with reversed-phase HPLC.

Figure 5.4 shows the elution curve of BC components from the foam outlet. The vertical axis indicates absorbance at 234 nm and the horizontal line

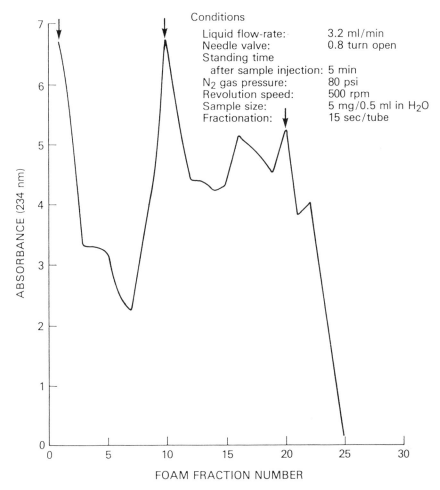

Figure 5.4. Elution curve of bacitracin components from foam line.

indicates the fraction number. This elution curve shows three major peaks indicated by arrows. The fractions corresponding to these peaks were subjected to HPLC analysis.

The HPLC chromatograms of BC components in foam fractions are shown in Fig. 5.5. Under the present reversed-phase HPLC conditions, BC was separated into more than 15 peaks. Generally, hydrophilic compounds show shorter retention time than hydrophobic compounds under the reversed-phase HPLC conditions. In this study, the HPLC elution time is used to estimate the polarity of BC components. Namely, BC-A, which elutes earlier in HPLC, is more hydrophilic than BC-F. The most hydrophobic compounds

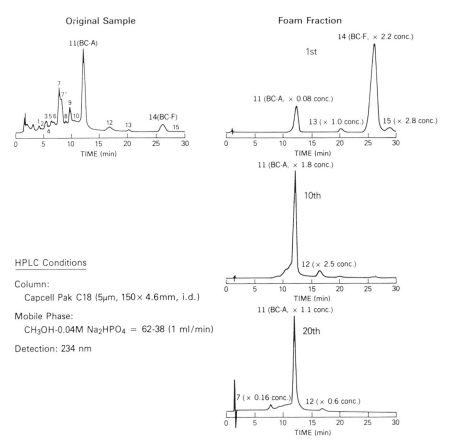

Figure 5.5. High-performance liquid chromatographic analyses of bacitracin components in foam fractions.

with the longest retention time in HPLC analysis, corresponding to peaks 14 and 15, were eluted in the first foam fraction with a small amount of peaks of less hydrophobic compounds corresponding to peaks 11 and 13. Peak 15 is hardly visible in the HPLC chromatogram of the original sample due to low concentration, but the same peak is clearly observed in the chromatogram of the first foam fraction. As peak 11, BC-A was almost isolated from other components in the 10th fraction. In the 12th fraction, peak 7 appeared on the HPLC chromatogram. Components with lower hydrophobicity than peak 7 did not appear in the foam fractions. These results clearly indicate that the bacitracin components are separated in the order of hydrophobicity of the molecule in the foam fractions, the most hydrophobic compounds being eluted first.

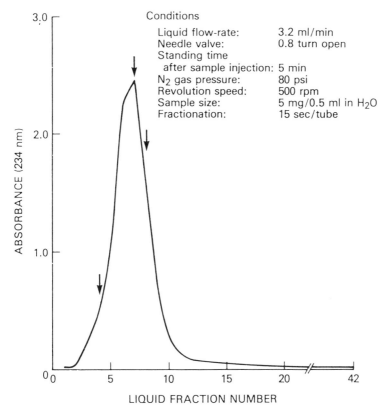

Figure 5.6. Elution curve of bacitracin components from liquid line.

The elution curve of BC components from the liquid outlet is shown in Fig. 5.6. It is quite different from that obtained from the foam fractions. There is only a single peak observed. The interesting HPLC chromatograms obtained from fractions 4, 6, and 9 at the three points indicated by arrows are shown in Fig. 5.7. The chromatogram of the fourth fraction shows, beside peaks 1 through 10, a group of more hydrophilic components, which are hardly visible in the chromatogram of the original sample due to the low concentration. In the sixth fraction, these hydrophilic compounds tend to disappear and peaks 2–10 dominate. In the ninth fraction, peaks 5 and 7' still remain as the main peaks while the intensity of peak 10 becomes much enhanced compared to that in the sixth fraction. Later fractions showed no additional peaks eluting. These results clearly indicate that the BC components elute in increasing order of their hydrophilicity in the liquid fractions.

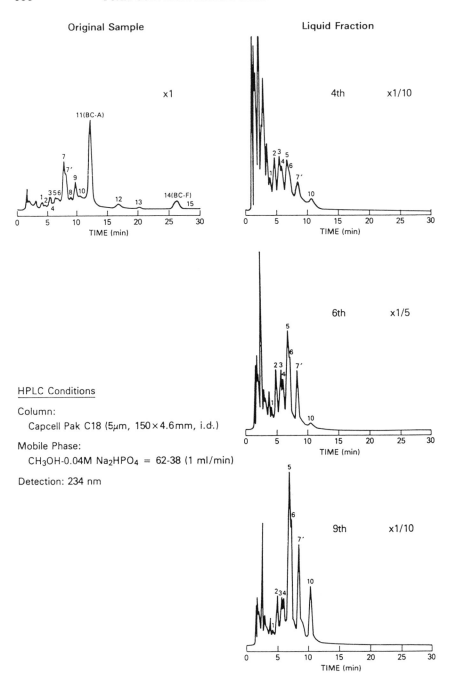

Figure 5.7. High-performance liquid chromatographic analyses of bacitracin components in liquid fraction.

5.5. CONTINUOUS ENRICHMENT AND STRIPPING OF BACITRACIN COMPONENTS

As described above, we were able to separate BC components in the order of hydrophobicity using foam CCC without any surfactant or other additives. Next, enrichment and concentration of BC components on continuous sample feeding are described below (8).

First, we conducted a series of preliminary studies to optimize the operational conditions for enrichment and concentration of BC components on continuous sample feeding while the sample size, sample feed rate, and nitrogen gas feed pressure were fixed as indicated at the bottom of Table 5.3. Liquid was not pumped from the liquid feed line, because liquid feeding dilutes the solutes and interferes with foam formation. All experiments were performed at a revolutional speed of 500 rpm as in above studies.

Effects of the needle valve opening at the liquid outlet and sample concentration on the enrichment was investigated with a 100-mL sample volume while varying the concentration from 10 to 100 ppm. Enrichment was initiated by introduction of nitrogen gas in the rotating column and the needle valve opening was adjusted to various levels while the sample was continuously introduced from the sample port using nitrogen gas pressure kept at 40 psi. Effluents from the foam and liquid outlets were each separately pooled in a container. Both effluents were analyzed by HPLC.

The results showed that opening the needle valve less than 1.0 turn or over 5.0 turns yielded no foam fraction. When the needle valve was opened between

Table 5.3. Optimization of Operational Conditions for Enrichment of Bacitracin on Continuous Sample Feeding

Conditions[a]	Enriched Concentration (times)			
	Peak 3	Peak 7	Peak 11	Peak 14
Needle valve				
< 1.0 turn open	—*	—	—	—
1 ∼ 3	13–77	52–180	180–4,000	220–11,670
> 5.0				
Sample concentration				
< 25 ppm	—	—	—	—
25	27–77	52–130	180–350	220–410
50	13–65	87–170	630–2,000	1,340–7,700
100	55–77	94–150	280–4,000	570–11,670

[a]Sample size, 100 mL in H_2O; sample feeding rate, 1.5 mL/min (40 psi); N_2 gas pressure, 80 psi; liquid feeding rate, 0; * is no foam eluted.

1 and 3 turns, HPLC peak 3 was enriched 13–77 times, peak 7 was enriched 52–180 times, peak 11 was enriched 180–4000 times and peak 14 was enriched 220 to over 11,000 times.

Studies on the sample concentration showed that no foam eluted at less than 25 ppm. As the concentration was increased to 25 ppm we were able to enrich the components. At 50 ppm sample concentration peak 11 was enriched 2000 times and peak 14 was enriched nearly 8000 times. Finally, at the 100 ppm sample concentration, peak 11 was enriched 4000 times and peak 14 was enriched over 11,000 times.

On the basis of these experimental results, we set foam CCC conditions for the large-scale sample feeding as shown in Table 5.4. Namely, needle valve, 2.0 turns open; sample concentration, 50 ppm; sample size, 2.5 L; sample feed rate, 1.5 mL/min at 40 psi; nitrogen gas feed pressure, 80 psi; liquid flow-rate, 0; sample collection, pooling the foam and the liquid effluents separately; and revolution speed, 500 rpm.

Figure 5.8 shows the results of HPLC analyses of bacitracin in the foam and liquid fractions obtained by large-scale continuous foam CCC. As demonstrated in the preliminary studies, the concentration in the foam fraction increases with the hydrophobicity of the components: Peak 3 was enriched 22 times; peak 7 was enriched 31 times; peak 11 was enriched 1400 times; peak 12 was enriched 1070 times; peak 13 was enriched 1380 times; and peak 14 was enriched 2260 times. In the liquid fraction, small amounts of peak 3 and 7 were detected.

Thus continuous enrichment and concentration in foam CCC is quite effective for the detection and isolation of a small amount of natural products with foaming capacity.

5.6. CONCLUSION

By using foam CCC, we were able to separate the components of bacitracin in the order of hydrophobicity and to continuously enrich the components

Table 5.4. Experimental Conditions for Large-Scale Foam CCC on Continuous Sample Feeding

Needle valve	2.0 turns open
Sample concentration	50 ppm
Sample size	2500 mL
Sample feeding rate	1.5 mL/min (40 psi)
N_2 gas pressure	80 psi
Liquid flow-rate	0
Sample collection	pooled
Revolution speed	500 rpm

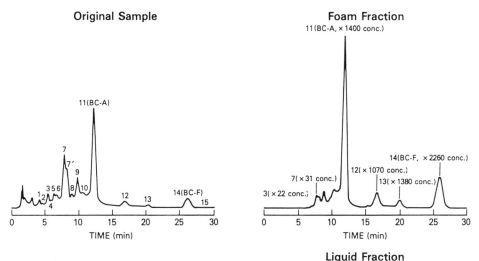

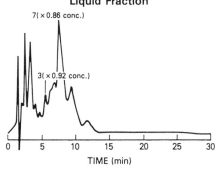

Figure 5.8. High-performance liquid chromatographic analyses of bacitracin components in foam and liquid fraction.

without any surfactant or other additives. The present method provides a number of advantages over other chromatographic methods as summarized below:

1. Enriched concentration of foam active sample.
2. Minimum decomposition or deactivation of biological samples.
3. No adsorptive sample loss onto the solid support matrix.
4. No risk of contamination.
5. Easy recovery of the sample after fractionation.
6. Low cost in operation.

Therefore, we believe that the present method has great potential in enrichment, stripping, and isolation of various natural and synthetic products in research laboratories and industrial plants.

REFERENCES

1. P. Somasudaran, *Sep. Purif. Methods*, **1**, 117 (1972).
2. Y. Ito, *J. Liq. Chromatogr.*, **8**, 2131 (1985).
3. M. Bhatnagar and Y. Ito, *J. Liq. Chromatogr.*, **11**, 21 (1988).
4. Y. Ito, *J. Chromatogr.*, **403**, 77 (1987).
5. H. Oka, Y. Ikai, N. Kawamura, M. Yamada, K.-I. Harada, Y. Yamazaki, and M. Suzuki, *J. Chromatogr.*, **462**, 315 (1989).
6. H. Oka, K.-I. Harada, M. Suzuki, H. Nakazawa, and Y. Ito, *Anal. Chem.*, **61**, 1988 (1989).
7. H. Oka, K.-I. Harada, M. Suzuki, H. Nakazawa, and Y. Ito, *J. Chromatogr.*, **482**, 197 (1989).
8. H. Oka, K.-I. Harada, M. Suzuki, H. Nakazawa, and Y. Ito, *J. Chromatogr.*, **538**, 213 (1991).

CHAPTER

6

pH-PEAK-FOCUSING AND pH-ZONE-REFINING COUNTERCURRENT CHROMATOGRAPHY

YOICHIRO ITO

Laboratory of Biophysical Chemistry, National Heart, Lung, and Blood Institute, National Institutes of Health, Bethesda, Maryland 20892

6.1. INTRODUCTION

This chapter is devoted to two closely related techniques: pH-peak-focusing and pH-zone-refining countercurrent chromatography. Inquiry about the cause of an abnormally sharp elution peak found in the course of purification of N-bromoacetyl-3,3',5-triiodo-L-thyronine (BrAcT$_3$) has led to the development of these two highly efficient preparative methods. These new techniques open a rich domain of applications for both analytical and preparative separations.

6.2. pH-PEAK-FOCUSING COUNTERCURRENT CHROMATOGRAPHY

6.2.1. Research for the Causative Agent Sharpening the BrAcT$_3$ Peak

Countercurrent chromatography (CCC) is a support-free liquid–liquid partition chromatography where the partition process takes place in an open column space free of a solid support matrix. Consequently, the method can eliminate an adsorptive effect, such as tailing of solute peaks, and usually yields symmetrical solute peaks where the peak width naturally increases with the retention time.

In the course of purifying BrAcT$_3$ by CCC fractionation (see Chapter 11, in this volume) we encountered an unusual phenomenon in that the product formed an extremely sharp elution peak with an efficiency of over 2000 TP (theoretical plates), while an earlier impurity peak showed normal broadness

High-Speed Countercurrent Chromatography, Edited by Yoichiro Ito and Walter D. Conway.
Chemical Analysis Series, Vol. 132.
ISBN 0-471-63749-1 © 1996 John Wiley & Sons, Inc.

of less than 500 TP (1, 2). Figure 6.1A shows the CCC elution profile of the reaction product obtained by bromoacetylation of T_3, monitored by the radioactivity of ^{125}I. The separation was performed with a two-phase solvent system composed of hexane, ethyl acetate, methanol, and 15 mM ammonium acetate (pH 4) at a volume ratio of 1:1:1:1 by eluting the lower aqueous phase

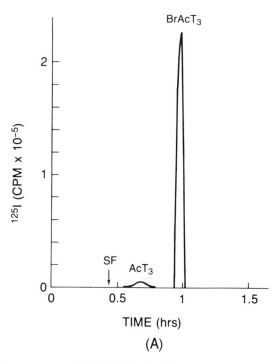

Figure 6.1. (A) Chromatogram of BrAcT$_3$ obtained with a standard two-phase solvent system (1). The BrAcT$_3$ is eluted in an abnormally sharp peak while the preceding AcT$_3$ peak has a normal profile. Experimental conditions: Apparatus consists of a HSCCC centrifuge equipped with a semipreparative multilayer coil separation column (1.6 mm i.d. and $\sim$315 mL capacity); solvent system is hexane/ethyl acetate/methanol/15 mM ammonium acetate (1:1:1:1), pH 4; sample is a crude bromoacetylation mixture (from 0.1 mmol of T$_3$ and a minute amount of $[^{125}I]T_3$) in 4 mL of solvent consisting of equal volumes of upper and lower phases; mobile phase is the lower aqueous phase (pH 5.2); flow-rate = 3 mL/min; revolution = 800 rpm; detection = ^{125}I radioactivity; retention of stationary phase is 69.8% of the total column capacity. (B) Chromatogram of BrAcT$_4$ obtained with a modified solvent system. The product peak formed a broad, skewed profile whereas the preceding impurity peak shows a sharp profile. Solvent system is hexane/ethyl acetate/methanol/15 mM ammonium acetate (4:5:4:5), pH 4; sample is the concentrate obtained after bromoacetylation of 0.1 mmol of T$_4$ diluted to 4 mL with the solvent consisting of equal volumes of each phase; other conditions are similar to those described in (A). Retention of the stationary phase was 60% (C): Chromatogram of BrAcT$_3$ obtained with a modified solvent system. Note that BrAcT$_3$ eluted in a broad peak, whereas AcT$_3$, which formed a broad peak in the standard solvent system (A), shows a sharpened peak profile (1). Experimental conditions are identical to those described in (A) except that the solvent/volume ratio was modified to 4:5:4:5. Retention of the stationary phase was 64.7% (SF = solvent front).

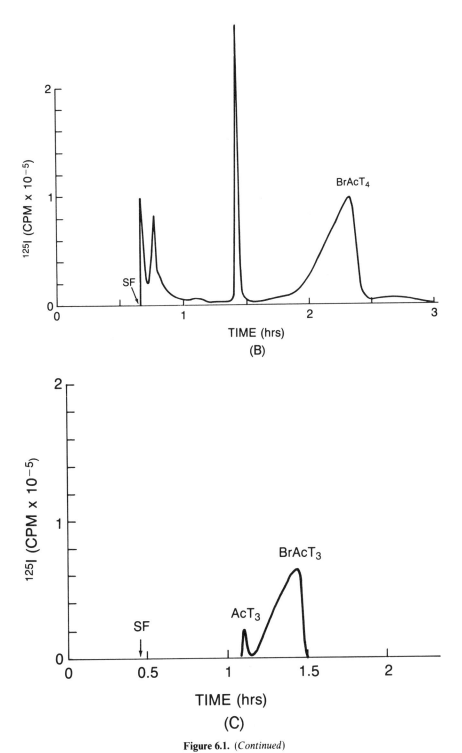

Figure 6.1. (*Continued*)

at a flow rate of 3 mL/min. In this chromatogram, the main $BrAcT_3$ peak shows a much sharper profile than the earlier eluting impurity peak, which was shown by mass spectromerty (2) to be N-acetyl-T_3 (AcT_3).

When N-bromoacetyl-L-thyroxine ($BrAcT_4$), an analog of $BrAcT_3$, was purified by CCC using a slightly modified solvent system, a normally broad, but unusually skewed peak of 300 TP was observed (1, 3). Figure 6.1B similarly illustrates the CCC elution profile of the reaction product obtained by bromoacetylation of T_4, and by using a modified solvent ratio of 4:5:4:5. In contrast to the chromatogram in Fig. 6.1A, the major peak ($BrAcT_4$) has a normally broad but somewhat skewed shape while a preceding impurity peak is unusually sharp, like the $BrAcT_3$ peak in Fig. 6.1A.

The possible cause of the sharp peaks may involve either the chemical nature of a particular solute or the effect of an exogenous agent in the solvent. Data obtained by modifying the experimental conditions supported the latter. For example, when the $BrAcT_3$ reaction mixture was eluted with the solvent system at a modified volume ratio of 4:5:4:5 (Fig. 6.1C), retention times of both $BrAcT_3$ and the preceding AcT_3 peaks were increased as expected due to an increase in the relative polarity of the stationary phase. However, the $BrAcT_3$ peak became much broader, skewing toward the solvent front, recalling the appearance of the $BrAcT_4$ peak in Fig. 6.1B. The preceding AcT_3 peak, on the other hand, was much sharper. This indicates that the sharp peak profile of $BrAcT_3$ in Fig. 6.1A is not an inherent property of the compound but instead depends on the retention time of the solute. One way in which this could occur is if an agent present in the sample solution, but invisible to the detector, affects the partitioning of the $BrAcT_3$ peak. That this is indeed the case is strongly supported by our finding that reduction of the size of the initial sample of bromoacetylation product or its dilution yielded a normally broad $BrAcT_3$ peak. Finally, rechromatography of a sample collected from the sharp peak also produced a normally broad peak.

In order to prove this point further, a sample was run using the same derivatization procedure but omitting T_3. Addition of this blank bromoacetylation solution to a small amount of crude $BrAcT_3$ substantially increased the sharpness of the $BrAcT_3$ peak compared to that obtained when the $BrAcT_3$ alone was analyzed (not shown).

A search for the substance responsible for sharpening the $BrAcT_3$ peak was undertaken by fractionating the blank bromoacetylation solution by CCC using the standard solvent system composed of hexane, ethyl acetate, methanol, and 15 mM ammonium acetate (pH adjusted to 4.0 with acetic acid) at a volume ratio of 1:1:1:1. The effluent from the CCC column was continuously monitored with a UV monitor at 206 nm and then collected in a fraction collector. The resulting chromatogram (not shown) revealed a flat baseline except for a small peak at the solvent front showing that the substance causing peak sharpening indeed has no chromophor. However, the pH of these collected fractions revealed a gradual drop starting near the solvent front to a minimum of pH 4.6 followed by an abrupt return to the original level of 5.3 at the retention time corresponding to the front of the sharp $BrAcT_3$ peak in Fig. 6.1A. This finding was reproduced in a run where a small amount of prechromatographed $BrAcT_3$ had been added to the blank bromoacetylation solution: The abrupt return to pH 5.3 again coincided with the beginning of the sharp $BrAcT_3$ peak (Fig. 6.2). These results clearly indicate that an acidic component(s) present in the bromoacetylation solution creates a pH gradient in the CCC effluent, strongly affecting the eluting peaks. Mass spectrometric analysis of the blank bromoacetylation solution showed the presence of much methyl bromoacetate (as expected from the

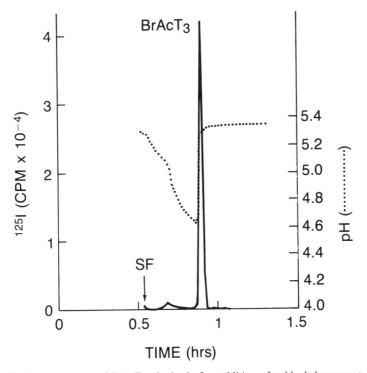

Figure 6.2. Chromatogram of $BrAcT_3$ obtained after addition of a blank bromoacetylation product (reaction mixture made without T_3). Note that the sharp $BrAcT_3$ peak coincides with the abrupt return point of the effluent pH shown by the dotted line. The sample solution was prepared by addition of 400 μL of blank bromoacetylation concentrate to $BrAcT_3$ concentrate corresponding to 0.025 mmol of T_3 and dissolution of this mixture in 4 ml of solvent. Other experimental conditions are the same as those described in Fig. 6.1A. Retention of the stationary phase was 56.8% (SF = solvent front).

methanol workup), in addition to a considerable amount of bromoacetic acid.

Each of these components and some other acids, such as HCl and trifluoroacetic acid (TFA), were then tested to observe their ability to reproduce the characteristic pH change in the CCC effluent as well as the sharpness of the BrAcT$_3$ peak. Inorganic acids, including HCl and HBr, eluting rapidly due to their high polarity, caused a sharp pH drop at the solvent front. Methyl bromoacetate failed to create any significant pH change in the CCC eluent. As expected, neither of these two groups of compounds produced a sharpened BrAcT$_3$ peak when added to a very dilute sample of the bromoacetylation product, that is, one that produced a normal, broad BrAcT$_3$ peak. On the other hand, organic acids, such as bromoacetic acid and TFA, did cause a characteristic pH change in the effluent, which was quite similar to that observed with the blank bromoacetylation solution. Furthermore, bromoacetic acid added to pure BrAcT$_3$ did indeed affect its elution profile. Figure 6.3 shows both this profile (solid line) detected by the radioactivity of ^{125}I along with the pH curve (dotted line). In a control experiment (Fig. 6.3A) the sample solution contained no acid additives, and BrAcT$_3$ is eluted in a relatively broad peak with a short retention time of less than 40 min. Addition of 1 mmol of bromoacetic acid to the sample solution (Fig. 6.3B) produced a sharpened BrAcT$_3$ peak with a delayed elution time at 70 min. The pH curve displays a pattern similar to that produced by the blank bromoacetylation product (Fig. 6.2). A close association between the abrupt return point of the pH curve and the sharp BrAcT$_3$ peak (Fig. 6.3B) shows that the bromoacetic acid is the agent responsible for sharpening the BrAcT$_3$ peak.

6.2.2. Chemohydrodynamic Mechanism of pH-Peak-Focusing

The above peak sharpening effect can be seen in detail in Fig. 6.4A, where a schematic view of the column is shown with the stationary and mobile phases arbitrarily separated. Thus, BrAcT$_3$ anions that happen to be at location 1 in the presence of bromoacetic acid, will rapidly protonate and enter the non-

Figure 6.3. Chromatogram of prepurified BrAcT$_3$ obtained with no acid additives (A) and with 1 mmol of bromoacetic acid (B) in the sample solution (1). The sharp peak profile of BrAcT$_3$ in Fig. 6.2 was reproduced by adding bromoacetic acid in the sample solution (B): Close association between the abrupt return point of the pH curve and the sharp BrAcT$_3$ peak strongly suggests that bromoacetic acid is the causative agent of the sharp peak. Sample solution A is CCC-purified-BrAcT$_3$ ($\sim$0.04 mmol) in 3.5 mL of the lower aqueous phase; sample solution B is CCC-prepurified BrAcT$_3$ ($\sim$0.04 mmol) and bromoacetic acid (1 mmol) in 3.5 mL of lower phase. Other experimental conditions are described in Fig. 6.1A. Retention of the stationary phase was 67% in A and 68% in B (SF = solvent front).

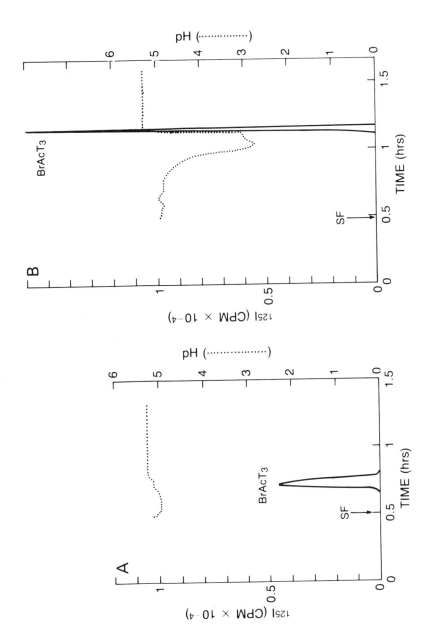

127

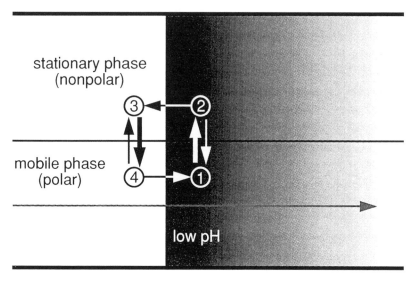

stationary phase
(nonpolar)

mobile phase
(polar)

low pH

(A)

REQUIREMENT FOR SHARP PEAK FORMATION

$K_b < K_a < K_{pH}$ Peak 1

$K_b \ll K_{pH} < K_a$ Peak 2

$K_b < K_{pH} \ll K_a$ Peak 3

$K_{pH} < K_b < K_a$ Peak 4

PARTITION COEFFICIENT
$(K = C_s/C_m)$

K_{pH} : acid agent

K_a : sample in low pH

K_b : sample in high pH

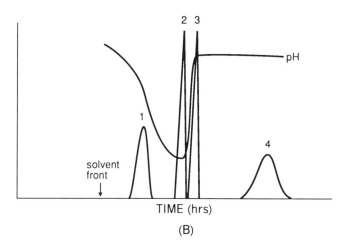

pH

2 3

1

4

solvent
front

TIME (hrs)

(B)

polar stationary phase (location 2). As the bromoacetic acid edge, sharply defined due to overload, moves forward, the compound finds itself in location 3, where the stationary phase is in contact with the higher pH mobile phase. Ionization of the carboxylic acid occurs and the $BrAcT_3$ enters the mobile phase (location 4), whereupon it rapidly accelerates to location 1 and the process is repeated. Thus, although elution of the sample is delayed, it undergoes the same partitioning action with distribution and diffusion lessened by the focusing action of the bromoacetic acid edge.

Clearly, for this focusing action to take place there must be a particular relationship between the partition coefficients of sample and causative agent. Figure 6.4B further details the general requirements for the effect to be manifest. Thus when partition coefficients (C_s/C_m) of both the free acid (K_a) and its salt (K_b) are less than the partition coefficient of the causative acid (K_{pH}), the solute will elute earlier (peak 1) than the pH gradient and there will be no focusing. On the other hand, when both K_a and K_b are greater than K_{pH}, the solute will elute after the gradient (peak 4), again without focusing. Peak focusing occurs only when K_{pH} falls between K_a and K_b as shown in peaks 2 and 3.

As shown in Fig. 6.1B, $BrAcT_4$, an analog of $BrAcT_3$, elutes as a broad peak in spite of the fact that the sample solution also contains bromoacetic acid. This can be explained on the basis of the different partition coefficients of these two compounds. The T_3 elutes before T_4 on C_{18} reversed-phase high-performance liquid chromatography (2). Similarly, $BrAcT_4$ is more hydrophobic than $BrAcT_3$ and therefore retained longer in the nonaqueous stationary phase so that it is eluted from the column considerably later than bromoacetic acid. However, the elution time of the pH gradient can be adjusted by injecting the pH-gradient-forming agent into the column through the sample port at a selected time during the CCC run so that it elutes just before $BrAcT_4$. This possibility was examined by using TFA as a pH-gradient-forming agent since it has a low-boiling point (72.4°C) and is easily eliminated from the collected fractions by evaporation under reduced pressure. The chromatogram in Fig. 6.5A was obtained by injecting $400\,\mu L$ of TFA into the column after $150\,mL$ of mobile phase had already been eluted. Now the $BrAcT_4$ peak is much sharper than that observed in Fig. 6.1B.

Figure 6.4. Chemohydrodynamic mechanism of pH-peak-focusing (1). (A) Schematic illustration of the peak-focusing process in the separation column. A portion of the column contains the nonpolar stationary phase in the upper half and the polar mobile phase in the lower half, where the solute molecules circulate at the sharp edge of the low pH region (shaded). For a more detailed description, see the text. (B) General requirements for pH-peak-focusing. Peak 1 is obtained when both K_a and K_b are smaller than K_{pH}, while peak 4 is obtained when both K_a and K_b are greater than K_{pH}. Peaks 2 and 3 are focused because K_{pH} falls between K_a and K_b.

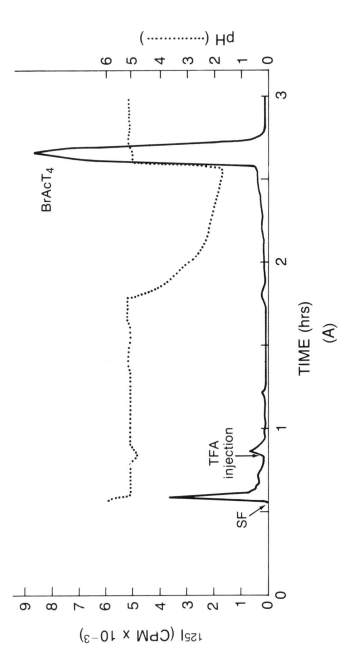

Figure 6.5. Chromatogram of BrAcT$_4$ obtained by two different peak-focusing methods. (A) A relatively sharp BrAcT$_4$ peak was produced from the solvent system composed of hexane/ethyl acetate/methanol/15 mM ammonium acetate (4:5:4:5), pH 4, by injecting TFA at the middle of the CCC run; 400 µL of TFA were injected into the column after 150 mL of the mobile phase were eluted. Note that the elution of the BrAcT$_4$ peak coincides with the abrupt return point of the eluate pH shown in the dotted line. The sample solution was prepared by dissolving 0.04 mmol of CCC-prepurified BrAcT$_4$ in 4 mL of the solvent. (B) A much sharper BrAcT$_4$ peak was produced by modifying the volume ratio of the solvent system to 1:1:1:1, pH 4, and adding 400 µL of TFA in the sample solution. Experimental conditions including flow rate, revolution speed, and so on, were identical to those described in Fig. 6.1B. Retention of the stationary phase was 39.5% (A) and 42.0% (B) of the total column capacity (SF = solvent front).

130

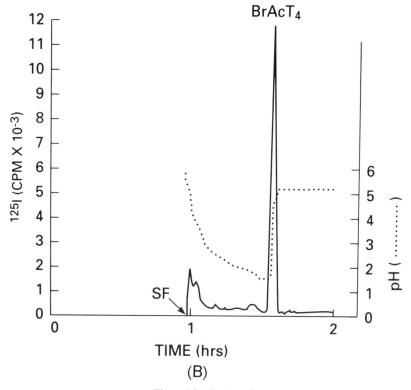

Figure 6.5. (*Continued*)

An alternative and more effective technique for bringing about the same effect would be to choose a suitable combination of the solvent system and gradient-forming organic acid in the sample solution so that the target solute is trapped into the low pH region during the fractionation, as illustrated in Fig. 6.4A. This scheme was effected by modifying the solvent ratio to 1:1:1:1 and introducing TFA in the sample solution. Figure 6.5B shows a chromatogram of CCC-prepurified $BrAcT_4$ (0.04 mmol) obtained by adding 400 μL of TFA in the sample solution. Further studies have shown that the amount of TFA can be reduced to 200 μL for microgram quantities of $BrAcT_4$.

The influence of the pH gradient described above may be observed with many other organic acids as solutes. When the pH of the two-phase solvent system is substantially higher than the pK_a of the target solute, the majority of these acids exist in an ionized form favoring the polar aqueous phase during partition. When the solvent pH is reduced below the pK_a of the compound, the un-ionized acids partition more to the stationary nonaqueous phase. If

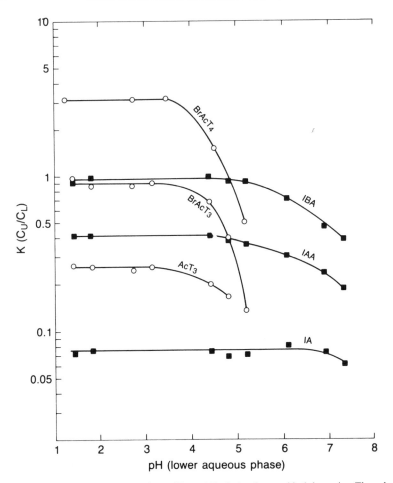

Figure 6.6. Effects of pH on the K values of T_3 and T_4 derivatives and indole auxins. The solvent system composed of hexane/ethyl acetate/methanol/15 mM ammonium acetate (1:1:1:1) was used and the solvent pH adjusted by addition of HCl. The K values of each compound is plotted against the pH in the lower aqueous phase. The two groups of compounds show different responses to the change of pH: The K values of T_3 derivatives show a sharp decrease between pH 4 and 5, whereas those of indole auxins display a gradual decrease after pH 5–6.

performed near the pK_a of the target solute, chromatography tends to produce a broad peak because more than one species is involved. In that case, a slight shift in the solvent pH in either direction would improve the sharpness of the solute involved.

A set of indole auxins, indole-3-acetamide (IA), indole-3-acetic acid (IAA), and indole-3-butyric acid (IBA), was selected for further experiments since

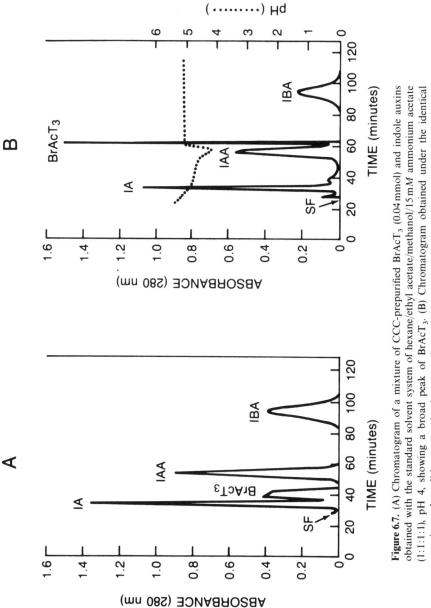

Figure 6.7. (A) Chromatogram of a mixture of CCC-prepurified BrAcT$_3$ (0.04 mmol) and indole auxins obtained with the standard solvent system of hexane/ethyl acetate/methanol/15 mM ammonium acetate (1:1:1:1), pH 4, showing a broad peak of BrAcT$_3$. (B) Chromatogram obtained under the identical experimental condition except that bromoacetic acid (380 mg) was added to the sample solution. Note that the focused BrAcT$_3$ peak was sharpened and its elution delayed. Retention of the stationary phase was 67.6% in both A and B (SF = solvent front).

133

these compounds have suitable partition coefficients in the above two-phase solvent system. Figure 6.6 shows the effect of pH on their partition coefficients compared to the effect on the K values of T_3 and T_4 derivatives. The pH value of the lower aqueous phase was adjusted by adding HCl. All components show constant partition coefficients up to pH 3.5. With further increase in pH, the K values of the T_3 derivatives, BrAcT$_3$ and AcT$_3$, start to fall and at the pH values between 4 and 5 decrease sharply while those of the indole auxins are not significantly altered before pH 5–7. The results show that the retention times and peak widths of the compounds in each group would be affected by a pH shift in those specific ranges.

Figure 6.7 shows a pair of chromatograms emphasizing this point. A mixture of BrAcT$_3$ (CCC prepurified) and indole auxins was chromatographed. The chromatogram in Fig. 6.7A was obtained from the sample mixture without acid additives. The three indole auxin peaks were well resolved and the broad BrAcT$_3$ peak eluted between the IA and IAA peaks. To observe the pH effect on the peak profile, bromoacetic acid (380 mg) was added to the sample and the column was eluted with the same mobile phase as used for Fig. 6.7A. The resulting chromatogram shown in Fig. 6.7B revealed a remarkable change in the BrAcT$_3$ peak, which became much sharper and was only eluted after IAA. In contrast, three indole auxin peaks are little affected by the pH shift.

In order to examine the effect of a pH change on the peak profile of the indole auxins, their fractionation by CCC was performed with the same solvent system adjusted to pH 7. Without added acids, the sample produced broad skewed peaks and a relatively short elution time for IAA and IBA. Addition of a trace amount (2 μL) of TFA (Fig. 6.8A) failed to improve the peak sharpness. However, 100 μL of TFA added to the sample solution (Fig. 6.8B) sharpened the IAA peak, which eluted immediately after the abrupt pH return point. The IBA peak eluting later showed little effect, maintaining a broad peak profile. A further increase of TFA (400 μL) (Fig. 6.8C) resulted in a considerable shift of the pH curve toward longer times to coincide with the

---▶

Figure 6.8. Chromatograms of indole auxins obtained with a neutral solvent system using sample mixtures containing TFA 2 μL of (A), 100 μL of (B), and 400 μL of (C) (1). Note that the amount of TFA in the sample solution determines the retention volume of the abrupt return point of the effluent pH, which coincides with the sharp peak of one of the auxins. Solvent system is hexane/ethyl acetate/methanol/15 mM ammonium acetate (1:1:1:1), pH 7; sample is indole auxins, IA, IAA, and IBA (2 mg each) + TFA 2 μL (A), 100 μL (B), and 400 μL (C) in 2 mL of solvent (1 mL of each phase); mobile phase is the lower aqueous phase (pH 7.4); flow-rate = 3 mL/min; revolution = 800 rpm. Retention of the stationary phase was 66.8% (A), 67.8% (B), and 28.1% (C). The low stationary phase retention in C is apparently due to introduction of a large volume of TFA in the sample solution (SF = solvent front).

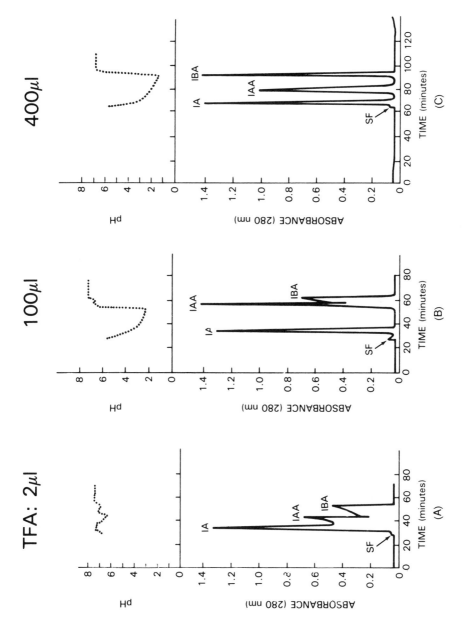

135

elution of the third peak (IBA). Consequently, the IBA peak was greatly sharpened while the second IAA peak returned to the normal profile, as seen in Fig. 6.7A.

As observed in Fig. 6.8 (upper diagrams), the profile of the pH elution curve is greatly influenced by the amount of TFA introduced in the sample solution: An increased amount of TFA results in a lower pH and longer persistence of the low pH. The higher TFA concentration produces relatively more un-ionized, less polar species, which go into the stationary organic phase. Hence, increased TFA concentration results in a longer retention time and more intense skewing of the pH elution curve.

6.2.3. Advanced Peak-Focusing Technique

6.2.3.1. Retainer Acid in the Stationary Phase and Eluent Base in the Mobile Phase

As previously described, the peak-focusing effect was originally produced by introducing a retainer acid such as bromoacetic acid or TFA in the sample solution, which was eluted with a mobile phase containing an ammonium acetate buffer. Under this condition, the pH elution curve shows a gradual drop starting near the solvent front followed by a sudden rise that coincides with elution of the sharp peak. However, it has been found that the sharp peak is similarly produced by adding the retainer acid to the stationary phase and eluent base to the mobile phase. In this case, the pH curve forms a stable flat line starting immediately after the solvent front. This modified method has several advantages over the original method, which occasionally produces some problems: Addition of a strong retainer acid, such as TFA, to the sample solution tends to expose the sample to a low pH and at the same time alters the two-phase composition in the sample zone, which may result in a lower reten-tion of the stationary phase in the separation column. In the modified method, a relatively large amount of the retainer acid can be introduced into the stationary phase with minimum complications. This method also facilitates the use of two or more retainer acids as spacers to improve the separation of the compounds. In this case, the retainers introduced into the stationary phase elute successively following the solvent front in the order of their pK_a or hydrophobicity forming a stepwise rising pH curve.

Figure 6.9 shows a chromatogram of retinoic acid (4 mg) in a solvent system composed of n-hexane/ethyl acetate/methanol/water (1:1:1:1). After the two solvent phases are equilibrated and separated, two retainer acids (TFA and octanoic acid) were added to the stationary phase, each at a concentration of $0.4\,\mu L/mL$, while ammonia was added to the mobile phase at 0.1% (v/v) as an eluent base. The run was performed at a flow-rate of 3 mL/min at 800 rpm

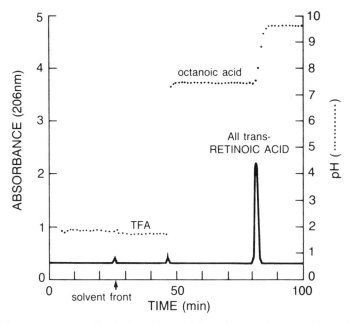

Figure 6.9. Chromatogram of retinoic acid obtained by the use of two retainer acids in the stationary phase. Both TFA and octanoic acid form a flat pH curve after the solvent front. Retinoic acid was eluted as a sharp peak focused at the end of the pH plateau of octanoic acid while some impurities were eluted at the transitional zone between the two retainer acids. Experimental conditions: Apparatus consists of a HSCCC centrifuge with a semipreparative multilayer coil (1.6 mm i.d. and ~315 mL capacity); solvent system is hexane/ethyl acetate/methanol/water (1:1:1:1); retainer acids are TFA and octanoic acid both 0.8 μL/mL in the upper organic stationary phase (pH 1.65); eluent base is 0.1% ammonia in the lower aqueous mobile phase (pH 10.01); sample is retinoic acid about 4 mg; flow-rate = 3 mL/min; revolution = 800 rpm; retention of stationary phase = 72.3%.

using a high-speed CCC (HSCCC) centrifuge with a 300 mL capacity multilayer coil. As expected, TFA eluted immediately after the solvent front followed by the elution of octanoic acid as indicated in the chromatogram. The retinoic acid in the sample solution eluted as a sharp peak at the end of the pH plateau formed by the octanoic acid. A sharp impurity peak is observed at the transition zone between the TFA and octanoic acid plateaus. The octanoic acid was selected as a retainer acid according to the hydrophobicity of retinoic acid. An advantage of using a retainer acid with a hydrophobicity close to that of the analyte is that impurities less hydrophobic than the retainer acid elute away from the analyte peak, since the retainer acid acts as a spacer between the impurities and the analyte.

A set of retainer acids can be used as spacers for effectively separating solutes with different hydrophobicity. Figure 6.10 shows such an example in the separation of dinitrophenyl (DNP) amino acids. The two-phase solvent system composed of methyl *tert*-butyl ether/acetonitrile/water (4:1:5, v/v/v) was equilibrated in a separatory funnel. After the two phases were separated, four retainer acids including TFA, acetic acid, propionic acid, and *n*-butyric acid, were added to the stationary phase, each at a concentration of $0.4\,\mu L/1\,mL$, while aqueous ammonia (eluent base) was added to the lower mobile phase at a concentration of 0.1% (v/v) to raise the pH to 10.77. The separation was performed under the identical condition employed for the separation of retinoic acid described above. Sharp peaks of three DNP–amino acids (each 1 mg) were widely separated by the plateaus of the retainer acids

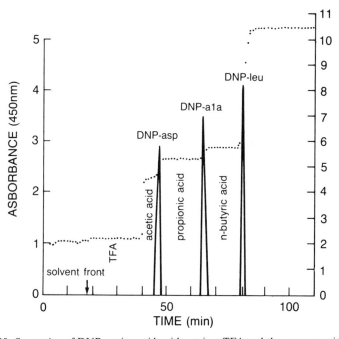

Figure 6.10. Separation of DNP–amino acids with retainer TFA and three spacer acids. Three DNP–amino acids were widely separated by the spacer acids added to the stationary phase. Experimental conditions: Apparatus consists of a HSCCC centrifuge with a semipreparative multilayer coil (1.6 mm i.d. and ~315 mL capacity); solvent system is methyl *t*-butyl ether–acetonitrile–water (4:1:5); retainer acids are TFA, acetic acid, propionic acid, and *n*-butyric acid, each $0.4\,\mu L/mL$ in the upper organic stationary phase; eluent base is 0.1% ammonia in the lower aqueous mobile phase (pH 10.77); sample is DNP-L-aspartic acid, DNP-L-alanine, and DNP-L-leucine, each 1 mg; flow-rate = 3 mL/min; revolution = 800 rpm; retention of stationary phase = 81.0%.

according to their hydrophobicities relative to those of the retainer acids: polar DNP-L-aspartic acid was eluted first at the transition zone between acetic acid and propionic acid; DNP-L-alanine with a moderate hydrophobicity between propionic acid and *n*-butyric acid, and hydrophobic DNP-L-leucine after *n*-butyric acid.

Note that, although not experimentally confirmed, when a set of retainer acids is introduced into the stationary phase, each pH plateau eluted after the solvent front represents one main retainer acid mixed with minute amounts of later eluting retainer acids except for the last eluting acid, which is supposed to be free of other retainer acids. For example, in the chromatogram shown in Fig. 6.10, the first pH plateau represents TFA mixed with minute amounts of all other retainer acids; the second pH plateau is mostly composed of acetic acid but also contains small amounts of propionic acid and *n*-butyric acid; the

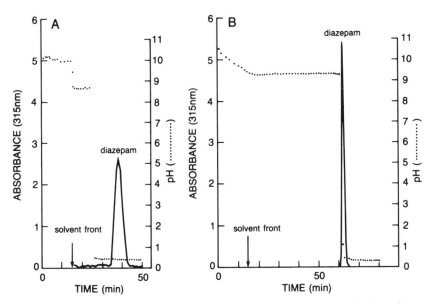

Figure 6.11. Chromatograms of diazepam obtained by adding a retainer base in the stationary phase. (A) diazepam formed a broad peak due to an insufficient amount of the retainer acid (triethylamine 1% in the stationary phase). (B) The diazepam peak was focused by increasing the concentration of the retainer acid five times (5%). Experimental conditions: Apparatus consists of a HSCCC centrifuge with a semipreparative multilayer coil (1.6 mm i.d. and ~315 mL capacity); solvent system is methyl *tert*-butyl ether–water; retainer base is triethylamine in the upper organic stationary phase at a concentration of 1% (A) and 5% (B); eluent base = 0.5 M HCl in the lower aqueous mobile phase; flow-rate = 3 mL/min; revolution = 800 rpm, retention of stationary phase = 78.0%.

third pH plateau is mostly *n*-propionic acid with a small amount of *n*-butyric acid; and the last plateau is composed entirely of pure *n*-butyric acid.

6.2.3.2.   *Application to the Separation of Basic Compounds*

All examples described above are separations of acidic compounds using a retainer acid(s). Recently, it has been found that the method can be equally well applied to basic analytes using a retainer base. A chromatogram of diazepam obtained by HSCCC in Fig. 6.11 illustrates such an example. The separation was performed as follows: A two-phase solvent system composed of methyl *tert*-butyl ether/water was equilibrated in a separatory funnel. After separation of the two phases, triethylamine (retainer base) was added to the upper organic stationary phase and the lower aqueous mobile phase was acidified with HCl (eluent acid) to make a concentration of 0.5 *M*. The chromatogram shown in Fig. 6.11A was obtained from the run using the stationary phase containing the retainer base at a concentration of 1% (v/v). The pH curve revealed that the retainer base was eluted at 25 min prior to the elution of the diazepam peak with a peak maximum at 38 min. Consequently, the diazepam was eluted as a broad peak. The experiment was repeated by increasing the concentration of the retainer base in the stationary phase by five times (5%, v/v). As shown in Fig. 6.11B, this resulted in a proportional increase of the retention time (after the solvent front) or the length of the pH plateau of the retainer base, and diazepam was eluted as a sharp peak at 62 min where the abrupt drop of pH occurred.

6.2.4.   Conclusion

This simple method uses an appropriate reagent (retainer acid or base) in the sample solution or the stationary phase to produce peak-focusing effects and delayed elution of the target compound(s). The method works for separation of both organic acids and bases by choosing suitable retainer acids and bases, respectively. The prime merit of the method is that increasing the sample concentration allows easier detection and it is often advantageous to collect samples in a minimum volume (4). Changing the elution time can often be helpful in avoiding coeluting impurities. Here this will occur as long as the impurities do not respond to the pH change in a similar fashion.

  While the peak-focusing technique has its major application in purification and concentration of a minor component present in a crude sample, it has been found that the retainer acid or base technique is also extremely useful in preparative-scale separations. Increasing the sample size produces a unique elution pattern comparable to that in displacement chromatography. Thus the method opens a rich domain of applications in preparative-scale separations.

This technique is named "pH-zone-refining CCC" and is described in Section 6.3.

6.3. pH-ZONE-REFINING COUNTERCURRENT CHROMATOGRAPHY

6.3.1. Characteristic Features of pH-Zone-Refining CCC

As mentioned earlier, the present method stemmed from the peak-focusing technique when a larger amount of sample was placed on the chromatographic column (5–9). Figure 6.12A illustrates a chromatogram of three DNP–amino acids obtained when the sample size was increased from 1 to 100 mg under otherwise identical experimental conditions to those described in Fig. 6.10. This increase in sample size produced rectangular individual peaks, which were quite different from those observed in the ordinary chromatogram. When the run was again repeated with a single retainer acid (TFA) eliminating the three spacer acids (i.e., acetic acid, propionic acid, and *n*-butyric acid), the three DNP–amino acid peaks were fused together while preserving their original rectangular shapes and partition coefficients as demonstrated by the plots in Fig. 6.12B. Further increase of the sample size only increases the width of the respective peaks without affecting the overall elution profile of the sample.

Figure 6.13 shows a chromatogram of a set of seven DNP–amino acids obtained by the same technique except that the retainer acid (TFA) was introduced in the sample solution (6). All components were separated as rectangular peaks arranged in the order of hydrophobicity with minimum overlapping, as shown by the partition coefficient values (K). The retainer acid introduced in the sample solution produced a gradual drop in pH starting at the solvent front followed by an abrupt rise at fraction 100, which coincided with the elution of the DNP–amino acids. During the elution of the DNP–amino acids, the pH rose stepwise according to the relative pK_a values of the DNP–amino acids eluted.

The elution pattern of pH-zone-refining CCC described above bears a remarkable resemblance to that observed in displacement chromatography (10–12) and isotachophoresis (13): The chromatogram consists of a train of concentrated rectangular peaks of major components each with minimum overlap. Impurities are separated and concentrated at the boundaries of the rectangular peaks. An increase of sample size only results in a proportional increase of the width of the respective peaks without affecting the overall elution pattern. Consequently, this method shares various advantages with displacement chromatography over other chromatographic methods. These advantages include (1) a large amount of sample can be purified in a short

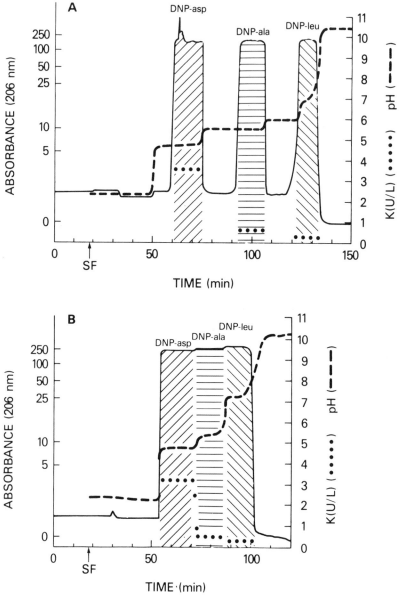

Figure 6.12. pH-zone-refining CCC of DNP–amino acids with (A) and without (B) spacer acids in the stationary phase. (A) Rectangular peaks of three DNP–amino acids were widely separated from each other by the spacer acids (acetic acid, propionic acid, and *n*-butyric acid). (B) Elimination of the spacer acids resulted in fusion of the rectangular peaks with minimum overlapping as demonstrated by the associated pH plateaus and partition coefficient values (K). Experiment conditions are identical to those in Fig. 6.10, except that the sample size was increased to 100 mg for each component. The K values were obtained by partitioning an aliquot of each fraction to the standard solvent system composed of chloroform/acetic acid/0.1 M HCl (2:2:1) (SF = solvent front).

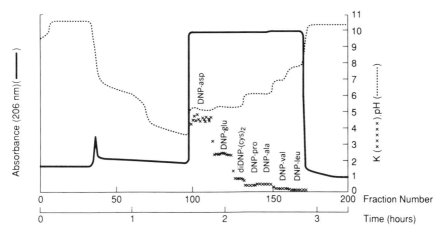

Figure 6.13. pH-zone-refining CCC of seven DNP–amino acids. The retainer acid (TFA) added to the sample solution produced a deep pH drop at fraction 100 followed by a train of rectangular solute peaks. Each pH plateau corresponds to the single component as indicated by the K values. Experimental conditions: Solvent system consists of methyl *tert*-butyl ether/acetonitrile/ water (4:1:5), NH_3 was added to the lower aqueous mobile phase at 0.1% (pH 10.5); sample consists of seven DNP–amino acids, each $50–100\,mg + 200\,\mu L$ of TFA in 10 mL of the upper stationary phase. Other conditions were the same as those described in Fig. 6.12.

column, (2) harvested fractions are highly concentrated, and (3) minute amounts of impurities may be isolated for characterization by conventional analytical methods. However, the chemohydrodynamic process in pH-zone-refining CCC is found to be quite different from that in displacement chromatography. The unique chemohydrodynamic mechanism of the present technique is described in Section 6.3.2.

6.3.2. Chemohydrodynamic Mechanism of pH-Zone-Refining CCC

Figure 6.14 schematically illustrates three model experiments to elucidate the chemohydrodynamic processes operating within the separation column together with the elution profile(s) of an acidic solute(s) in pH-zone-refining CCC (7).

Figure 6.14A shows the preparation of the mobile and stationary phases to initiate the CCC expeirments: First, the two-phase solvent system, composed of an ether (e.g., methyl *tert*-butyl ether) and distilled water, is thoroughly equilibrated in a separatory funnel and the two phases are separated. The upper organic phase is then mixed with a proper amount of the retainer acid, such as TFA, and used as the stationary phase (shaded). A proper amount of an eluent base, such as ammonia, is added to the lower aqueous phase and this is used as the mobile phase. In each experiment the separation is initiated by filling the entire column space with the acidified stationary organic phase followed by introduction of the sample, which is dissolved in the stationary phase. Then the mobile phase is pumped through the column while the apparatus is run at optimum speed. Model experiments are illustrated with three different samples: a small quantity of solute S (R-COOH) (Fig. 6.14B); a larger quantity of the same solute (Fig. 6.14C); and large quantities of three different solutes, S_1 (R_1-COOH), S_2 (R_2-COOH), and S_3 (R_3-COOH) (Fig. 6.14D). In each figure, the upper diagram shows the chemohydrodynamic process within the separation column and the lower diagram shows the elution profile of the solute peak(s).

In Fig. 6.14B (upper diagram), a portion of the separation column shows a chemohydrodynamic mechanism similar to that previously described in Fig. 6.4A. The retainer acid (TFA) forms a sharp trailing border that corresponds to an abrupt pH drop as indicated above the separation column. On the left side of this retainer border, the solute molecule is mostly distributed in the mobile phase as an ionized form ($R-COO^-$) associated with a counterion of the eluent base (NH_4^+). As soon as the solute anion crosses the retainer border and becomes exposed to a low pH, it is protonated by TFA (CF_3COOH) forming the hydrophobic nonionized form (R-COOH), which is then quickly partitioned into the stationary phase. As the TFA border is pushed forward, the protonated solute molecule is again exposed to a high pH, loses a proton to the eluent base (NH_3), and is partitioned back into the mobile phase. This process is repeated until the retainer acid is eluted from the column. Consequently, the solute elutes as a sharp peak at the point where an abrupt rise of the pH occurs in the eluate, as shown in the lower diagram (Fig. 6.14B). Note that mobile phase flow is from left to right in the upper diagram but from right to left in the lower diagram.

When the sample size is increased, the solute forms a broad band (zone S) with a specific pH behind the sharp trailing border of the retainer acid (the upper diagram in Fig. 6.14C). At the front end of the solute zone, the mobile phase constantly carries the solute ions past the TFA border into the low pH zone where they are protonated and quickly partitioned into the stationary phase. This constant supply of the solute molecules to the stationary phase at the front end of the solute zone is compensated at the rear end where the solute molecules are returned to the mobile phase by the action of NH_3 molecules

Figure 6.14. Chemohydrodynamic mechanism of pH-zone-refining CCC. (A) Preparation of solvent phases for initiating the imaginary experiment and B–D: The chemohydrodynamic process within the column (upper diagram) and the elution profile (lower diagram) obtained from a small sample size of single component (B); a large sample size of the same component (C); and large quantities of three major components (D). (see text for details).

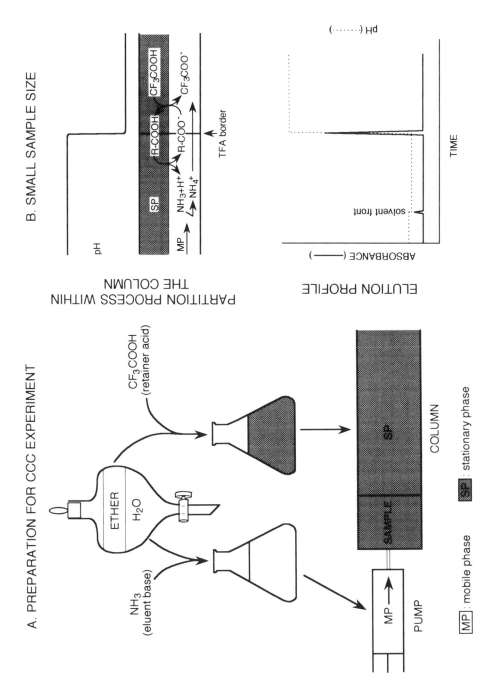

A. PREPARATION FOR CCC EXPERIMENT

B. SMALL SAMPLE SIZE

145

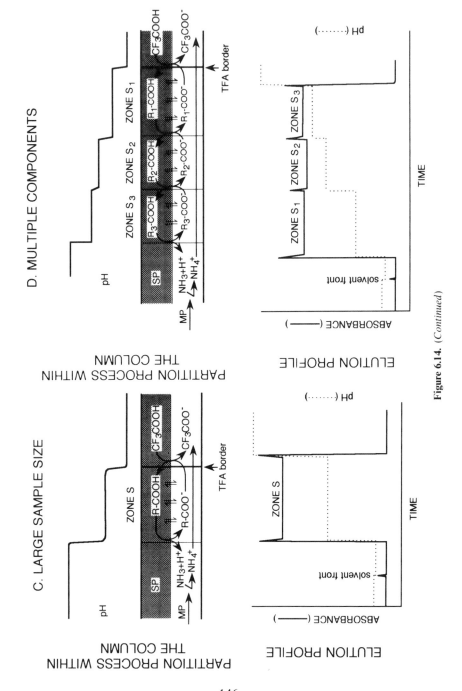

Figure 6.14. (*Continued*)

146

that deprotonate the solute molecules forming the hydrophilic ionized form. Consequently, once the steady-state chemohydrodynamic equilibrium is established, the solute zone maintains its width and remains at constant pH. Impurities present within the solute zone, provided they possess different hydrophobicities or pK_a values, are subjected to a steady partition process between the two phases and are eliminated from the solute zone toward the boundaries on either side, where they accumulate to form sharp bands. This efficient partition process continues until the entire solute zone is eluted from the column. Consequently, the solute forms a rectangular peak associated with sharp impurity peaks at its edges, as shown in the lower diagram (Fig. 6.14C).

Introduction of three different solutes, S_1 (R_1-COOH), S_2 (R_2-COOH), and S_3 (R_3-COOH), each in large quantities (Fig. 14D, upper diagram) results in competition of the solutes to occupy the column space immediately behind the sharp border of the retainer acid. Among those, solute S_1 which has the lowest pK_a value and hydrophobicity, soon establishes its own zone by protonating other components that are in turn partitioned into the stationary phase delaying their forward movement. The competition continues among the other two solutes in which solute S_2 with a lower pK_a value and hydrophobicity drives out S_3 to establish the second zone. Finally, solute S_3 occupies the end of the zone train forming a sharp trailing border, as shown in the upper diagram (Fig. 6.14D). In practice, the time sequence of the above zone formation is not in this order and the third and last solute may form the sharp trailing border earlier than the second solute. As indicated by curved arrows, proton transfer takes place at each zone boundary from the solute molecules in the preceding zone (stationary phase) to those in the following zone (mobile phase) as governed by the difference in pH between these two neighboring zones, while the NH_4^+ counterions continuously move with the mobile phase. After chemohydrodynamic equilibrium is reached, all zones move at the same rate as that of the trailing TFA border while constantly maintaining their own widths and pH. Within each zone, an efficient partition process takes place to eliminate impurities that are driven out to form narrow bands at the zone boundaries. Consequently, the solutes are eluted as a train of rectangular peaks with sharp impurity peaks at their boundaries, as shown in the lower diagram of Fig. 6.14D.

6.3.3. Simple Mathematical Model of pH-Zone-Refining CCC

An attempt has been made to formulate the two-phase liquid–liquid partition process in pH-zone-refining CCC (7).

An imaginary experiment is initiated by thoroughly equilibrating methyl *tert*-butyl ether and distilled water in a separatory funnel at room temperature. After the two phases are separated, a retainer acid (TFA) is added to the upper

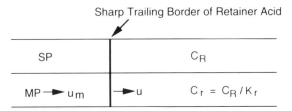

Figure 6.15. Sharp trailing border of retainer acid in the column. As the retainer acid present in the stationary phase is partitioned into the flowing mobile phase, a sharp trailing border of the retainer acid is formed. After the partition equilibrium is established between the two phases, the retainer border travels at a uniform rate that is lower than that of the mobile phase.

organic phase at a concentration of C_R while an eluent base (NH$_3$) is added to the lower aqueous phase at a concentration of C_E. The column is then completely filled with the acidified organic (stationary) phase and rotated while the basic aqueous (mobile) phase is pumped into the column at a flow rate of u_m (L/min). The eluate is collected in test tubes and the pH of the fractions subsequently measured.

Figure 6.15 schematically illustrates the longitudinal cross section through a portion of the coiled column, which arbitrarily shows the stationary organic phase in the upper half and the mobile aqueous phase in the lower half. As the mobile phase flows through the column by partially displacing the stationary phase, the retainer acid in the stationary phase is partitioned into the mobile phase mostly by forming a salt with the eluent base. This causes a depletion of the retainer acid from the stationary phase to form a sharp trailing boundary of the retainer acid that moves through the column at a rate lower than that of the mobile phase. The rate of travel of this retainer boundary through the column finally becomes constant (u L/min through the column space occupied by the mobile phase) after the concentration of the retainer acid in the mobile phase reaches an equilibrium with the concentration in the stationary phase (C_R). At this point, the concentration of the retainer acid in the mobile phase is expressed as $C_r = C_R/K_r$, where K_r is the partition coefficient of the retainer acid at this equilibrium state.

6.3.3.1. *Retention Volume and Rate of Travel of the Trailing Border of the Retainer Acid*

We assume that the establishment of the above chemohydrodynamic equilibrium is instantaneous. Then, the retention volume of the retainer acid (V_r) (the volume of the mobile phase required to elute the retainer acid completely from the column) and the moving rate of the sharp retainer border (u) may be obtained by the following equations:

$$C_R V_s = C_r(V_r - V_m) \tag{6.1}$$

$$V_m/u = V_r/u_m \tag{6.2}$$

where V_s and V_m indicate the volumes of the stationary and mobile phases within the column, respectively, after the solvent front is eluted. Equation (6.1) shows that the net amount of the retainer acid in the stationary phase retained in the column is equal to the amount eluted from the column. In Eq. (6.2), the left term indicates the time required for the retainer acid to travel through the entire length of the column, which equals the retention time of the retainer acid in the chromatogram indicated in the right term.

From Eqs (6.1) and (6.2), the retention volume of the retainer acid (V_r), the traveling rate of the retainer border (u), and the partition coefficient of the retainer acid at the equilibrium state (K_r) are given in the following equations:

$$V_r = K_r V_s + V_m \tag{6.3}$$

$$u = u_m/[(V_s/V_m)K_r + 1] \tag{6.4}$$

$$K_r = [(u_m/u) - 1](V_m/V_s) \tag{6.5}$$

6.3.3.2. Formation of the Solute Zone Behind the Trailing Border of the Retainer Acid

The experiment is continued by similarly filling the column with the stationary organic phase acidified with the retainer acid. This is followed by injection of the sample solution containing an acidic solute with a pK_a and hydrophobicity higher than that of the retainer acid. Then, the mobile phase containing an eluent base (NH_3) at a concentration of C_E is eluted through the column at a flow-rate of u_m (L/min).

As described earlier, the retainer acid (such as TFA) quickly forms a sharp trailing border that moves through the column space occupied by the mobile phase at a uniform rate of u (L/min). Since the solute molecule forms a hydrophilic ion in the high pH zone behind the sharp trailing border of the retainer acid, it quickly moves with the mobile phase. However, as soon as the solute molecule passes through the sharp retainer border, it is exposed to a low pH, protonated to form a hydrophobic nonionized acid, and quickly partitioned into the stationary phase. As a result, movement of the solute molecule through the column is delayed. As the mobile phase flows forward, the solute molecule again finds itself in a high pH zone where it is deprotonated and

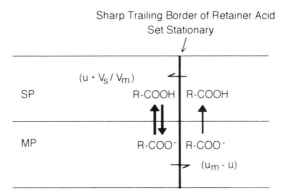

Figure 6.16. Partition processes at the vicinity of the retainer border. In order to ease comprehension of the hydrodynamic process, the sharp trailing border of the retainer acid is set stationary. Then, the observer sees the two solvent phases countercurrent through the retainer border. Since the solute molecules migrating into the right side of the border are returned by the stationary phase, the solute concentrates on the left side of the border until the partition equilibrium is reached. The further supply of solute molecules results in increasing the width of the solute zone toward the left (see text for details).

partitioned mostly into the mobile phase. As this process continues with an abundant supply of the solute molecules in the mobile phase, the solute concentration behind the sharp border of the retainer acid gradually rises and alters both the pH and the partition coefficient of the solute until a partition equilibrium is established between the two phases.

In order to facilitate mathematical treatment of the above chemohydrodynamic process, the moving trailing border of the retainer acid (Fig. 6.15) is set stationary as indicated in Fig. 6.16. From this modified viewpoint, the mobile and stationary phases move countercurrently by passing through the retainer border at the relative rate of $(u_m - u \text{ L/min})$ and $(-uV_s/V_m \text{ L/min})$, respectively, as indicated in the diagram.

Now let us compare the partition process on both sides of the retainer border at the steady-state chemohydrodynamic equilibrium. On the right side where pH is low, all solute molecules present in the flowing mobile phase are immediately protonated to form the hydrophobic nonionic acid, quickly transferred into the stationary phase, and sent back to the left side of the retainer border. On the left side of the retainer border, where pH is higher, the solute molecules are initially partitioned into the mobile phase. As this chemohydrodynamic process continues, however, both solute concentration and the partition coefficient (due to the nonlinear isotherm) increase on the left side of the retainer border until the partition equilibrium is reached between the solute concentration in the stationary phase (C_s) and that in the mobile phase (C_m). These solute movements are indicated by three thick arrows across the interface in the diagram.

At this equilibrium state, the rate of the solute transfer from the mobile to the stationary phase on the right side of the retainer border is given by the solute concentration in the mobile phase (C_m) multiplied by the rate of travel of the mobile phase through the trailing border of the retainer acid $(u_m - u)$. The amount of the solute transferred from the mobile phase is then uniformly distributed in the stationary phase at the concentration (C_s) which is determined by the relative flow-rate of the stationary phase through the trailing border of the retainer acid (uV_s/V_m). Therefore.

$$C_m(u_m - u) = C_s uV_s/V_m \qquad (6.6)$$

which becomes

$$K_S = C_s/C_m = [(u_m/u) - 1](V_m/V_s) \qquad (6.7)$$

Comparison between Eqs. (6.5) and (6.7) implies that under the above equilibrium state the solute partition coefficient (K_S) on the left side of the retainer border is equal to that of the retainer acid itself (K_r) on the right side of the retainer border, that is, $K_S = K_r$. This clearly indicates that the concentration ratio between the stationary and the mobile phases in the equilibrated solute zone is determined by that of the retainer acid. Once this partition equilibrium is reached, the further supply of the solute through the mobile phase results in the development of a solute zone behind the sharp border of the retainer acid, the width of the solute zone being proportional to the amount of the solute introduced into the separation column.

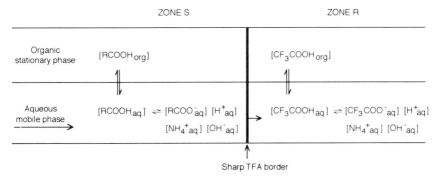

Figure 6.17. Chemohydrodynamic equilibrium of retainer acid (in zone R) and solute S (in zone S) on respective sides of the sharp TFA border in separation column.

Figure 6.17 shows the simplified chemodynamic equilibrium of TFA (CF_3COOH) and solute S (RCOOH) between the stationary organic (org) phase and the flowing aqueous (aq) phase on an assumption that the concentration of ionized components in the organic phase is negligible. The pH of the mobile phase in the solute zone (zone S) on the left side of the sharp TFA border is given from the following three equations:

$$K_{D-S} = [RCOOH_{org}]/[RCOOH_{aq}] \tag{6.8}$$

$$K_S = [RCOOH_{org}]/([RCOOH_{aq}] + [RCOO^-_{aq}]) \tag{6.9}$$

$$K_{a-S} = [RCOO^-_{aq}][H^+_{aq}]/[RCOOH_{aq}] \tag{6.10}$$

where K_{D-S} is the partition ratio of solute S (RCOOH) and K_{a-S}, the dissociation constant of the solute. These equations are reduced to

$$pH_{z-S} = pK_a + \log\{(K_{D-S}/K_S) - 1\} \tag{6.11}$$

where pH_{z-S} is the pH of the mobile phase in the solute zone. As shown in Eq. (6.11), the pH of the solute zone is determined by the pK_a and hydrophobicity (K_{D-S}) of the solute as well as its partition coefficient (K_S). Since $K_S = K_r$. pH_{z-S} can be computed using Eq. (6.16) described below.

The chemodynamic equilibrium of TFA (CF_3COOH) between the stationary organic phase and the flowing aqueous phase in front of its sharp trailing border is shown in zone R in Fig. 6.17. From this diagram the partition coefficient of TFA (K_r) is obtained by solving the following four simultaneous equations:

$$
\begin{aligned}
K_r &= [CF_3COOH_{org}]/([CF_3COOH_{aq}] + [CF_3COO^-_{aq}]) \\
&= C_R/([CF_3COOH_{aq}] + [CF_3COO^-_{aq}])
\end{aligned} \tag{6.12}
$$

$$
\begin{aligned}
K_{D-r} &= [CF_3COOH_{org}]/[CF_3COOH_{aq}] \\
&= C_R/[CF_3COOH_{aq}]
\end{aligned} \tag{6.13}
$$

$$[CF_3COO^-_{aq}][H^+_{aq}]/[CF_3COOH_{aq}] = K_{a-r} \tag{6.14}$$

$$[CF_3COO^-_{aq}] + [OH^-_{aq}] = [H^+_{aq}] + [NH^+_{4\,aq}] \tag{6.15}$$

where K_{D-r} indicates the partition ratio of TFA, C_E is the concentration of eluent base, and K_{a-r} is the dissociation constant of TFA.

From Eqs. (6.12–6.15), the partition coefficient of the retainer acid is given by the following equation:

$$K_r = \{(C_E/2C_R) + 1/K_{D-r} + [(C_E/2C_R)^2 + K_{a-r}/C_R K_{D-r}]^{1/2}\}^{-1} \quad (6.16)$$

which indicates that K_r becomes greater as C_R increases and/or C_E decreases.

6.3.3.3. Multiple Solute Zones behind the Retainer Border

When two or more different solutes are introduced in larger quantities, each solute tries to occupy the column space behind the sharp border of the retainer acid. In this competition, the solute (S_1), which has the lowest pK_a and lowest hydrophobicity, yields the lowest pH at the equilibrium concentration, drives out all other components, and establishes its equilibrium zone immediately behind the retainer acid. As previously analyzed, the partition coefficient of solute S_1 within the equilibrium zone is equal to that of the preceding retainer acid, that is $C_{s-1}/C_{m-1} = K_r$. As does the retainer acid, S_1 forms a sharp trailing border that is then followed by the second solute (S_2) with the second lowest pK_a and second lowest hydrophobicity, which produces the next higher pH at the equilibrium concentration. Similarly, the partition coefficient of solute S_2 within the equilibrium zone is equal to those of the preceding zones, that is, $C_{s-2}/C_{m-2} = C_{s-1}/C_{m-1} = K_r$. The sharp trailing boundary of this S_2 zone is again followed by the third solute (S_3) with the next higher pK_a and next higher hydrophobicity, and so forth.

Consequently, the solutes form a train of equilibrium zones that are arranged in an increasing order of pK_a and hydrophobicity behind the sharp border of the retainer acid, as shown in Fig. 6.18. As mentioned earlier, the partition coefficient of the solute in each zone is equal to that of the retainer acid, therefore,

$$K_r = C_{s-1}/C_{m-1} = C_{s-2}/C_{m-2} = C_{s-3}/C_{m-3} = \cdots \quad (6.17)$$

$$pH_{z-r} < pH_{z-1} < pH_{z-2} < pH_{z-3} < \cdots \quad (6.18)$$

The sharp trailing border of the retainer acid and the following solute zones move through the column at the same rate and elute in this order to form a succession of pH zones each corresponding to a single solute with a specific pK_a and a characteristic hydrophobicity (K_{D-s}).

Figure 6.19 elucidates the relationship between the mobile phase pH and K for the solutes within the column (left) and the profile of their pH zones eluted (right). These curves may be drawn from Eq. (6.11) by inserting K_{D-s} and pK_a of the solutes. If these parameters are not available, each curve can be

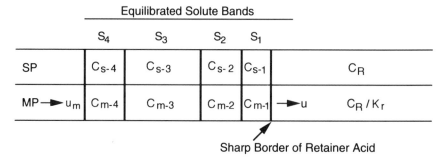

Figure 6.18. A train of solute zones formed behind the retainer border in the column. All solute zones travel together at the same rate determined by that of the retainer acid. The partition coefficient of the solute (C_s/C_m) in each zone is equal to that of the retainer acid (K_r) (see text for details).

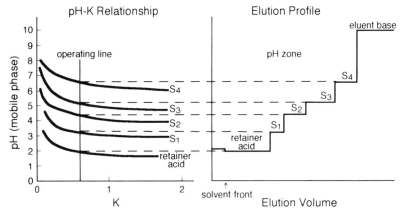

Figure 6.19. Relationship between pH versus K curves and eluted pH-zones. The operating line vertically drawn at the critical K value intersects each curve to indicate the pH level of the respective solute zone (see text for details).

experimentally obtained by equilibrating increasing amounts of the solute or retainer acid with the solvent system (containing the eluent base but no retainer acid), and measuring both the pH of the lower aqueous phase and the solute concentration in the upper and lower phases. Then, the diagram is constructed by plotting the pH on the ordinate against K (the solute concentration in the upper organic phase divided by that of the lower aqueous phase) on the abscissa. Due to the nonlinear isotherm, K increases with increasing amounts of solute accompanied by decreasing pH. In Fig. 6.19 (left) five pH–K curves are arranged from the top to the bottom in the order of decreasing pK_a of the solutes where the lowest curve represents that of the retainer acid with the lowest pK_a. The vertical line drawn through the critical K value (K_r) is called the operating line, which intersects each pH curve and determines the pH level in the corresponding solute zone eluted as shown in the right diagram.

The pH–K diagram is useful for predicting the experimental results including the order of solute elution, the pH level of each solute zone, and feasibility of the separation. A good separation is expected from a set of pH–K curves, which show even distribution of each curve with a great distance apart.

6.3.4.   Displacement pH-Zone-Refining CCC

The overall results described above indicate that the present technique bears a distinct resemblance to displacement chromatography (10–12). These two chromatographic methods share common features, such as the formation of a chain of rectangular elution peaks, which travel together at the same rate, and the concentration of minor components at the peak boundaries. However, careful comparison of these two techniques reveals that the key reagents, that is, a retainer in pH-zone-refining CCC and a displacer in displacement chromatography, act in an opposite manner. The retainer transfers the solute from the mobile phase to the stationary phase at the front end of the solute band (Fig. 6.14D), whereas the displacer transfers the solute from the stationary phase to the mobile phase at the back of the solute bands. From this point of view, pH-zone-refining CCC may be more properly considered as "reverse displacement pH-zone-refining CCC."

Recently, it was found that pH-zone-refining CCC can be performed in a manner identical to displacement chromatography simply by interchanging the role of the mobile and stationary phases (8). In this way, the original eluent base becomes a retainer for analytes in the stationary phase while the original retainer acid becomes a displacer, transferring the analytes from the stationary phase to the mobile phase at the back of the solute bands. The chemohydrodynamic mechanism of displacement pH-zone-refining CCC is illustrated in Fig. 6.20.

A model experiment is initiated by filling the entire column with the stationary aqueous phase that contains a retainer, NH_3. This is followed by injection of a sample solution containing three major components [solutes S_1 (R_1-COOH), S_2 (R_2-COOH), S_3 (R_3-COOH)]. The column is then eluted with an organic mobile phase containing a displacer, TFA. As the mobile phase moves through the column by partially displacing the stationary phase, it distributes TFA anions to the stationary phase retained in the column according to its partition coefficient, which is determined by the pH at a given point in the column. This partition process quickly depletes the TFA from the flowing mobile phase front resulting in formation of a sharp TFA front border. The sharpness of this TFA border is maintained by the concentration-dependent partition behavior of TFA or its nonlinear isotherm. After equilibrium is

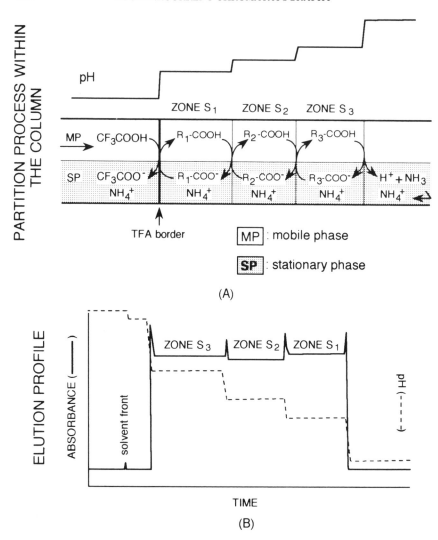

Figure 6.20. Chemohydrodynamic mechanism of displacement mode of pH-zone-refining CCC. (A) Partition process within the column. (B) Elution profile.

established, this sharp TFA border travels through the column at a uniform rate lower than that of the mobile phase.

Figure 6.20A schematically shows a portion of the separation column that contains the stationary aqueous phase in the lower half (shaded) and the mobile organic phase in the upper half. The sharp TFA border is indicated by a thick line across the column.

In the early stage of the experiment, all solute molecules present on the right side of the TFA border are exposed to a high pH, hence they are mostly deprotonated to become a hydrophilic form (R-COO$^-$) and partitioned into the aqueous stationary phase. As the TFA front advances, these molecules are exposed to a low pH, protonated into a hydrophobic form (R-COOH) and transferred into the flowing organic phase. In other words, TFA in the mobile phase gradually displaces all solute molecules present in the stationary phase in a manner analogous to the action of the displacer in displacement chromatography.

As this process continues, the solute concentration at the immediate front of the TFA border increases causing the pH to fall. In this situation, solute S_1 with a lower pK_a and hydrophobicity among the three components acts as a displacer of the other two solutes to occupy the column space immediately after the sharp TFA border, where it forms the first zone (zone S_1) by making a sharp front border. The competition continues among the other two solutes in which solute S_2 with a lower pK_a and hydrophobicity will form the second zone (zone S_2), which is in turn preceded by the third zone (zone S_3) consisting of solute S_3. In practice the time sequence to form second and the third sharp borders may not be in this order, as described earlier.

When this partition process is completed a train of solute zones is formed in front of the TFA border as in displacement chromatography (Fig. 6.20A). Each zone consists of a single species, is equipped with self-sharpening boundaries, and has the solute partition coefficient (K_S) equal to that of TFA (K_r) in the succeeding zone. As indicated earlier, all zones are arranged in a decreasing order of pK_a and hydrophobicity and move together at the same rate determined by that of the succeeding TFA border.

As indicated by curved arrows, proton transfer and displacement of the solute molecules takes place between the two neighboring species at each zone boundary. Ammonium ion created at the far front border (solute S_3) remains permanently in the stationary phase and serves as the counterion for all species. Charged impurities present in each solute zone are quickly eliminated toward the zone boundaries on either side where they accumulate to form narrow bands as seen in displacement chromatography. Consequently, the solute is eluted as successive rectangular peaks with minimum overlap and with sharp impurity peaks at their boundaries, as shown in Fig. 6.20B. Each zone shows a distinct pH plateau arranged in a downward staircase fashion (dotted line). As described earlier for the reverse pH-zone-refining CCC technique, net pH values of these plateaus may be predicted from three parameters, that is pK_a, K_D (partition ratio), and K_S (partition coefficient) of each solute (see Eq. 6.11).

The overall features of displacement pH-zone-refining CCC is quite similar to those in the reverse displacement mode described earlier and the method is used for the separation of organic bases by adding a retainer acid, such as HCl, to the aqueous stationary phase and displacer base, such as triethylamine, to the organic mobile phase. However, there are some advantages and disadvantages for each mode of pH-zone-refining CCC. The displacement mode has a problem in monitoring the pH of the effluent, which is largely composed of organic solvent, although this difficulty is obviated by the use of ternary or quaternary two-phase solvent systems. On the other hand, the displacement mode has an advantage over the reverse displacement mode in that it yields salt-free fractions in the organic phase, which are easily evaporated. The most important advantage of the displacement mode, however, derives from the fact that the retainer is permanently held in the stationary phase within the column instead of being eluted. Consequently, the displacement mode is better adapted to a ligand-affinity separation, which may be applied to nonionizable compounds as in displacement chromatography.

6.3.5. pH-Zone-Refining CCC versus Displacement Chromatography

As described above, the present technique closely resembles displacement chromatography in many aspects. Similarities and differences between these two techniques are summarized in Table 6.1.

Comparison between these two chromatographic methods, however, shows that pH-zone-refining CCC has some important advantages over displacement chromatography. One of these advantages is that suitable chromatographic conditions can be found easily for a given set of samples. Table 6.2 lists the analytes, two-phase solvent systems, and key reagents that have been successfully applied to the present methods. These examples may provide a useful guide for the choice of experimental conditions in practical applications.

As key reagents for separating acidic analytes, TFA is used as a retainer acid (displacer acid) and ammonia as an eluent base (retainer base) for the reverse displacement mode (the displacement mode). If desirable, common organic acids, such as formic acid, acetic acid, propionic acid, butyric acid, and octanoic acid, may be used individually or in combination as spacers. The two-phase solvent systems can be selected according to the hydrophobicity and solubility of the sample. For solutes with a moderate hydrophobicity, binary solvent systems composed of an ether (diethyl ether or methyl *tert*-butyl ether) and water is successful in many cases. If solubility of the sample is a problem, the solvent system may be modified to a ternary system, such as ether/acetonitrile/water (4:1:5). For the separation of hydrophobic analytes, one may choose a series of solvent systems composed of hexane/ethyl acetate/methanol/water, where the hydrophobicity of the system is conveniently adjusted by changing the ratio of hexane to ethyl acetate from 5:5:5:5 to

Table 6.1. Comparison between pH-Zone-Refining CCC and Displacement Chromatography

	pH-Zone-Refining CCC		Displacement Chromatography
	Reverse Displacement Mode	Displacement Mode	
Key reagents	Retainer		Displacer
Solute transfer	$MP^b \rightarrow SP^c$		$SP \rightarrow MP$
Acting location	Front of solute bands		Back of solute bands
Formation of solute bands	A train of individual solute bands with minimum overlapping		
Traveling rate of solute bands	All solute bands move together at the same rate as that of the key reagent		
K^a in solute band	Same as and determined by that of the key reagent		
Impurities	Concentrated at the boundaries of the solute bands		
Peak profile	A train of rectangular peaks associated with sharp impurity peaks at their boundaries		
Solute concentration in mobile phase is determined by	Concentration of counterions in aqueous phase		Solute affinity to stationary phase
Elution order is determined by	Solute pK_a and hydrophobicity		Solute affinity to stationary phase

[a] Partition coefficient expressed by solute concentration in the stationary phase divided by that in the mobile phase = K.
[b] Mobile phase = MP.
[c] Stationary phase = SP.

10:0:5:5 (through 6:4:5:5, 7:3:5:5, 8:2:5:5, and 9:1:5:5) to optimize the partition coefficient of the sample (17, 18). For the separation of polar compounds, methyl *tert*-butyl ether/*n*-butanol/acetonitrile/water systems may be useful. The partition coefficient of the solutes in the solvent system is also adjusted by changing the concentration of the retainer acid(s), such as TFA, in the stationary phase. On the other hand, the separations of basic analytes were all successfully performed using a combination of triethylamine in the organic phase and hydrochloric acid in the aqueous phase.

Another advantage of pH-zone-refining CCC is that the system can retain a large volume of the stationary phase, from 50 to 80% of the total column

Table 6.2. Samples and Solvent Systems Applied to the Present Methods

Sample[a]	Solvent Systems[b] (Volume Ratio)	Key Reagents[c] Retainer	Eluent	Elution Profile Peak	Plateau	References
DNP–amino acids (0.01–1 g)	MBE/H$_2$O	TFA (0.04%/SP)	NH$_3$ (0.1%/MP)	+	+	27
	MBE/AcN/H$_2$O (4:1:5)	TFA (200 μL/SS)	NH$_3$ (0.1%/MP)	+	+	6, 7, 27
	MBE/AcN/H$_2$O (4:1:5)	TFA (0.04%/SP)	NH$_3$ (0.1%/MP)	+	+	5, 9, 27
	MBE/AcN/H$_2$O (4:1:5)	TFA + spacer acids (each 0.04%/SP)	NH$_3$ (0.1%/MP)	+	+	5, 9, 27
	MBE/H$_2$O (DPCCC)	NH$_3$ (22 mM/SP)	TFA (10.8 mM/MP)		+	8
	MBE/H$_2$O (DPCCC)	NH$_3$ (44 mM/SP)	TFA (10.8 mM/MP)		+	8
	MBE/H$_2$O (DPCCC)	NH$_3$ (22 mM/SP)	TFA (10.8 mM/MP) (spacer acids/MP or SS)		+	8
N-CBZ-amino acids (0.15g)	MBE/H$_2$O	TFA (0.04%/SP)	NH$_3$ (0.1%/MP)		+	27
N-t-BOC-amino acids (0.6 g)	MBE/H$_2$O	TFA (0.04%/SP)	NH$_3$ (0.1%/MP)		+	27
Proline (OBzl) (1 g)	MBE/H$_2$O	TEA (10 mM/SP)	HCl (10 mM/MP)		+	7
Amino acid(OBzl) (0.7 g)	MBE/H$_2$O	TEA (10 mM/SP)	HCl (10 mM/MP)		+	19
Amino acid(OBzl) (10 g)	MBE/H$_2$O	TEA (5 mM/SP)	HCl (20 mM/MP)		+	19
N-CBZ-Dipeptides (0.8 g)	MBE/AcN/H$_2$O (2:2:3)	TFA (16 mM/SP)	NH$_3$ (5.5 mM/MP)		+	20
N-CBZ-Dipeptides (3 g)	MBE/AcN/H$_2$O (2:2:3)	TFA (16 mM/SP)	NH$_3$ (5.5 mM/MP)		+	20
N-CBZ-Tripeptides (0.8 g)	n-BuOH/MBE/AcN/H$_2$O (2:2:1:5)	TFA (16 mM/SP)	NH$_3$ (2.7 mM/MP)		+	20
Dipeptide-βNA (0.3 g)	MBE/AcN/H$_2$O (2:2:3)	TEA (5 mM/SP)	HCl (5 mM/MP)		+	20

Compound (amount)	Solvent system	Acid	Mobile phase			Ref.
Bacitracin (0.2 g)	MBE/AcN/n-BuOH/0.1%NH$_3$ (1:1:1:3)	TFA (0.04%/SP)	MP	+		27
Bovine insulin (0.2 g)	n-BuOH/H$_2$O	TFA (0.04%/SP)	NH$_3$ (0.1%/MP)	+	+	27
Retinoic acid (0.004–0.036 g)	Hex/EtOAc/MeOH/H$_2$O (1:1:1:1)	TFA + octanoic acid (each 0.04%/Sp)	NH$_3$ (0.1%/MP)	+	+	27
BrAcT$_3$ (trace–1 mmol)	Hex/EtOAc/MeOH/15 mM AcONH$_4$ (1:1:1:1)	TFA (200 µL/SS)	MP	+	+	1, 2, 4, 27
BrAcT$_4$ (trace–1 mmol)	Hex/EtOAc/MeOH/15 mM AcONH$_4$ (1:1:1:1)	TFA (0.04–0.08%/SP)	MP	+	+	1, 3, 4, 27
Indole auxins (1.6 g)	MBE/H$_2$O	TFA (0.04%/SP)	NH$_3$ (0.1%/MP)	+	+	7
Nile blue A (0.1 g)	EtOAc/H$_2$O	TFA (200 µL/SS)	MP		+	27
TCF (0.01–1 g)	DEE/AcN/10 mM AcONH$_4$ (4:1:5)	TFA (200 µL/SS)	MP		+	9, 14
Red No. 3 (0.5 g)	DEE/AcN/10 mM AcONH$_4$ (4:1:5)	TFA (200 µL/SS)	MP		+	9, 15
Orange No. 5 (0.01–5 g)	DEE/AcN/10 mM AcONH$_4$ (4:1:5)	TFA (200 µL/SS)	MP	+	+	5, 7, 9
Orange No. 10 (0.35 g)	DEE/AcN/10 mM AcONH$_4$ (4:1:5)	TFA (200 µL/SS)	MP		+	9
Red No. 28 (0.1–6 g)	DEE/AcN/10 mM AcONH$_4$ (4:1:5)	TFA (200 µL/SS)	MP	+	+	9, 16
Eosin YS (0.3 g)	DEE/AcN/10 mM AcONH$_4$ (4:1:5)	TFA (200 µL/SS)	MP	+	+	27
Diazepam (0.005 g)	MBE/H$_2$O	TEA (5%/SP)	HCl (0.5 M/MP)	+		27

161

Table 6.2. Samples and Solvent Systems Applied to the Present Methods

Sample[a]	Solvent Systems[b] (Volume Ratio)	Key Reagents[c] Retainer	Key Reagents[c] Eluent	Elution Profile Peak	Elution Profile Plateau	References
Amaryllis alkaloids (3 g)	MBE/H₂O	TEA (5 mM/SP)	HCl (5 mM/MP)		+	21
	MBE/H₂O (DPCCC)	HCl (10 mM/SP)	TEA (10 mM/MP)		+	21
Vinca alkaloids	MBE/H₂O (DPCCC)	HCl (5 mM/SP)	TEA (5 mM/MP)		+	27
Structural isomers (15 g)	MBE/AcN/H₂O (4:1:5)	TFA (0.32%/SP)	NH₃ (0.8%/MP)		+	22
Stereoisomers (0.4 g)	Hex/EtOAc/MeOH/H₂O (1:1:1:1)	TFA + octanoic acid (each 0.04%/SP)	NH₃ (0.025%/MP)		+	7, 9, 23
(±)-DNB-leucine (2 g)	MBE/H₂O	TFA + chiral selector (each 40 mM/SP)	NH₃ (20 mM/MP)		+	26

[a]Dinitrophenyl = DNP; carbobenzoxy = CBZ; *tert*-butoxycarbonyl = *t*-BOC; benzylesters = OBzl; naphthyl amide = βNa; tetrachlorofluorescein = TCF; amaryllis alkaloids = crinine, powelline, and crinamidine; *Vinca* alkaloids = vincamine and vincine; structural isomers = 2- and 6-nitro-3-acetamido-4- chlorobenzoic acid; stereoisomers; 4-methoxymethyl-1-methyl-cyclohexane carboxylic acid; dinitrobenzoyl = DNB.

[b]The upper organic phase was used as the stationary phase (SP) and the lower aqueous phase, the mobile phase (MP), except in DPCCC, where the above relationship is reversed. Methyl *tert*-butyl ether = MBE; acetonitrile = AcN; butanol = BuOH; hexane = hex; ethyl acetate = EtOAc; methanol = MeOH; ammonium acetate = AcONH₄; diethyl ether = DEE; displacement mode = DPCCC.

[c]Trifluoroacetic acid = TFA; acetic acid = AcOH; in stationary phase = SP; in mobile phase = MP; in sample solution = SS; triethylamine = TEA; *N*-dodecanoyl-L-proline-3,5-dimethylanilide = chiral selector. Note that in DPCCC eluent becomes displacer.

capacity for preparative-scale separations. Furthermore, the lack of the solid support matrix in the column space permits introduction of partially dissolved samples in a suspension with minimum risk of clogging the separation column.

On the other hand, displacement chromatography has at least one important advantage over pH-zone-refining CCC in that the method covers a broad spectrum of samples including both ionizable and nonionizable compounds, whereas the application of the present method has so far been limited to ionizable compounds. In principle, however, the displacement mode operation of pH-zone-refining CCC will be applied to ligand-affinity separations for nonionizable analytes.

6.3.6. Applications of pH-Zone-Refining CCC

Since 1993, the pH-zone-refining CCC technique has been applied to a variety of samples in a broad spectrum of hydrophobicity (Table 6.2). Extensive applications of reverse displacement pH-zone-refining CCC to acidic hydroxyxanthene dyes are described in Chapter 12 in this volume. Several examples of typical applications are presented in Sections 6.3.6.1–5 below.

6.3.6.1. Separation of Basic Amino Acid Derivatives

Application of pH-zone-refining CCC to basic compounds can be carried out by a combined use of an organic retainer base in the organic stationary phase and an inorganic eluent acid in an aqueous mobile phase. In the present examples, triethylamine was used as a retainer base and HCl as an eluent acid (19).

Figure 6.21A shows a chromatogram of seven amino acid benzylesters obtained by reverse displacement pH-zone-refining CCC. The separation was performed with a two-phase solvent system of methyl *tert*-butyl ether/water. Triethylamine (10 mM) was added to the upper stationary phase and HCl (10 mM) was added to the lower aqueous phase, which was then used as the mobile phase. All components (each 100 mg) were separated in 3 h. Sharp peak boundaries drawn in the chromatogram each contained no more than several milliliters of a mixing zone. Some irregularities in the pH curve was apparently caused by carryover of the stationary phase while the partition coefficient values (K_{std}) determined by a standard solvent system indicate clear separation of each component. A larger scale preparative separation of these amino acid benzylesters is shown in Fig. 6.21B. The separation was performed with 5-mM triethylamine in the organic stationary phase and 20 mM HCl in the aqueous mobile phase. A 10-g sample mixture (3.3 g for each component) was separated in 7 h. Despite an over 10-fold increase in sample size, the mixing zones at peak

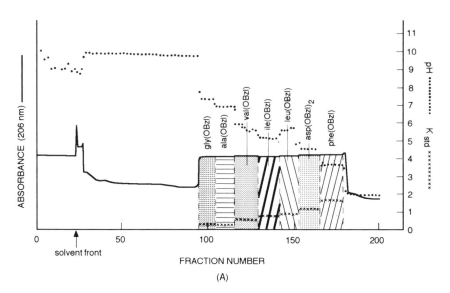

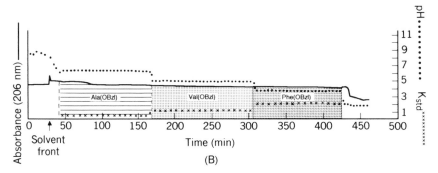

Figure 6.21. Separation of amino acid benzylesters by pH-zone-refining CCC. Identification of the CCC fraction was made by standard partition coefficients (K_{std}) using a two-phase solvent system composed of hexane/ethyl acetate/methanol/0.1 M NaOH (1:1:1:1, v/v). Experimental conditions: Apparatus consists of a HSCCC centrifuge; column is a semipreparative multilayer coil, 1.6 mm i.d., 160 m long, and 325 mL capacity; solvent system is methyl *tert*-butyl ether/water, triethylamine at 10 mM (A) and 5 mM (B) in the upper organic stationary phase and hydrochloric acid at 10 mM (A) and 20 mM (B) in the lower aqueous mobile phase; sample is (A) a mixture of seven amino acid benzylesters as labeled in the chromatogram, each 100 mg, dissolved in 20 mL of solvent, and (B) a mixture of three amino acid benzylesters as labeled in the chromatogram, each 3.3 g, dissolved in 100 mL of solvent; flow-rate = 3 mL/min; revolution = 800 rpm; retention of stationary phase = 71.2% of the total column capacity.

boundaries remained no more than several milliliters. Here the pH curve shows three flat zones, each corresponding to the specific analyte as indicated by the partition coefficient curve.

6.3.6.2. *Separation of Peptide Derivatives*

Two groups of oligopeptide derivatives, acidic N-CBZ(carbobenzoxy)-peptides and basic peptide-β-NA(naphthylamides), were each successfully separated by pH-zone-refining CCC (20). In these separations the composition of the two-phase solvent system was optimized according to the polarity of the analytes. Relatively hydrophobic peptides are separated using a ternary solvent system composed of methyl *tert*-butyl ether/acetonitrile/ water (2:2:3 by volume), while more polar peptides are partitioned with a quaternary solvent system composed of methyl *tert*-butyl ether/*n*-butanol/acetonitrile/water (2:2:1:5).

Figure 6.22A shows a chromatogram of eight different N-CBZ(carbobenzoxy)-dipeptides obtained with the above ternary two-phase solvent system. A 100 mg amount of each peptide, totaling 800 mg, was separated within 4 h. Figure 6.22B shows a similar chromatogram of five N-CBZ-tripeptides in which only the terminal amino acid is different. The separation was performed with the above quaternary solvent system. The present system was also applied to the separation of basic peptide βNA, as shown in Fig. 6.22C. The separation was performed using the ternary system where triethylamine was added to the organic stationary phase and HCl to the aqueous mobile phase. The pH plateau (dotted line) displays a downward staircase pattern, each level corresponding to the respective analyte.

The separations of free amino acids and peptides have not been reported. These zwitterions have a high polarity in either acidic or basic condition and need more careful selection of the solvent system and the retainer or displacer analyte.

6.3.6.3. *Separation of Alkaloids*

One important application of the present technique is to isolate biologically active compounds such as alkaloids from the plant extract. Figure 6.23 shows the separation of three basic alkaloids from a crude extract of *Crinum moorei* using both reverse displacement (A) and displacement (B) modes (21). In both separations, a binary solvent system composed of methyl *tert*-butyl ether/ water was used where triethylamine (5–10 mM) was added to the organic phase and HCl (5–10 mM) to the aqueous phase. Each alkaloids was identified by thin-layer chromatography (TLC) as well as by comparison of mass spectroscopy (MS) and nuclear magnetic resonance (NMR) spectrum analyses with

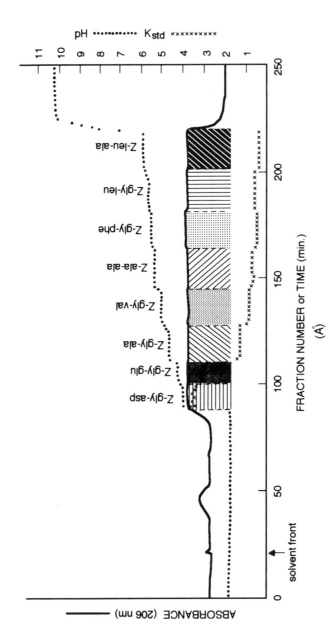

Figure 6.22. Separations of N–CBZ-peptides by pH-zone-refining CCC. Experimental conditions: Apparatus consists of a HSCCC centrifuge; column is a semipreparative multilayer coil, 1.6 mm i.d., 160 m long, and 325 mL capacity; solvent system is (A) methyl *tert*-butyl ether/acetonitrile/water (2:2:3), TFA at 16 m*M* in the organic stationary phase and ammonia at 5.5 m*M* in the aqueous mobile phase, (B) methyl *tert*-butyl ether/*n*-butanol/acetonitrile/water (2:2:1:5), TFA at 16 m*M* in the organic stationary phase and ammonia at 2.7 m*M* in the aqueous mobile phase, (C) methyl *tert*-butyl ether/acetonitrile/water (2:2:3), triethylamine at 5 m*M* in the organic stationary phase and hydrochloric acid at 5 m*M* in the aqueous mobile phase; sample is (A) a mixture of eight N–CBZ-dipeptides as labeled in the chromatogram, each 100 mg; (B) a mixture of five N–CBZ-tripeptides as labeled in the chromatogram, each 100 mg; (C) a mixture of three dipeptide-*β*NA as labeled in the chromatogram, each 100 mg; flow-rate = 3.3 mL/min; revolution = 800 rpm; retention of stationary phase is (A) 65.1%, (B) 59.4%, and (C) 67.0%.

166

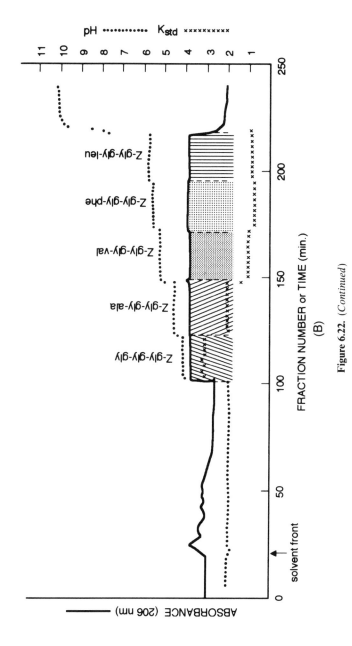

Figure 6.22. (*Continued*)

167

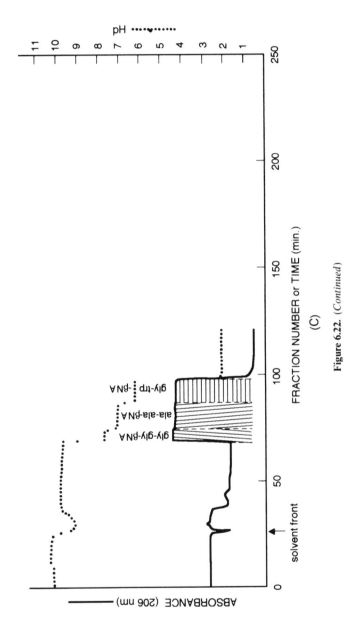

Figure 6.22. (*Continued*)

168

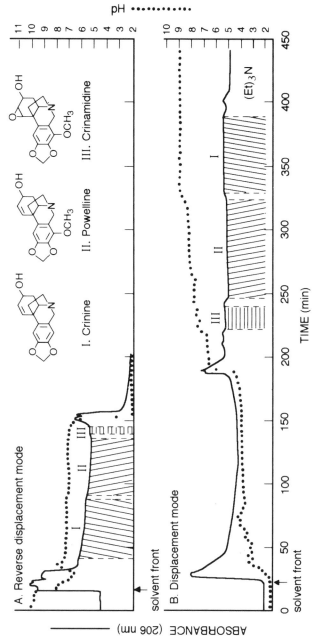

Figure 6.23. Chromatograms of crude alkaloid extract of *Crinum moorei* obtained by reverse displacement mode (A) and displacement mode (B) of pH-zone-refining CCC. Experimental conditions: Apparatus consists of a HSCCC centrifuge; column is a semipreparative multilayer coil, 1.6 mm i.d, 160 m long, and 325-mL capacity; solvent system is methyl *tert*-butyl ether–water; stationary phase is (A) upper organic phase with 5 mM triethylamine and (B) lower aqueous phase with 10 mM HCl; mobile phase is (A) lower aqueous phase with 5 mM HCl and (B) upper organic phase with 10 mM triethylamine; flow = 3.3 mL/min in head-to-tail-elution mode (A) and tail-to-head elution mode (B); sample is a crude alkaloid extract of *Crinum moorei*, 3 g dissolved in 30 mL of solvent system consisting of equal volumes of each phase; revolution (A) is 800 rpm and (B) is 600 rpm.

169

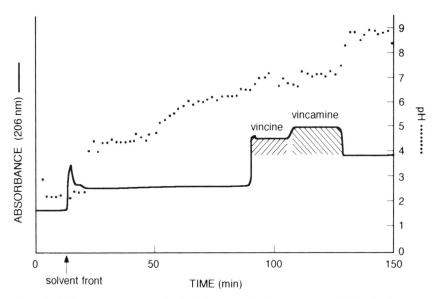

Figure 6.24. Chromatogram of crude alkaloid extract of *Vinca minor* obtained by displacement mode of pH-zone-refining CCC. Experimental conditions: Apparatus consists of a HSCCC centrifuge; column is a semipreparative multilayer coil, 1.6 mm i.d., 160 m long, and 325 mL capacity; solvent system is methyl *tert*-butyl ether/water, 5 mM triethylamine in upper organic mobile phase and 5 mM HCl in lower aqueous stationary phase; flow-rate is 3.3 mL/min in tail-to-head elution mode; sample is a crude alkaloid extract of *Vinca minor*, 300 mg dissolved in 30 mL of solvent system consisting of equal volumes of each phase; revolution = 800 rpm; retention of stationary phase: 90.4%.

standard compounds. In the reverse displacement mode the alkaloids were eluted as HCl salt, while in the displacement mode they are eluted as free bases. Figure 6.24 shows another example, in which vincine and vincamine were separated from a crude extract of *Vinca minor*. Here again a binary solvent system of methyl *tert*-butyl ether/water was used for separation (Table 6.2).

6.3.6.4. *Separation of Structural Isomers*

One important domain of applications in pH-zone-refining CCC is to purify the desired component(s) from a crude reaction mixture. Isolation of a large amount of isomers from a synthetic mixture is sometimes very difficult in the conventional chromatographic method. Figure 6.25 illustrates a gram-quantity separation of structural isomers from a crude reaction mixture (22). A 15 g mixture of 2- and 6-nitro-3-acetamido-4-chlorobenzoic acids was separated by the present technique using a ternary solvent system composed of methyl *tert*-butyl ether/acetonitrile/water (4:1:5). As indicated in the chromatogram,

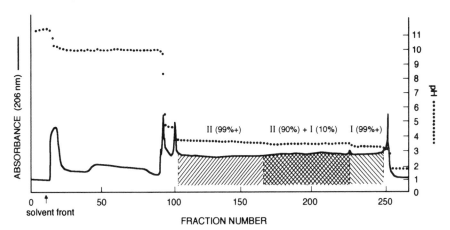

Figure 6.25. Chromatogram of isomeric mononitro-derivatives of 2- and 6-nitro-3-acetamido-4-chlorobenzoic acid by displacement mode of pH-zone-refining CCC. Experimental conditions: Apparatus consists of a HSCCC centrifuge; column is a semipreparative multilayer coil, 1.6 mm i.d., 160 m long, and 325 mL capacity; solvent system is methyl *tert*-butyl ether/acetonitrile/ water (4:1:5), 0.8% aqueous ammonia in lower aqueous stationary phase and 0.32% TFA in upper organic mobile phase; flow = 3.3 mL/min in tail-to-head elution mode; sample is a nitration product of 3-acetamido-4-chlorobenzoic acid (15 g) dissolved in 100 mL of solvent system consisting of equal volumes of each phase; revolution = 800 rpm.

the main plateau contains a mixing zone of the two isomers covering over 40% of the total area, while the rest of the area yields gram-quantities of each component. If desired, the fractions corresponding to the mixing zone may be rechromatographed to increase the yield of pure components. More closely related cis-trans stereoisomers of 4-methoxymethyl-1-methyl-cyclohexane-carboxylic acid have been successfully resolved using hexane/ethyl acetate/ methanol/water system (1:1:1:1) with TFA and octanoic acid as a retainer and a spacer, respectively (23).

6.3.6.5.   *Separation of Enantiomers*

With an increasing demand for chiral separations, more than 100 chiral stationary phases are now commercially available for analytical-scale separations (24). However, few preparative-scale separations (25) have been reported because of the high cost of solid stationary phases required for large columns. High-speed CCC is an excellent alternative since the column requires no solid support and can be used for separating a variety of enantiomers simply by introducing suitable chiral selectors to the stationary liquid phases.

Recently, gram quantities of racemates were successfully separated by pH-zone-refining CCC in our laboratory using *N*-dodecanoyl-L-proline-3,5-

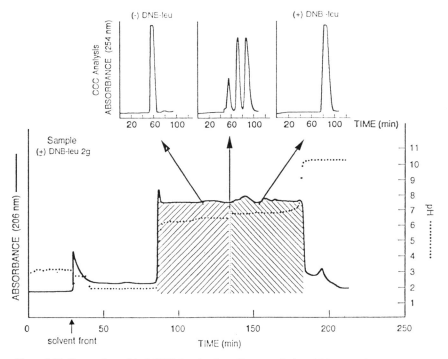

Figure 6.26. Separation of (±)-DNB-leucine by pH-zone-refining CCC. Experimental conditions: Apparatus consists of a HSCCC centrifuge; column is a semipreparative multilayer coil, 1.6 mm i.d., 160 m long, and 325 mL capacity; solvent system is (A) methyl *tert*-butyl ether/water; stationary phase is the upper organic phase to which TFA (40 mM) and chiral selector (40 mM) were added; mobile phase is the lower aqueous phase to which aqueous ammonia was added at 20 mM; sample is (±)-DNB-leucine 2 g; flow-rate = 3 mL/min; revolution = 800 rpm; analysis consists of the optical rotation and CD and analytical CCC (upper diagram). Analytical CCC were carried out using the same column by the standard CCC technique under the following conditions: Solvent system is hexane/ethyl acetate/methanol/10 mM HCl (8:2:5:5); stationary phase is the organic phase containing the same chiral selector (20 mM); flow = 3 mL/min in head-to-tail elution mode; revolution = 800 rpm.

dimethylanilide as a chiral selector. Figure 6.26 (lower diagram) shows a chromatogram obtained from 2 g of (±)-dinitrobenzoyl-leucine using a two-phase solvent system composed of methyl *tert*-butyl ether/water, where TFA and chiral selector were added to the organic stationary phase each at 40 mM and ammonia at 20 mM to the aqueous mobile phase (26). The sample was eluted as a rectangular peak that was divided into nearly equal parts by two pH zones. Fractions taken from each zone and from the zone boundary were analyzed by the standard analytical-scale CCC technique (26) using the same chiral selector at 10 mM in the organic stationary phase (the upper diagram).

As indicated in the chromatogram, the two enantiomers are efficiently separated with a minimum mixing zone of less than 5% of each peak. The impurity was also concentrated at the mixing zone and detected by the analytical chromatogram (middle peak). The present method may be extended to the separation of various other kinds of racemates by dissolving suitable chiral selectors in the liquid stationary phase.

6.3.7. Conclusion

pH-zone-refining CCC is a new preparative separation method that provides a rich field of applications for the separation of ionizable compounds. This method offers various advantages over conventional CCC methods, such as large sample capacity, high concentration of eluted fractions, and enrichment and detection of minor components present in a large quantity of the crude sample. The method has been successfully applied to the separation of natural and synthetic products including acidic and basic derivatives of amino acids and oligopeptides; hydroxyxanthene dyes; alkaloid; indole auxins; and structural, geometrical and optical isomers; and so on. Although the practical application of the present method is currently limited to organic acids and bases, future application will cover amphoteric compounds and nonionizable analytes.

ACKNOWLEDGMENT

I am indebted to Dr. Henry M. Fales of the National Institutes of Health for reviewing the manuscript with valuable suggestions.

Glossary

V_m	Volume of the mobile phase (MP) in the column (L)
V_s	Volume of the stationary phase (SP) in the column (L)
u_m	Flow-rate of MP (L/min)
u	Traveling rate of the trailing border of the retainer acid through MP (L/min)
V_r	Volume of MP required to elute the retainer acid completely from the column (L)
K_r	Partition coefficient of the retainer acid in the equilibrium zone
K_s	Partition coefficient of solute S in the equilibrium zone
K_{D-r}	Partition ratio of the retainer acid
K_{D-s}	Partition ratio of solute S

K_{a-r} Dissociation constant of the retainer acid
K_{a-s} Dissociation constant of solute S
pH_{z-r} pH of MP in the equilibrium zone of the retainer acid
pH_{z-s} pH of MP in the equilibrium zone of solute S
C_r Molar concentration of the retainer acid in MP
C_R Molar concentration of the retainer acid in SP
C_E Molar concentration of the eluent base in MP
C_m Molar concentration of solute S in MP in the equilibrium zone
C_s Molar concentration of solute S in SP in the equilibrium zone

REFERENCES

1. Y. Ito, Y. Shibusawa, H. M. Fales, and H. J. Cahnmann, *J. Chromatogr.*, **625**, 177 (1992).
2. H. J. Cahnmann, E. Gonçalves, Y. Ito, H. M. Fales, and E. A. Sokoloski, *J. Chromatogr.*, **538**, 165 (1991).
3. H. J. Cahnmann, E. Gonçalves, Y. Ito, H. M. Fales, and E. A. Sokoloski, *Anal. Biochem.*, **204**, 344 (1992).
4. Y. Ito and H. J. Cahnmann, Method for Concentrating a Solute by Countercurrent Chromatography, *U.S. Patent* **5, 354, 473**.
5. A. Weisz, A. L. Scher, K. Shinomiya, H. M. Fales, and Y. Ito, *J. Am. Chem. Soc.*, **116**, 704 (1994).
6. Y. Ito, K. Shinomiya, H. M. Fales, A. Weisz, and A. L. Scher, *the 1993 Pittsburgh Conference and Applied Spectroscopy*, Atlanta, GA, March 7–12, 1993, Abstract 54P.
7. Y. Ito, K. Shinomiya, H. M. Fales, A. Weisz, and A. L. Scher, Abstract for *the 206th National ACS Meeting*, Chicago, IL, Aug. 22–27, 1993.
8. Y. Ito and Y. Ma, *J. Chromatogr.*, **672**, 101 (1994.)
9. Y. Ito and A. Weisz, pH-Zone Refining Countercurrent Chromatography, *U.S. Patent* **5, 332, 504**, July 26, 1994.
10. A. Tiselius, *Arkiv. Kem. Mineral Geol.*, **14B**, 22 (1940).
11. S. Cleasson, *Ann. N. Y.. Acad. Sci.*, **49**, 183 (1948).
12. C. Horváth, A. Nahum, and J. H. Frenz, *J. Chromatogr.*, **218**, 365 (1981).
13. P. Boček, M. Deml, P. Gebauer, and V. Dolnik, *Analytical Isotachophoresis*, B. J. Radola (Ed.), VCH, 1988.
14. A. Weisz, K. Shinomiya, and Y. Ito, *The 1993 Pittsburgh Conference and Applied Spectroscopy*, Atlanta, GA, March 7–12, 1993, Abstract 865.
15. A. Weisz, D. Andrzejewski, R. J. Highet, and Y. Ito, Abstract for *the 10th International Symposium on Preparative Chromatography (PREP 93)*, Arlington, VA, June 14–16, 1993.

16. A. Weisz, D. Andrzejewski, and Y. Ito, Abstract for *the 10th International Symposium on Preparative Chromatography (PREP 93)*, Arlington, VA, June 14–16, 1993.

17. F. Oka, H. Oka, and Y. Ito, *J. Chromatogr.*, **538**, 99 (1991).

18. Y. Ito, *Countercurrent Chromatography in Chromatography, V,* Jounral of Chromatography Library, Part A. Chapter 2 E. Heftmann (Ed.) Elsevier Scientific, Amsterdam, 1992.

19. Y. Ma and Y. Ito, *J. Chromatogr.* A, **678**, 233 (1994).

20. Y. Ma and Y. Ito, *J. Chromatogr.* A, **702**, 197 (1995).

21. Y. Ma, Y. Ito, E. Sokoloski, and H. M. Fales, *J. Chromatogr.* A, **685**, 259 (1994).

22. Y. Ma, Y. Ito, D. Torok, and H. Ziffer, *J. Liq. Chromatogr.*, **17(16)**, 3507 (1994).

23. C. Denekamp, A. Mandelbaum, A. Weisz, and Y. Ito, *J. Chromatogr.*, A, **685**, 253 (1994).

24. C. J. Welch, *J. Chromatogr.* A, **666**, 3 (1994).

25. E. Francotte and A. Junker-Buchheit, *J. Chromatogr.*, **576**, 1 (1992).

26. Y. Ma, Y. Ito, and A. Foucault, *J. Chromatogr.*, A., **704**, 75 (1995).

27. Y. Ito (unpublished data).

APPLICATIONS

CHAPTER

7

HIGH-SPEED COUNTERCURRENT CHROMATOGRAPHY OF NATURAL PRODUCTS

MARC MAILLARD, ANDREW MARSTON, AND
KURT HOSTETTMANN

*Institut de Pharmacognosie et Phytochimie, Ecole de Pharmacie, Université de Lausanne,
CH 1015 Lausanne, Switzerland*

7.1. INTRODUCTION

The isolation and separation of products from biological material (plants, microorganisms, or animal sources) are often very tedious and time-consuming operations. The scale of the challenge is enormous, particularly when one considers that a plant may contain up to 10,000 different chemical constituents.

With the development of increasingly sophisticated apparatus and techniques for the isolation and the structure elucidation of compounds of biological origin, the natural products chemist can now solve numerous problems that were unsoluble only a few years ago. Thus, the isolation of previously inaccessible minor components, present only in very small amounts, or the separation of closely related compounds from complex mixtures can today be achieved with considerable success.

However, it should be pointed out that while physicochemical methods for the elucidation of molecular structure become more and more refined, enabling structure determination to become a routine step, the successful isolation of pure substances from biological material remains largely empirical.

The nature of the separation problem is generally directly related to the type of isolation that is performed. Techniques and approaches may vary entirely according to the amount of sample that is to be separated. The methodology of isolation of small quantities (milligram or less) required for structure determination purposes is different from the one used for larger amounts (hundred milligram to kilogram quantities) needed for comprehen-

High-Speed Countercurrent Chromatography, Edited by Yoichiro Ito and Walter D. Conway.
Chemical Analysis Series, Vol. 132.
ISBN 0-471-63749-1 © 1996 John Wiley & Sons, Inc.

sive biological testing, semisynthesis, or even for production of therapeutic agents.

Since chromatographic methods play the major role in obtaining a pure compound from complex mixtures, any new addition to the arsenal of chromatographic tools is highly welcome. Furthermore, techniques that are rapid and do not lead to decomposition, material loss, or artifact formation are needed. For example, the rediscovery of countercurrent chromatography (CCC) in the form of droplet CCC (DCCC) about 20 years ago (1) introduced a very good alternative to other preparative and semipreparative chromatographic methods, such as column chromatography using various solid stationary phases (silica gel, alumina, polyamide, cellulose, and Sephadex), flash chromatography, preparative liquid chromatography, and all the other techniques that use a solid or a liquid stationary phase fixed on an inert support. In contrast to these latter techniques, CCC is a liquid–liquid separation method that does not require a solid sorbent. There are numerous points in favor of this kind of chromatographic technique, some of which are listed below.

Since there is no solid support on which irreversible adsorption can occur, it is possible to totally recover the introduced sample. Indeed, all the compounds that can interact chemically with solid supports (and then be irreversibly retained or denatured) are recovered either in the mobile phase or by removing them from the stationary phase after the chromatographic run. In addition, such adsorption effects are often accompanied by band tailing, a phenomenon that is not observed in liquid–liquid separations. Finally, when compared to the highly specialized and costly solid supports used in bonded phase liquid chromatography, CCC represents a very interesting and low-cost alternative, especially for preparative scale isolation processes. High-speed countercurrent chromatography (HSCCC) is one of the most recent developments of the liquid–liquid separation techniques, and although it shares certain characteristics with its other relatives, such as DCCC (2) or RLCCC (3), HSCCC also presents some advantages. Unlike DCCC, it is not necessary for the mobile phase to form droplets as it passes through the stationary phase. In fact, there is no restriction on the choice of solvents as long as their mixture forms a stable two-phase system. Another benefit of the HSCCC technique is related to the fact that it is possible at any stage of the run to switch the phase or elution mode, that is, at any point during the separation the flow direction can be reversed and elution commenced with stationary phase. Furthermore, the use of centrifugal force for retention of the stationary phase allows an increase in the speed of performed separations, without loss of resolution. This finding is of crucial importance, particularly in the isolation and separation of unstable or highly sensitive compounds.

With these advantages, there is no doubt that HSCCC techniques will become increasingly used by chemists involved in the separation of bioactive components from crude extracts of plants, marine organisms, or microbial fermentation media.

7.1.1. Instruments

Basic principles of HSCCC have already widely been described elsewhere (4, 5) and in other chapters of this book. However, a brief description of the design of the two alternative types of HSCCC apparatus currently available on the market will be given here.

> *Rotating Coil Instruments.* One (or two to three) separation coil(s), normally composed of a single Teflon tube(s) wrapped around a spool, describing a planetary movement around a central axis.
>
> *Cartridge Instruments.* A system of cartridges assembled in a spinning rotor.

Both methods require the use of centrifugal force to retain the stationary phase in the separator. This results in a very effective way of using the column space, since under these conditions, almost 100% of the efficient column space is used for the mixing of the two phases. Thus the interfacial area of the phases is maximized (6).

As there is no solid support, retention of injected samples depends uniquely on the respective partition coefficients of each constituent of the mixture between the phases of the two-phase solvent system. The number of theoretical plates of the column system can be easily modified by increasing the number (or the length) of coils, by reducing the internal diameter of the coiled column, or by reducing the helical diameter. In the most recent development of HSCCC, mobile phase is pumped at a very high flow-rate, while increased rotation speed allows retention of stationary phase inside the separation unit. Naturally this setup leads to much bigger constraints, which call for some drastic technical modifications in designing a functional centrifugal apparatus.

Finally, both rotating coil and cartridge systems allow a great flexibility in the choice of the solvent system or flow direction. This latter system can even be easily switched during the run by a simple manipulation. Modifications of the pumping procedure allow variation of phase ratios in the coil or cartridge, gradient operation, and particularly easy reversal of the mobile and the stationary phase (7).

7.1.1.1. Rotating Coil Instruments

Over the last few years, a variety of prototypes have been constructed, most of which have found very little practical application. However, two chromatographs have been commercialized, and the majority of the published applications have been performed on these instruments. The Multilayer Coil Planet Centrifugal Countercurrent Chromatograph, also known as the "multilayer coil separator–extractor," developed by Ito and marketed by P.C. Inc., Potomac, MD, consists of a single length of 2.6-mm i.d. polytetrafluoroethylene (PTFE) tubing, with a capacity of about 350 mL. Rotation speeds for separation are generally of the order of 700–800 rpm. The coil is balanced by a counterweight and describes a planetary motion around a central axis (see Fig. 7.1) (8). The second commercially available rotating coil instrument is the Multicoil Centrifugal Countercurrent Chromatograph, which has been introduced by Pharma-Tech Research Corp., Baltimore, MD. This is a development of the multilayer coil planet centrifuge (MLCPC) described above and, depending on the model, consists of two or three coils mounted symmetrically on each side of the rotary frame of the centrifuge (9). This geometry avoids use of a counterweight to balance the system. Each coiled column undergoes synchronous planetary movement and is equipped with flow-tubes arranged in such a fashion that they do not twist, enabling seal-free operation of the instrument (Fig. 7.2), thus avoiding leakages. Increasing the number of the coils results in an increased column capacity and enhances the partition

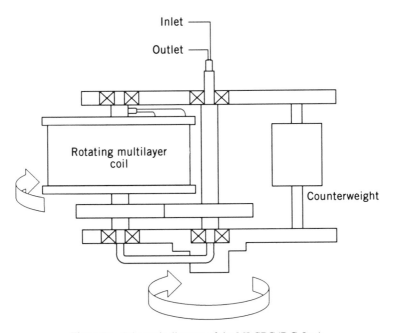

Figure 7.1. Schematic diagram of the MLCPC (P.C. Inc.).

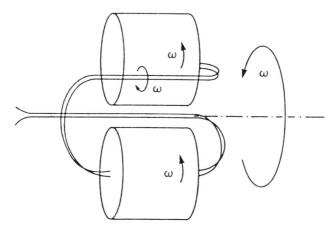

Figure 7.2. Principle of operation of the twin-CPC (ω = rate of rotation).

efficiency (10). The total solvent volume depends on the diameter and length of Teflon tubing used, while rotational speeds of up to 1500 rpm are possible. Recently, a two-coil apparatus known as the Kromaton (SEAB, Villejuif, France) has appeared on the market.

When we are concerned with the principle of rotating coil instruments, it is important to understand that the motion of the coil(s) causes a vigorous agitation of the two phases. A repetitive mixing and settling process, ideal for solute partitioning, occurs at over 13 times per second (Fig. 7.3) (11). These rapid and successive partitions explain how it is possible to have efficient separations with such a small volume of solvent. Unfortunately, the mechanism of hydrodynamic distribution of the solvent phases in the coil(s) is not known. Retention of the stationary phase is strongly dependent on interfacial tension, density difference, and the viscosity of the two phases. Back pressures are generally low during the run (2–6 bar), which avoids technical problems such as leakages.

7.1.1.2. *Cartridge Instruments*

These chromatographs, constructed by Sanki Engineering Ltd., Kyoto, Japan, were first described in 1982 (12). The columns in such instruments are actually a series of channels drilled into a block of PCTFE and connected (top to bottom) with each other by a small passageway parallel to the columns (Fig. 7.4). These columns are arranged in a block, called a cartridge. Each cartridge contains the equivalent of 400 separation channels. Twelve cartridges, which contain a total of 250 mL of solvent, are connected by a narrow-bore Teflon tube and arranged around the circumference of a rotor in such

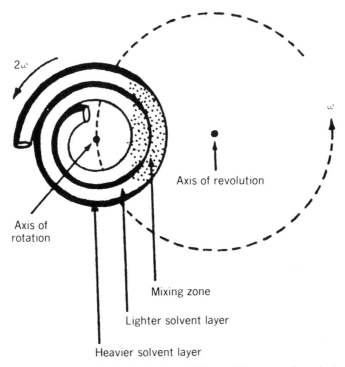

Figure 7.3. Solvent distribution in the multilayer coil (ω = rate of rotation).

a way that the longitudinal axes of the separation channels are parallel to the direction of the centrifugal force (Fig. 7.5). Since the rotor is spinning during the chromatographic run, the system is provided with a special connection, a rotary seal joint, which allows the leak-free passage of solvent into the apparatus under pressure.

Back pressure tends to be high (20–60 bar) during elution. This phenomenon is directly related to the flow-rate of the mobile phase, and also increases quadratically with the spin rate (13). Under these conditions most of the separations are achieved at speeds of about 1000 rpm.

7.1.2. Solvent Systems

The correct choice of a solvent system is of crucial importance to a successful separation. Several approaches for the selection of the most judicious solvent system can be used when undertaking a HSCCC chromatographic separation.

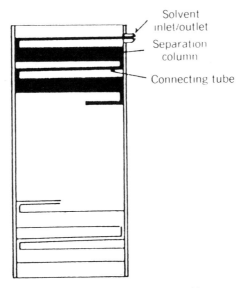

Figure 7.4. Cross section of a cartridge.

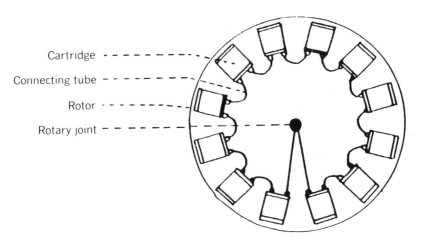

Figure 7.5. Disposition of cartridges in the Sanki instrument.

7.1.2.1. Consideration of Known Systems

A survey of the specialized literature reveals numerous examples of solvent systems used in different CCC separations (14–16). Consultation of these references may give some indications of possible systems useful for the separation in question. For example, the classical chloroform/methanol/ water system or the less polar n-hexane/ethyl acetate/methanol/water system can be chosen as the starting point, and by modifying the relative proportions of each individual solvent, it is possible to finally obtain the required distribution of sample between the two phases. Chloroform-based solvent systems provide large density differences and relatively high interfacial tensions between the two solvent phases: they are consequently frequently employed in HSCCC. Furthermore, their short settling times allow a reduction in the amount of displaced stationary phase and normally produce satisfactory phase retention. However, due to their physicochemical properties, these chloroform/methanol/water systems can easily lead to overpressure problems with the cartridge instruments; separation with certain proportions of this solvent system are troublesome with the Sanki apparatus. On the other hand, ethyl acetate/methanol/water solvent combinations are not always compatible with the MLCPC, since the equilibrium state is difficult to reach and there is a continual leakage of stationary phase.

7.1.2.2.   Thin-Layer Chromatography

By analogy with the method of solvent search previously described for DCCC (17), the selection of appropriate biphasic solvent systems can be performed by thin-layer chromatography (TLC). Indeed, this method of screening, using silica gel plates, is also applicable for centrifugal partition chromatographic techniques. However, some modifications have to be introduced into the original method. Thus, when migrating the TLC plate in the organic layer of a two-phase aqueous solvent system, the ideal R_f values of the sample for chromatography with the organic phase, which are used as the mobile phase, should lie between 0.2 and 0.4 for the Sanki apparatus. For the P.C. Inc. MLCPC, best results are obtained if the R_f values are a little lower.

This method only gives an approximate indication of the utility of a particular solvent system because TLC involves both partition and adsorption mechanisms, whereas the centrifugal partition chromatography (CPC) instruments are based on purely liquid–liquid partition phenomena. Okuda et al. (18) used TLC on cellulose plates for selection of solvent systems applicable to the separation of tannins.

7.1.2.3.   High-Performance Liquid Chromatography

An alternative method of selecting suitable solvents is to monitor the distribution of the sample between the two phases constituting the solvent system. Indeed, it is theoretically possible in CPC to predict the locations of eluted solute peaks once their partition coefficients are known. Thus the knowledge

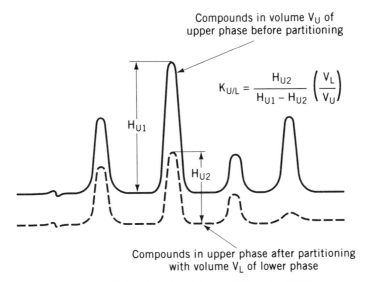

Figure 7.6. Estimation of partition coefficients by HPLC.

of these parameters is of great benefit. An analytical high-performance liquid chromatography (HPLC) method has been described to determine the partition coefficients of the components of a mixture (19). Reversed-phase HPLC (RP-HPLC) is used to determine the respective concentrations of each component distributed between the two immiscible liquid phases. The partition coefficient K is calculated from the detector response following injection of a solution of the compounds before and after extraction with an immiscible solvent (Fig. 7.6). This procedure is particularly useful for the analysis of natural product mixtures, since all of the individual components may not be fully identified.

7.1.2.4. Partition Ratio of Biological Activity

Another method, limited to bioactive products, for undertaking a solvent search, is based on the distribution of biological activity of the sample to be separated. The sample is first shaken with the two-phase solvent system. Both phases are then screened for the activity in question. The aim is to find a solvent system that gives a good distribution of activity between the two phases. This method has found several applications, mainly in the isolation of antibiotics (20).

7.1.2.5. *Analytical HSCCC*

Utilization of analytical HSCCC can, in some cases, be very interesting for the choice of solvent systems for preparative-scale separation. For example, a Pharma-Tech Model CCC-3000 analytical HSCCC instrument (coil capacity 43 mL) has been used to separate an artificial mixture of the three flavonoid aglycones hesperetin (**1**), kaempferol (**2**), and quercetin (**3**). At a flow-rate of 0.7 mL/min and a rotational speed of 2000 rpm, the analysis of a 3 × 0.5-mg mixture was complete within 60 min (Fig. 7.7a). Chloroform/ methanol/water (33:40:27) was employed as the solvent system, with the lower phase as the mobile phase. Direct transposition of this solvent system to a Pharma-Tech preparative multicoil centrifugal CCC (70 m × 2.6 mm i.d., total coil capacity 650 mL, 1000 rpm), with a flow-rate of 3 mL/min allowed the separation of a 15-mg sample of the same flavonoid mixture (Fig. 7.7b) (21).

However, use of an analytical apparatus for choosing preparative-scale conditions is not always very reliable. Differences between the two installa-

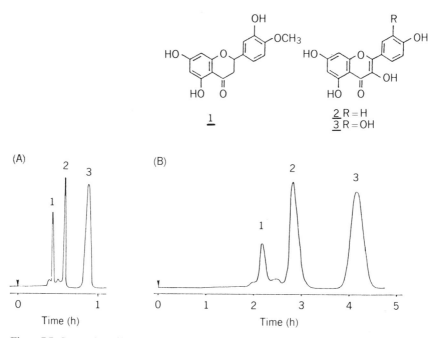

Figure 7.7. Separation of hesperitin (**1**), kaempferol (**2**), and quercetin (**3**). Solvent system consists of $CHCl_3$–MeOH–H_2O (33:40:27) (mobile phase = lower phase); Detection 254 nm. (a) Pharma-Tech CCC-3000 (coil capacity: 43 mL); flow-rate 0.7 mL/min. Rotational speed is 2000 rpm. (b) Pharma-Tech CCC-1000 (coil capacity: 650 mL); flow-rate is 3 mL/min. Rotational speed is 1000 rpm.

tions, such as the coil i.d., mean that solvent systems that are very efficient at the preparative scale cannot always be efficiently used with an analytical apparatus, since their viscosity causes high carry-over of the stationary phase. On other hand, in a small coil (< 40 mL), leakage of only a very minor amount of stationary phase results in a dramatic decrease in the retention percentage of the stationary phase.

7.2. PREPARATIVE APPLICATIONS

An extensive range of applications of the various centrifugal countercurrent instruments is now available and several reviews have appeared on the subject (4, 5, 8, 16). High-speed CCC is being employed routinely in many research laboratories, both in industry and in universities, for the preparative separation of natural products on anything from a microgram to a gram scale. Aqueous and nonaqueous solvent systems are used and the separation of either very polar or nonpolar compounds poses no particular problems, as long as the mixture to be injected is soluble in at least one of the phases.

The advantage of HSCCC over DCCC in the separation of natural products lies in the speed of operation (Fig. 7.8), while relative to methods employing solid supports (sorbents), irreversible adsorption is avoided and decomposition is kept to a minimum. Resolution is not always very high but both crude extracts and semipure fractions can be successfully chromatographed. This technique is particularly suitable for polar substances.

Although certain separations have been performed on prototypes or home-made installations, most of the applications reported have used one of

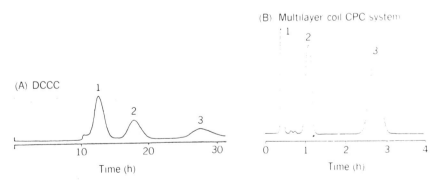

Figure 7.8. Comparison of DCCC and HSCCC (multilayer coil separator–extractor) for the separation of hesperitin (**1**), kaempferol (**2**), and quercetin (**3**). Solvent system consists of $CHCl_3$–$MeOH$–H_2O (33:40:27) (mobile phase = lower phase); detection = 254 nm. (a) Flow-rate = 48 mL/h. (b) Flow-rate = 3 mL/min; rotational speed is 700 rpm.

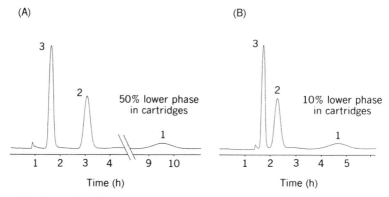

Figure 7.9. Separation of flavonoids 1–3 on a Sanki cartridge CCC instrument with different phase ratios. Solvent system is $CHCl_3$–MeOH–H_2O (5:6:4) (mobile phase = upper phase); flow-rate = 2.5 mL/min; rotational speed = 400-rpm; detection = 254 nm. (a) 50% lower phase in cartridges. (b) 10% lower phase in cartridges.

the following three systems: the multilayer coil separator–extractor (P.C. Inc.), the CCC-1000 (Pharma-Tech.), or the Sanki cartridge instruments.

The upper and lower phases of a given solvent system can be pumped independently by connecting two pumps to a CCC. Hence, the proportion of the two phases in the instrument can be changed at will. This finding has important implications for gradient operation, for reversed phase operation, and also for simply changing elution times. This modification of basic CCC operation has been applied to both rotating coil (7) and cartridge instruments (22). In the example shown in Fig. 7.9a for a cartridge instrument, running the upper and lower phase pumps simultaneously at 5 mL/min gives 50% of each phase in the instrument. The sample, a mixture of three flavonoids, is eluted in 10h. Adjustment of pump flow-rates to give initially 10% of the lower phase allows complete separation of the mixture after 5h (Fig. 7.9b) (22).

7.2.1. Flavonoids

A large number of polar natural products have been separated by CCC; many of them are polyphenolics. Flavonoid aglycones and glycosides are well suited to liquid–liquid chromatography because there is no tendency to "tail," as found on the usual sorbents (silica gel, polyamide, etc.). Furthermore, the different countercurrent methods give quantitative recovery of injected samples, which is never the case in liquid–solid chromatography of polyphenols.

For example, the first step of the separation of flavonol glycosides from *Epilobium parviflorum* (Onagraceae) involved introduction of a methanol

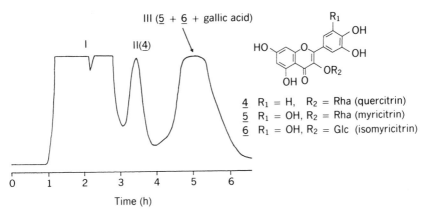

Figure 7.10. HSCCC of a methanol extract from *Epilobium parviflorum* (Onagraceae). Solvent system is $CHCl_3$–MeOH–H_2O (7:13:8) (mobile phase = lower phase); flow-rate = 3.5 mL/min; rotational speed = 700 rpm; Detection = 254 nm.

extract (2 g) onto a multilayer coil separator–extractor eluted with the lower phase of the solvent system chloroform/methanol/water (7:13:8) (Fig. 7.10). Three major bands were obtained, starting with lipophilic compounds (I), followed by band II, containing quercitrin (**4**). A gel permeation chromatographic step on Sephadex LH-20 (MeOH) was sufficient to obtain the glycoside in the pure state. Band III contained myricitrin (**5**), isomyricitrin (**6**), and gallic acid. After chromatography on LH-20 (MeOH), the three substances were obtained in 35-, 18-, and 26-mg yields, respectively (23). For HSCCC, the coil of the instrument was filled with 80% upper and 20% lower phase by the two-pump modification (7).

Extracts of *Gingko biloba* (Gingkoaceae) are extremely successful commercially on the European pharmaceuticals market at the present time. Flavonoids from the leaves have been isolated by a combination of HSCCC and semipreparative HPLC. The liquid–liquid step made use of gradient elution, starting with water as the stationary phase and eluting with ethyl acetate. Increasing amounts of *i*-butanol were then added to the ethyl acetate, reaching the proportion ethyl acetate/*i*-butanol (6:4) at the end of the elution. A total of seven flavonol glycosides was obtained from 500 mg of leaf extract (24) (Table 7.1).

Flavonoid glycosides have been separated from the aerial parts of *Oxytropis ochrocephala* (Leguminosae) by a horizontal flow-through CPC, consisting of four columns (total capacity 120 mL) around a holder, balanced by a counterweight. The two-phase solvent system used was chloroform/ethyl acetate/methanol/water (2:4:1:4), with the lower phase as the mobile phase. Ethyl acetate was added to the ternary system to increase the polarity. When

Table 7.1. Separations of Plant-Derived Natural Products by Centrifugal Countercurrent Chromatography

Sample	Instrument[a]	Solvent System	References
Flavonoids	Sanki LLN	$CHCl_3-MeOH-H_2O$ (33:40:27)	76
	Ito–HSCCC	$CHCl_3-MeOH-H_2O$ (33:40:27)	7
	Ito–HSCCC	$CHCl_3-MeOH-H_2O$ (7:13:8)	23
	Ito–HSCCC	$CHCl_3-MeOH-H_2O$ (4:3:2)	77
	Triple-coil HSCCC	$CHCl_3-MeOH-H_2O$ (4:3:2)	78
	Sanki L-90	$CHCl_3-MeOH-H_2O$ (7:13:8)	26
Flavonoid glycosides	Horizontal Flow-through CPC	$CHCl_3-EtOAc-MeOH-H_2O$ (2:4:1:4)	25
	Sanki LLN	$EtOAc-94\%\ EtOH-H_2O$ (2:1:2)	76, 79
	Sanki LLN	$EtOAc-nBuOH-H_2O$ (2:1:2)	79
	Ito–HSCCC	$EtOAc-H_2O-EtOAc-iBuOH-H_2O$	24
Xanthone glycosides	Sanki LLN	$CHCl_3-MeOH-H_2O$ (4:4:3)	27, 28
Lignan glycosides	Ito–HSCCC	$CHCl_3-MeOH-H_2O$ (7:13:8)	80
	Ito–HSCCC	$CHCl_3-MeOH-H_2O$ (5:5:3)	81
	Ito–HSCCC	$nC_6H_{14}-CH_2Cl_2-MeOH-H_2O$ (2:4:5:2)	81
	Ito–HSCCC	$nC_6H_{14}-CH_2Cl_2-MeOH-H_2O$ (1:5:4:3)	89
Lignans	Dual CCC	$nC_6H_{14}-EtOAc-MeOH-H_2O$ (10:5:5:1)	34
	Ito–HSCCC	$nC_6H_{14}-CH_3CN-EtOAc-H_2O$ (8:7:5:1)	33
Phenolic glycosides	Ito–HSCCC	$C_6H_{12}-Me_2CO-EtOH-H_2O$ (7:6:1:3)	82
	Ito–HSCCC	$CHCl_3-MeOH-H_2O$ (7:13:8)	80
	Ito–HSCCC	$nC_6H_{14}-EtOAc-MeOH-H_2O$ (3:7:5:5)	84
	Ito–HSCCC	$nC_6H_{14}-EtOAc-MeOH-H_2O$ (3:7:5:5)	76
Polyphenols	Sanki LLN	$C_6H_{12}-EtOAc-MeOH-H_2O$ (7:8:6:6)	22
	Sanki L-90	$CHCl_3-MeOH-H_2O$ (7:13:8)	76

Compound	Instrument	Solvent system	Reference
Chalcones	Sanki LLN	$CHCl_3$–MeOH–H_2O (7:13:8)	29, 30
Tannins	Sanki LLN	nBuOH–nPrOH–H_2O (4:1:5)	29
	Sanki LLN	nBuOH–nPrOH–H_2O (2:1:3)	18, 85
	Sanki LLN	nBuOH–HOAc–H_2O (4:1:5)	31
	Ito-HSCCC	nBuOH–0.1 M NaCl (1:1)	116
Coumarins	Sanki LLN	nC$_6$H$_{14}$–EtOAc–MeOH–H_2O (3:7:5:5)	76
	Ito-HSCCC	nC$_6$H$_{14}$–EtOAc–MeOH–H_2O (3:7:5:5)	84
Coumarin glycosides	Ito-HSCCC	$CHCl_3$–MeOH–H_2O (13:23:16)	86
Anthranoids	Sanki LLN	nC$_6$H$_{14}$–CH$_3$CN–MeOH (8:5:2)	76
	Ito-HSCCC	nC$_6$H$_{14}$–CH$_3$CN–MeOH (8:5:2)	7
Naphthoquinones	Sanki LLN	nC$_6$H$_{14}$–CH$_3$CN–MeOH (8:5:2)	76
Iridoid glycosides	Sanki LLN	$CHCl_3$–MeOH–H_2O (9:12:8)	22
	Ito-HSCCC	$CHCl_3$–MeOH–iPrOH–H_2O (5:6:1:4)	35
	CCC-1000	$CHCl_3$–MeOH–H_2O (9:12:8)	28
Sesquiterpenes	Ito-HSCCC	iC$_8$H$_{18}$–EtOAc–MeOH–H_2O (7:3:6:4)	87
Norditerpenes	Sanki LLN	$CHCl_3$–MeOH–H_2O (5:6:4)	88
Quassinoids	Ito-HSCCC	$CHCl_3$–MeOH–H_2O (5:6:4)	89
Triterpenes	Ito-HSCCC	nC$_6$H$_{14}$–EtOAc–MeOH–CH$_3$CN (5:2:4:5)	90
	Triple-coil HSCCC	nC$_6$H$_{14}$–94% EtOH–H_2O (6:5:2)	78
Saponins	Sanki LLN	$CHCl_3$–MeOH–H_2O (7:13:8)	76
	Ito-HSCCC	$CHCl_3$–MeOH–iBuOH–H_2O (7:6:3:4)	37
	Ito-HSCCC	$CHCl_3$–MeOH–H_2O (7:13:8)	36
	Ito-HSCCC	$CHCl_3$–MeOH–iPrOH–H_2O (5:6:1:4)	38
Cardiac glycosides	Ito-HSCCC	$CHCl_3$–MeOH–HOAc–H_2O (5:3:1:3)	91
Alkaloids	Multilayer HSCCC	nC$_6$H$_{14}$–EtOH–H_2O (6:5:5)	39
	Multilayer HSCCC	$CHCl_3$–0.07 M sodium phosphate (1:1)	83
	Ito-HSCCC	$CHCl_3$–MeOH–H_2O (10:10:1)	93
	Ito-HSCCC	$CHCl_3$–MeOH–0.5% HBr/H_2O (5:5:3)	94

(*Continued*)

193

Table 7.1. (*Continued*)

Sample	Instrument[a]	Solvent System	References
	Ito-HSCCC	nBuOH–Me$_2$CO–H$_2$O (8:1:10)	95
	Ito-HSCCC	nBuOH–0.1 M NaCl (1:1)	96
	Ito-HSCCC	CHCl$_3$–0.07 M sodium acetate (1:1)	97
	Ito-HSCCC	CCl$_4$–MeOH–H$_2$O (10:10:1)	98
Retinals	Sanki LLN	C$_6$H$_6$–nC$_5$H$_{12}$–CH$_3$CN–MeOH (500:200:200:11)	99
Carotenoids	Ito-HSCCC	CCl$_4$–MeOH–H$_2$O (5:4:1)	40, 97
Gingerols (phenolic ketones)	Ito-HSCCC	nC$_6$H$_{14}$–EtOAc–MeOH–H$_2$O (3:2:3:2)	41
Cyclohexadienone derivatives	Sanki LLN	nC$_6$H$_{14}$–94% EtOH–EtOAc–H$_2$O (83:67:33:17)	100
	Ito-HSCCC	nC$_6$H$_{14}$–94% EtOH–EtOAc–H$_2$O (83:67:33:17)	100
Humulone derivatives	Ito-HSCCC	nC$_6$H$_{14}$–CH$_3$CN–$tert$BuOMe (10:10:1)	101
Polyacetylenic alcohols	Ito-HSCCC	nC$_6$H$_{14}$–CH$_3$CN–$tert$BuOMe (10:10:1)	42, 101

[a] Ito-HSCCC = High-speed countercurrent chromatograph/multilayer coil separator–extractor (P.C.Inc., Potomac, MD); CCC-1000: Pharma-Tech.

compared with the same separation by semipreparative HPLC, centrifugal CCC had a lower column efficiency and a longer separation time, but a higher loading capacity. In addition, fractions obtained by CCC were claimed to be of higher purity (25).

Among several new phenolic compounds isolated from licorice, originating from northwest China, is an isoflavone glicoricone (**7**) with a phloroglucinol-type substitution pattern in the B ring. For its separation, 20 g of the ethyl acetate extract was chromatographed on a Sanki L-90 instrument equipped with twelve 1000-mL cartridges (capacity 850 mL), using the solvent system chloroform/methanol/water (7:13:8). The final purification was done by chromatography on MCI gel CHP-20P and preparative TLC (26).

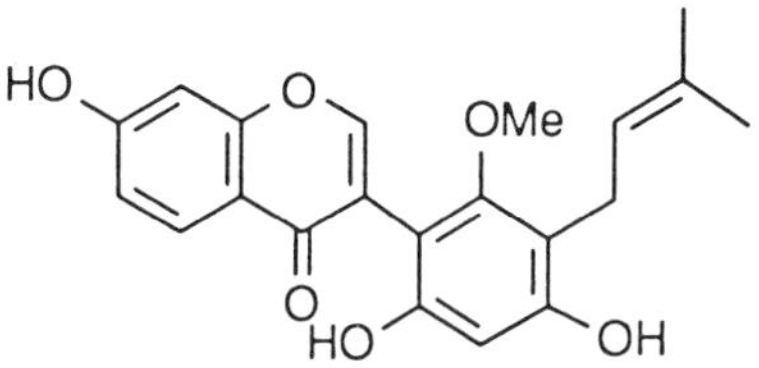

7

7.2.2. Xanthones

The xanthones are another class of polyphenolics that lends itself well to liquid–liquid chromatography applications. These compounds have potential therapeutic use as inhibitors of monoaminooxidases. A three-step isolation procedure involving a cartridge CCC instrument enabled the three new xanthone glycosides **8**–**10** to be obtained from *Halenia campanulata*, a South American species of the Gentianaceae (Fig. 7.11). Initial gel filtration was followed by centrifugal partition chromatography with the solvent system chloroform/methanol/water (4:4:3), using the lower phase as the mobile phase, at 3.2 mL/min and 300 rpm. As xanthone glycosides **9** and **10** eluted together, and **8** was still contaminated with a small amount of impurity (Fig. 7.12), final purification by semipreparative HPLC was necessary (27, 28).

7.2.3. Tannins

Tannins include a wide variety of phenolic compounds, ranging from single glycosides of gallic acid to complex condensed and polymerized derivatives of catechin, epicatechin, and related compounds. Their separation poses special problems since there is often irreversible adsorption and even hydrolysis on solid supports (18, 29). Preparative HPLC is also accompanied by sample loss

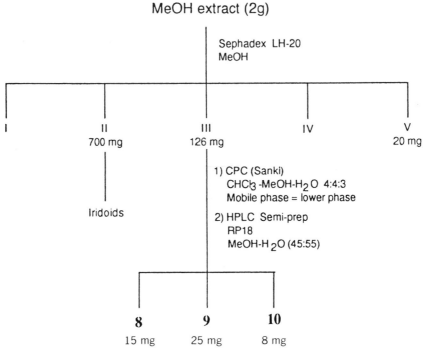

Figure 7.11. Isolation of xanthone glycosides from *Halenia campanulata* whole plant (Gentianaceae).

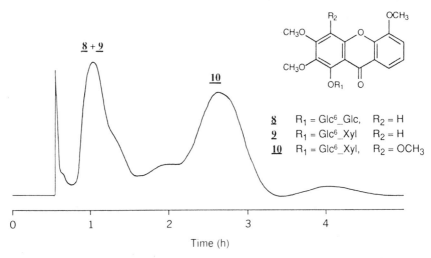

Figure 7.12. Separation of xanthone glycosides from *Halenia campanulata*. (Gentianaceae). Sanki LLN (12 cartridges). Solvent is $CHCl_3-MeOH-H_2O$ (4:4:3) (mobile phase = lower phase); flow rate = 3.2 mL/min; rotational speed = 300 rpm; detection = 254 nm.

and deterioration or contamination of the column (30). Centrifugal CCC has proved to be an ideal technique for the resolution of these particular problems.

Among the examples of successful separations performed on a Sanki cartridge instrument are the isolation of the water-soluble tannin geraniinic acid A from *Geranium thunbergii* (Geraniaceae) (31) and the isolation of the ellagitannin liquidambin from *Liquidambar formosana* (Hamamelidaceae) (32). In the latter case, up to 3 g of sample were loaded onto the chromatograph. While *n*-butanol/*n*-propanol/water (4:1:5) was used for the above-mentioned separation, different proportions (2:1:3) were employed for the isolation of tannins from *Stachyurus praecox* (Stachyuraceae) leaves, with the upper phase as the mobile phase (18).

One quite remarkable separation is that of castalagin (**11**) from vescalagin (**12**), since they are diastereoisomers differing only in the configuration of the hydroxyl group of the central carbohydrate moiety. These compounds were extracted from *Lythrum anceps* (Lythraceae) leaves and chromatographed on a Sanki cartridge instrument, with the solvent system *n*-butanol/*n*-propanol/water (4:1:5) (mobile phase = upper phase) (18).

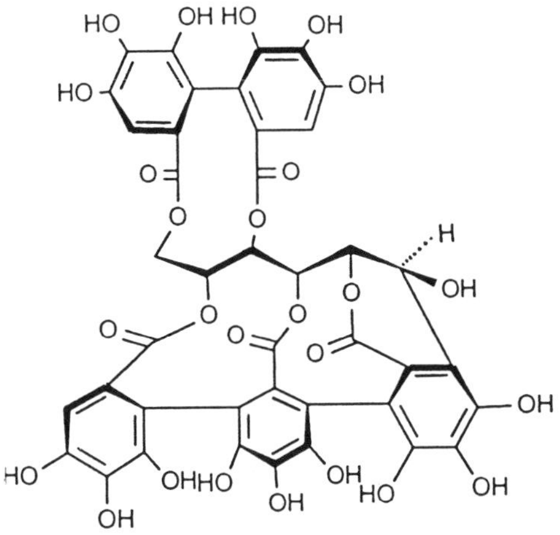

11

A readily hydrolyzable trimeric tannin, nobotanin J, and a tetrameric tannin, nobotanin K, from *Heterocentron roseum* (Melastomataceae) were obtained in the pure state by chromatography with the solvent system *n*-butanol/*n*-propanol/water (4:1:5). When purification of nobotanin J was attempted by gel filtration, extensive hydrolysis occurred (29, 30).

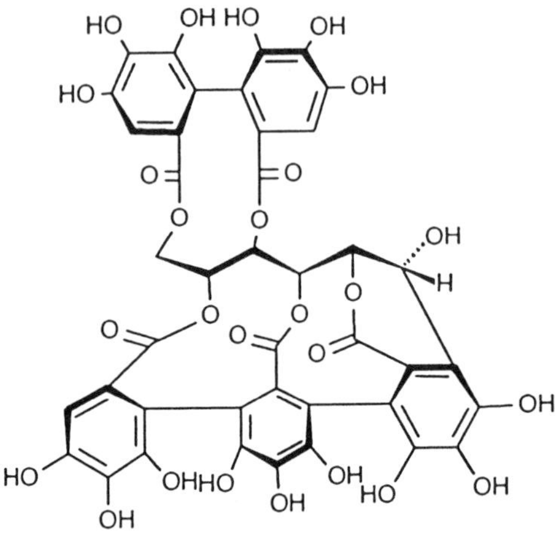

12

Elaeocarpusin, which decomposes to form geraniin and corilagin on simply standing in water for several hours, was obtained from *Geranium thunbergii* (Geraniaceae) leaves by centrifugal cartridge chromatography with the solvent system *n*-butanol/*n*-propanol/water (4:1:5), using the upper phase as the mobile phase (18).

7.2.4. Lignans

Several applications of centrifugal CCC have been reported for the separation of lignans and their glycosides from natural products (see Table 7.1). In one case, the immunomodulatory compound eleutheroside E (**13**) was isolated from the roots of Siberian ginseng (*Eleutherococcus senticosus*, Araliaceae) by a strategy involving HSCCC. An initial open-column chromatography step on silica gel of the methanol extract gave 1.3 g of a fraction containing the desired lignan glycoside. A portion of this fraction (730 mg) was chromatographed on a multilayer coil separator–extractor filled with 50% of each phase of the chloroform/methanol/water solvent system, to give 35 mg of pure eleutheroside E (**13**) (Fig. 7.13) (22).

Initial purification by centrifugal chromatography of insecticidal neolignans from *Magnolia virginiana* (Magnoliaceae) was claimed to be better, less expensive, and more efficient than traditional open-column or more recent flash chromatographic methods (33). A hexane extract of the leaves was chromatographed using the lower layer of the solvent system hexane/acetonit-

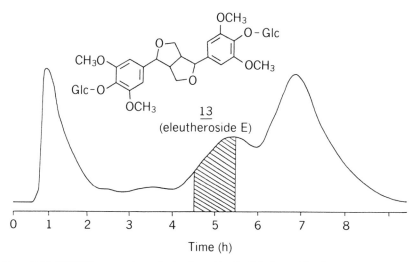

Figure 7.13. HSCCC separation of eleutheroside E (**13**) from *Eleutherococcus senticosus* (Araliaceae) roots. Solvent is $CHCl_3$–$MeOH$–H_2O (7:13:8) (mobile phase = lower phase); 50% lower phase in coil; flow-rate = 3.5 mL/min; rotational speed = 700 rpm; detection = 254 nm.

rile/ethyl acetate/water (8:7:5:1) as the mobile phase. This very interesting solvent system contains only a small proportion of water, probably to provide compatibility with the very lipophilic extract. Subsequent purification of the fractions provided a biphenyl ether (**14**) and two biphenyls (**15, 16**), which were not only insecticidal to *Aedes aegypti*, the vector of yellow fever, but also fungicidal, bactericidal, and toxic to brine shrimp (33).

Dual CCC is very useful for the separation of crude extracts and fractions that are composed of many constituents with a wide range of polarities. The sample is fed into the middle portion of a multilayer coil and the polar and nonpolar components are eluted from appropriate ends of the column, that is, this is true CCC. By this means, normal and reverse phase elutions can be performed simultaneously. Separation of schisanhenol (**17**) (32 mg) and schisanhenol acetate (**18**) (4 mg) from a crude ethanol extract (125 mg) of *Schisandra rubriflora* (Schisandraceae) kernels was possible by dual CCC on a 400 mL, 2.6-mm i.d. PTFE coil with the solvent system hexane/ethyl acetate/methanol/water (10:5:5:1) at a flow-rate of 2 mL/min (34).

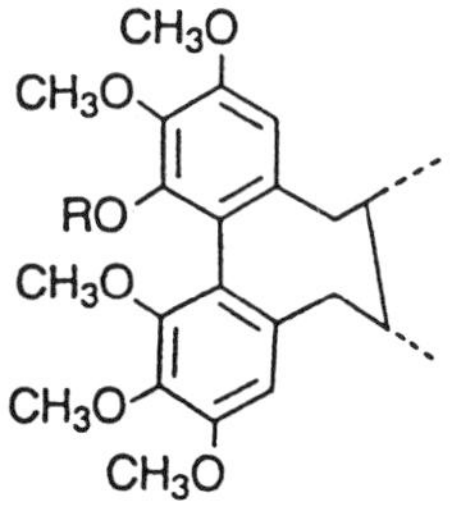

| **17** | R = H |
| **18** | R = Ac |

7.2.5. Monoterpene Glycosides

A useful one-step separation of a constituent of a crude plant extract is shown in Fig. 7.14. This involves a seco-iridoid glycoside from the roots of gentian (*Gentiana lutea*, Gentianaceae). The plant drug is very bitter and is often employed for its tonic properties. The roots were extracted with methanol and after evaporation of solvent, the residue was taken up in water and partitioned with petroleum ether, followed by ethyl acetate and *n*-butanol. Countercurrent chromatography of the butanol extract (500 mg) gave the pure bitter principle gentiopicrin (**19**) (113 mg) after an elution time of approximately 6 h, with 20% of lower phase of the solvent system chloroform/ methanol/water (9:12:8) in the coil (22).

A separation scheme involving silica gel chromatography, low-pressure liquid chromatography (LC), semipreparative HPLC, and centrifugal CCC has been used to isolate a whole series of iridoid glycosides from *Rogeria adenophylla* (Pedaliaceae). After open-column chromatography on silica gel of a methanol extract of the aerial parts, one fraction (796 mg) was injected onto a multilayer coil separator–extractor [solvent system:chloroform/ methanol/ isopropanol/water (5:6:1:4), mobile phase = lower phase]. From the nine

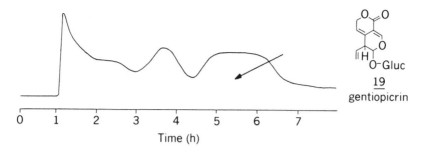

Figure 7.14. High-speed CCC separation of a *Gentiana lutea* (Gentianaceae) methanol root extract. Solvent system is $CHCl_3$–MeOH–H_2O (9:12:8) (mobile phase = lower phase); flow-rate = 3mL/min; rotational speed = 700 rpm.

resulting fractions, low-pressure LC (RP-8) with methanol/water or acetonit-rile/water allowed the isolation of 8-O-*trans*-coumaroylharpagide (**20**) (12 mg), procumbide (**21**) (99 mg), 8-O-*cis*-cinnamoylharpagide (**22**) (38 mg), and 6′-O-*p*-coumaroylharpagide (**23**) (8 mg) (35).

	R_1	R_2
20	*trans-p*-coumaroyl	H
22	*cis-p*-coumaroyl	H
23	H	*p*-coumaroyl

Two secoiridoid glycosides (**24, 25**) were obtained from *Halenia campan-ulata* (Gentianaceae) after LH-20 gel filtration and CCC on a Pharma-Tech CCC-1000 instrument (700 rpm, 3 mL/min) with the solvent system chloro-form/methanol/water (9:12:8) (mobile phase = lower phase). These two isomers differ only in the configuration at C7 but were well separated by liquid–liquid chromatography (28).

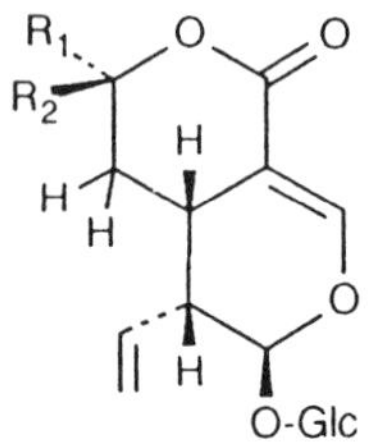

24   R$_1$ = OCH$_3$, R$_2$ = H
25   R$_1$ = H   R$_2$ = OCH$_3$

7.2.6. Saponins

As in the case of DCCC, centrifugal CCC is well suited to the separation of saponins. In one example, a 2-g portion (4 × 0.5 g) of an 80% methanol extract of leaves from *Abrus fruticulosus* (Leguminosae) was chromatographed by HSCCC using the solvent system chloroform/methanol/water (7:13:8) (mobile phase = lower phase) at a flow-rate of 1.5 mL/min. Final purification by flash chromatography or over-pressured LC allowed the rapid identification of the sweet-tasting glycosides abrusosides A–D (36).

Two saponins, asiaticoside (26) and madecassoside (27), which differ only in the presence of one hydroxyl group, were separated from an extract of *Centella asiatica* (Umbelliferae) by HSCCC. This medicinal plant is used to aid wound healing and in the treatment of leprosy. The two-phase solvent system used was chloroform/methanol/*i*-butanol/water (7:6:3:4), with the lower phase as the mobile phase, at a flow-rate of 4 mL/min. A 400-mg sample of the evaporated mother liquors of *C. asiatica* was injected. Detection of the non-UV-active saponins was achieved by direct coupling to TLC (37).

High-speed CCC was employed as a separation step in the isolation of saponins from the African plant *Sesamum alatum* (Pedaliaceae). After chromatography of a methanol extract of the aerial parts on a silica gel column, a fraction (1.25 g) was injected onto a multilayer coil separator–extractor [chloroform/methanol/isopropanol/water (5:6:1:4), with the organic phase as the mobile phase] before final purification by low-pressure LC on a RP-8 column. A novel 18,19-secoursane disaccharide (28) was obtained (38).

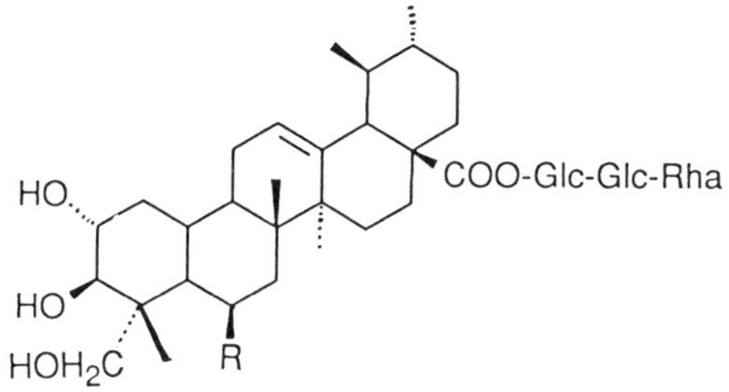

26   Asiaticoside   R = H
27   Madecassoside   R = OH

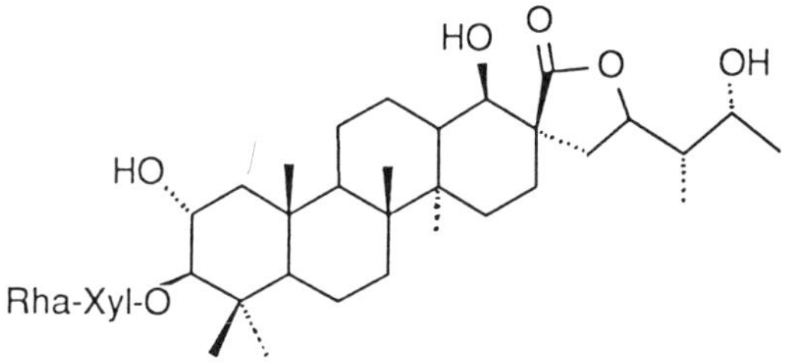

28

7.2.7. Alkaloids

Repetitive sample injections are possible for the separation of close-running compounds on rotating coil instruments. This has been shown by the separation of vincamine and vincine, alkaloids from *Vinca minor* (Apocynaceae). After 20 repetitive injections (at 42-min intervals) of 1.7 mg of sample mixture, 16.5 mg of **29** and 14 mg of **30** were obtained on a 230-mL instrument. The solvent system used was hexane/ethanol/water (6:5:5), with the lower phase as the mobile phase.

Interestingly, resolution of the HSCCC system was not changed when it was shutdown overnight and restarted the next day with the same stationary phase in the column (39).

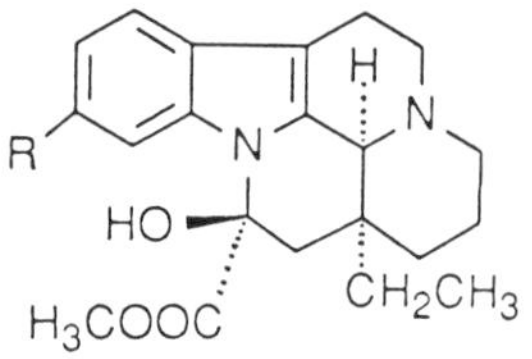

| 29 | Vincamine | R = H |
| 30 | Vincine | R = OCH₃ |

7.2.8. Carotenoids

The pigments cochloxanthin (**31**) and dihydrocochloxanthin (**32**) from *Cochlospermum tinctorium* (Cochlospermaceae) could neither be isolated by HPLC nor by preparative TLC. However, with a multilayer coil separator–extractor, they were obtained in one step from a methanol extract of the roots. The solvent system was carbon tetrachloride/methanol/water (5:4:1) (upper phase as mobile phase), the flow-rate 4 mL/min and the rotational speed 800 rpm. The extract (800 mg) was first dissolved in a 1:1 mixture (10 mL) of the two solvent phases before injection. Separation was achieved within 2 h (40).

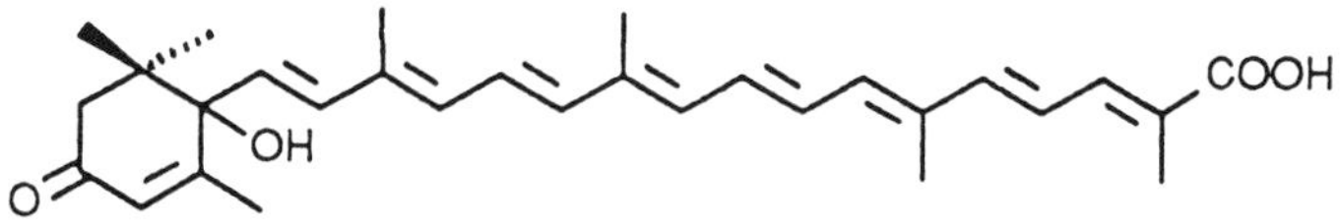

31

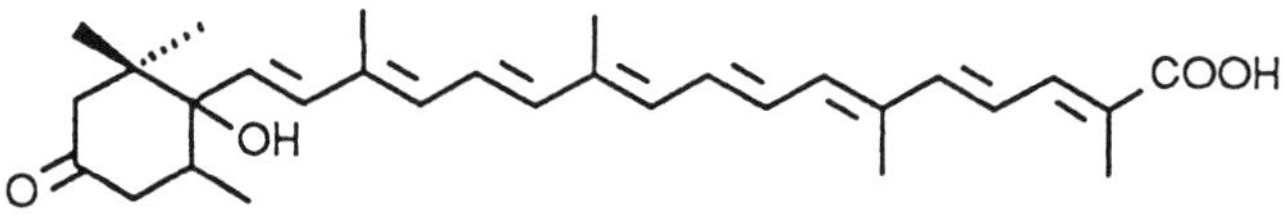

32

7.2.9. Gingerols

Gingerols are homologous series of phenolic ketones found in ginger (*Zingiber officinale*, Zingiberaceae) rhizomes. Three of these homologs (**34–36**) have been separated by a combination of silica gel column chromatography and

CCC. The second and final chromatographic step was performed on a multi-layer coil instrument using the upper phase of hexane/ethyl acetate/methanol/water (3:2:3:2) as mobile phase. By changing the composition to 2:3:2:3, a further compound, [4]-gingerol (33), was eluted (41).

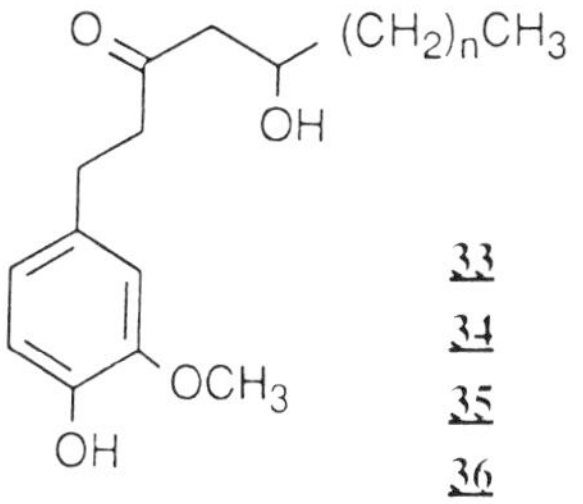

33	[4]-gingerol	n = 2
34	[6]-gingerol	n = 4
35	[8]-gingerol	n = 6
36	[10]-gingerol	n = 8

7.2.10. Polyacetylenic Alcohols

For the separation of apolar compounds, as has been reported for anthranoids and naphthoquinones (16), it is sometimes advantageous to use nonaqueous solvent systems, particularly when solubility is a problem. This approach has also been adopted for the isolation of polyacetylenic alcohols from parsley root (*Petroselinum crispum*, Umbelliferae). When 350 mg of pure diethyl ether extract was injected onto a HSCCC instrument, the antifungal compounds falcarinol (37, 17 mg) and falcarindiol (38; 4 mg) were obtained. As mobile phase, the acetonitrile phase of hexane/acetonitrile/*t*-butylmethyl ether (10:10:1), was employed. This one-step procedure minimized the problems that occur during the usual multistep separation of these oxygen-, light- and heat-sensitive compounds (42).

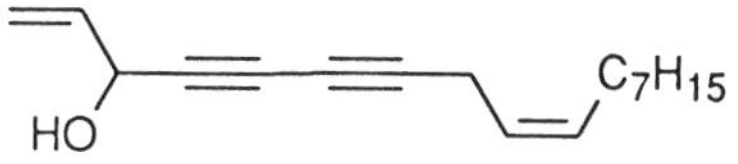

37

38

7.2.11. Marine Natural Products

Centrifugal partition chromatography is ideal for the separation of this group of delicate natural products (Table 7.2). Attempts at purification of antitumor ecteinascidins from the colonial tunicate *Ecteinascidia turbinata*, for example, by normal or reversed-phase chromatography usually led to extensive loss of activity. On the other hand, CPC proved to be a very effective means of separating these light- and acid-sensitive alkaloids (43).

The variety of different metabolites isolated from marine organisms by HSCCC is large but many contain nitrogen and most have been obtained from sponges. Among these nitrogen-containing products is a series of pyr-roloquinoline alkaloids, isobatzellines A (**39**), B (**40**), and C (**41**) from a *Batzella* sponge. The isobatzellines exhibited *in vitro* cytotoxicity against P-388 leukemia cell line and antifungal activity against *Candida albicans*. They were isolated from an extract of the sponge after solvent partition and a CCC step. This procedure involved eluting with the upper phase of the solvent system heptane/chloroform/methanol/water (2:7:6:3) (44).

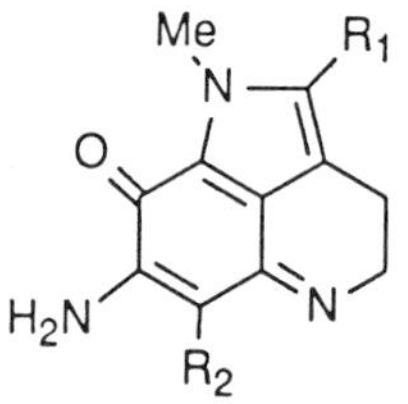

$$
\begin{array}{ll}
\underline{39} & R_1 = SMe, R_2 = Cl \\
\underline{40} & R_1 = SMe, R_2 = H \\
\underline{41} & R_1 = H, R_2 = Cl
\end{array}
$$

A polycyclic aromatic alkaloid, meridine (**42**), has been isolated from the ascidian *Amphicarpa meidiana*, collected in South Australia. The alkaloid was obtained directly from a methanol/chloroform extract after chromatography on a multilayer coil separator–extractor with chloroform–methanol–5% aqueous hydrochloric acid (5:5:3; mobile phase = lower phase) (45).

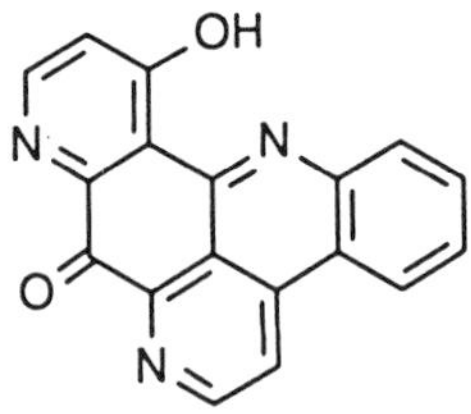

42

The same instrument (capacity 370 mL) was used for the separation of pyrrolimidazoles **43–45** from the Indo-Pacific marine sponges *Axinella* sp.

Table 7.2. Separations of Marine Natural Products by Countercurrent Chromatography

Organism	Class of Compound	Solvent System	References
Sponge *Tedania ignis*	Indoles, carbazole, β-carboline, phenolics	$CHCl_3$–MeOH–H_2O (25:34:20)	104
Sponge *Tedania ignis*	Diketopiperazine	$CHCl_3$–MeOH–H_2O (25:34:20)	105
Sponge *Calyx podatypa*	N-methylpyridinium salts	nC_7H_{16}–CH_3CN–CH_2Cl_2 (10:7:3)	106
Sponge *Xestospongia wiedenmayeri*	Methoxylamine pyridines	nC_7H_{16}–CH_3CN–CH_2Cl_2 (50:30:15)	50
Ascidian *Amphicarpa meridiana*	Aromatic alkaloids	$CHCl_3$–MeOH–5% aq. NH_3 (5:5:3)	45
Sponge *Batzella* sp.	Pyrroloquinoline alkaloids	nC_7H_{16}–$CHCl_3$–MeOH–H_2O (2:7:6:3)	44, 111
		nC_7H_{16}–EtOAc–MeOH–H_2O (4:7:4:3)	44, 111
		$CHCl_3$–iPr_2NH–MeOH–H_2O (7:1:6:4)	111
Sponges *Axinella* sp. and *Hymeniacidon* sp.	Pyrrololactams	$nBuOH$–0.01 M K_3PO/0.01 M K_2HPO_4 (1:1)	46
Bryozoan *Bugula neritina*	Bryostatin (macrocyclic lactone)	nC_6H_{14}–EtOAc–MeOH–H_2O (14:6:10:7)	107
Sponges *Dercitus* sp. and *Stelletta* sp.	Acridine alkaloids	CH_2Cl_2–MeOH–H_2O (5:5:3)	108
Tunicate *Clavelina picta*	Quinolizidines	nC_6H_{14}–CH_3CN–CH_2Cl_2 (10:7:3)	109
Sponge *Spongosorites ruetzleri*	Imidazolediylbis[indoles]	nC_7H_{16}–EtOAc–MeOH–H_2O (4:7:4:3)	48
		nC_7H_{16}–EtOAc–MeOH–H_2O (5:7:4:3)	48
Sponge *Discodermia polydiscus*	(Aminoimidazolinyl)indole	$CHCl_3$–MeOH–H_2O (5:10:6)	110
Sponges *Theonella* sp.	Cyclic peptide	nC_6H_{14}–EtOAc–MeOH–H_2O (3:7:5:5)	47
Sponges *Theonella* sp.	Arginine derivative	$CICH_2CH_2Cl$–$CHCl_3$–MeOH–H_2O (2:3:10:6)	113
Sponges *Plakortis lita*	Cyclic peroxides	nC_7H_{16}–CH_2Cl_2–CH_3CN (5:1:4)	112
Tunicate *Ascidia nigra*	Tunichromes	$iAmOH$–$nBuOH$–$nPrOH$–H_2O–HCOOH–*tert*butyl sulfide (32:48:40:120:1:4)	115
Sponges *Dercitus* sp.	Indole derivative	$CHCl_3$–MeOH–H_2O (5:10:6)	114

and *Hymeniacidon* sp. Following gel filtration on Sephadex LH-20, the crude fractions were subjected to HSCCC. For the solvent system, K$_3$PO$_4$ (2.1 g) and K$_2$HPO$_4$ (1.7 g) were dissolved in of 1 L water, and 1 L of *n*-butanol was added. The pH 11.2 phosphate buffer was selected because the products were more soluble in aqueous alkaline solutions. The chromatograph was filled with stationary (upper) phase and after sample introduction, the mobile (lower) phase was pumped through the coil. Samples were dissolved in 1:1 *n*-butanol/ 0.01 *M* K$_3$PO$_4$, 0.01 *M* K$_2$HPO$_4$. The separated compounds were precipitated and then purified by polyamide CC6 and Sephadex LH-20 chromatographic steps. Pyrrololactams **43** (16 mg) and **45** (28 mg) were isolated from 140 mg of an *Axinella* fraction in about 5 h, while **44** and **45** were isolated from *Hymeniacidon*. Compounds **44** and **45** inhibited PS leukemia cell growth (46).

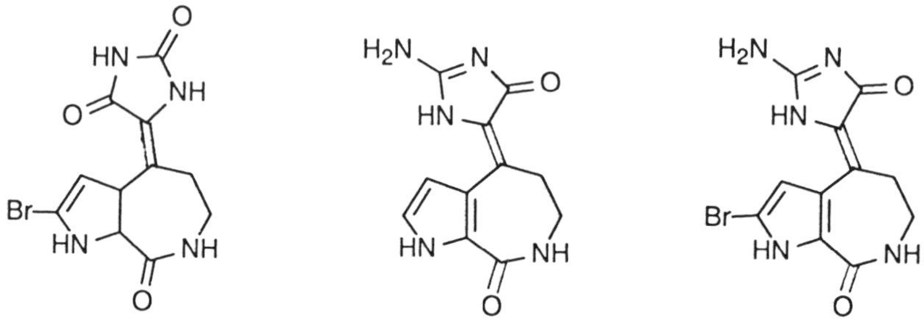

43   **44**   **45**

One step in the isolation of a cytotoxic peptide from a *Theonella* species of sponge (Japan) involved centrifugal CCC with hexane/ethyl acetate/methanol/ water (3:7:5:5) (47), while a very similar solvent system was used for the isolation of imidazolediylbis[indoles] from a Caribbean deep-sea sponge (48).

Following reports on the separation of a cytotoxic sesquiterpene/methylene quinone from a deep water sponge by HSCCC employing a *nonaqueous* heptane/dichloromethane/acetonitrile (10:3:7) solvent system (49), a whole variety of marine natural products have been isolated by the same means. Some of these products are listed in Table 7.2, in some instances with hexane and others with heptane as the least polar solvent. For example, xestamine A (**46**), a long-chain methoxylamine pyridine has been isolated from the

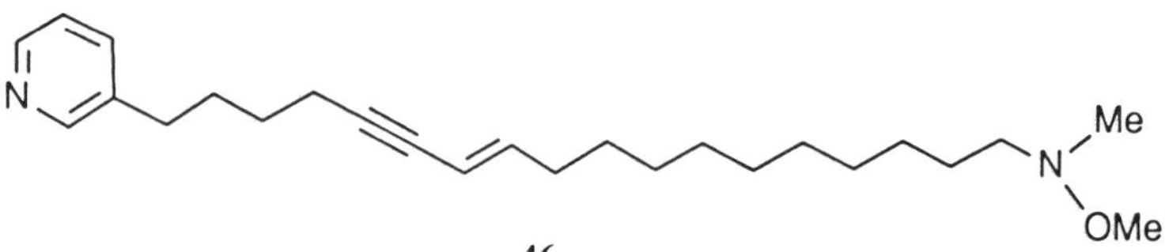

46

Table 7.3. Separations of Antibiotics by Centrifugal Countercurrent Chromatography

Sample	Instrument	Solvent System	References
Polyene antibiotics	Synchronous coil planet centrifuge	$CHCl_3 - MeOH - H_2O$ (4:4:3)	102
Glycoside antibiotics	Shimadzu HSCCC	$nBuOH - Et_2O - H_2O$ (10:4:12)	56
Siderochelin A	Ito-HSCCC	$CHCl_3 - MeOH - H_2O$ (7:13:8)	20
Efrotomycin	Ito-HSCCC	$CCl_4 - CHCl_3 - MeOH - H_2O$ (5:5:6:4)	20
Pentalenolactone	Ito-HSCCC	$CHCl_3 - MeOH - H_2O$ (1:1:1)	20
Bu 2313B	Ito-HSCCC	$nC_6H_{14} - CH_2Cl_2 - MeOH - H_2O$ (5:1:1:1)	20
Tirandamycin	Ito-HSCCC	$nC_6H_{14} - EtOAc - MeOH - H_2O$ (70:30:15:6)	20
Dunaimycins	Ito-HSCCC	$nC_6H_{14} - EtOAc - MeOH - H_2O$ (70:30:15:6)	52
(Macrolide antibiotics)		$nC_6H_{14} - EtOAc - MeOH - H_2O$ (8:2:10:5)	52
		$nC_6H_{14} - EtOAc - MeOH - H_2O$ (8:2:5:5)	52
Actinomycins	Ito-HSCCC	$nC_6H_{14} - Et_2O - MeOH - H_2O$ (1:5:4:5)	53
Valinomycin	Ito-HSCCC	$nC_6H_{14} - MeOH - H_2O$ (10:9:1)	51
Pipericidin A1	Ito-HSCCC	$nC_6H_{14} - Et_2O - MeOH - H_2O$ (4:1:4:1)	51
Concanamycin	Ito-HSCCC	$nC_6H_{14} - EtOAc - MeOH - H_2O$ (1:1:1:1)	51
Tomaymycin	Ito-HSCCC	$nC_6H_{14} - EtOAc - MeOH - H_2O$ (1:3:1:3)	51
Acetoxycycloheximide E73	Ito-HSCCC	$CHCl_3 - C_6H_5CH_3 - MeOH - H_2O$ (5:4:5:4)	51
Tiacumicin	Ito-HSCCC	$CCl_4 - CHCl_3 - MeOH - H_2O$ (7:3:7:3)	103
Benzanthrins	Ito-HSCCC	$CCl_4 - CHCl_3 - MeOH - H_2O$ (4:1:4:1)	54
Coloradocin	Ito-HSCCC	$CHCl_3 - MeOH - H_2O$ (1:1:1)	55

sponge *Xestospongia wiedenmayeri*, with the aid of a heptane-containing mixture (50).

7.2.12. Antibiotics

One of the areas in which HSCCC has been particularly well exploited is the field of antibiotics (Table 7.3). This exploitation is not immediately evident from the published literature because many of the applications involve small-scale separations of microbial metabolites from industrial laboratories or development plants, details of which are kept secret for obvious reasons. Liquid–liquid partition techniques are particularly suitable for the separation of antibiotics because these bioactive microbial metabolites are often produced in very small quantities and have to be removed from other secondary metabolites and unmetabolized media ingredients. Antibiotics are normally biosynthesized as mixtures of closely related congeners and many are labile molecules, thus requiring mild separation techniques with a high resolution capacity. When searching for solvent systems, a preliminary TLC screening is helpful, but another possibility is to subject the two phases directly to a bioassay when an antibiotic mixture is partitioned in a small amount of the two immiscible components (20). A screening of the following solvent systems has been recommended (51) for antibiotics (and, incidently for other natural products):

Hexane/methanol/water 10:9:1
Hexane/ethyl acetate/methanol/water 1:1:1:1
Chloroform/methanol/water 5:8:5

Partitioning into the organic layer will be increased in the second system when the ethyl acetate/hexane ratio is increased. The third solvent mixture is one of the most polar chloroform/methanol/water systems.

Reported work from the Abbott laboratories in Chicago included the isolation of siderochelin A, efrotomycin, pentalenolactone, Bu2313B, and tirandamycins A and B from the crude extracts of *Bacteroides fragilis* and *Staphylococcus aureus*. Charges of up to 600 mg were introduced for the different separations (20). Recently, the dunaimycins, spiroketal 24-membered macrolides, were isolated from *Streptomyces diastatochromogenes* by procedures that involved HSCCC with different compositions of the solvent system hexane/ethyl acetate/methanol/water (52).

Actinomycins have been separated on HSCCC instruments. Pure actinomycins C_1 (2.3 mg), C_2 (18.9 mg), and C_3 (22.3 mg) were obtained from an actinomycin C complex. Centrifugal CCC is thus capable of separating very closely related products: an ethyl group in C_2 replaces a methyl substituent in C_1, while a second ethyl group in C_3 replaces a methyl group in C_2 (53).

Two isomeric benz[a]anthraquinones, benzanthrins A (**47**) and B (**48**), have been isolated from *Nocardia lurida* by a strategy that involved a final HSCCC step (54). For the purification of coloradocin, successive 400-mg samples were injected into the multilayer coil, until 4.7 g had been employed (55).

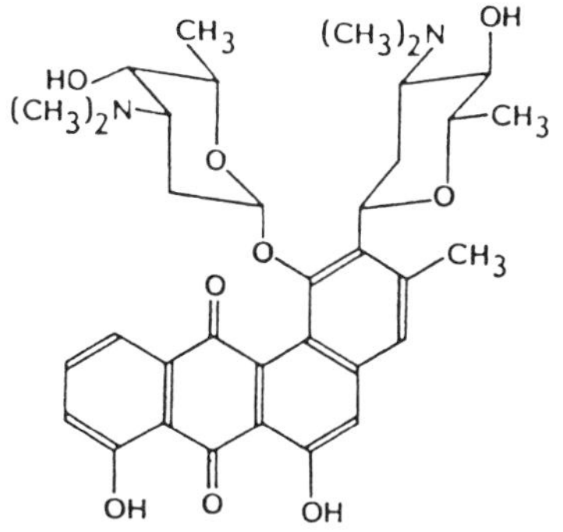

47 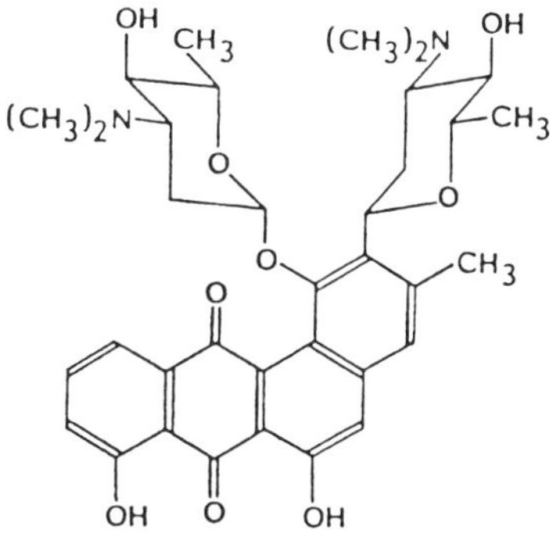   48

The sporaviridins, water-soluble basic glycoside antibiotics, have complex structures, are unstable under basic conditions, and exist as mixtures of closely related compounds. A sample of six sporaviridins was resolved on a Shimadzu HSCCC prototype instrument (325-mL total capacity; 800 rpm). Selection of the solvent system was based on partition coefficient data from chloroform/methanol/water, chloroform/ethanol/methanol/water and *n*-butanol/diethyl ether/water mixtures. After HPLC analysis, the final system adopted was *n*-butanol/diethyl ether/water (10:4:12). Sample was introduced by so-called "sandwich" injection (sample solution injected after filling the stationary phase, and then the mobile phase is eluted). The six components were separated within 3.5 h, employing a total elution volume of 500 mL (56).

7.3. ANALYTICAL APPLICATIONS

As seen from the first part of this chapter, HSCCC is a rapid high-efficiency separation method used in the last two decades for the isolation of natural products on a preparative and semipreparative scale.

Curiously, it is only in the last few years that analytical applications of this technique have been described. In fact, several problems first had to be solved. For example, reduction of the Teflon tube internal diameter results in a steady carryover of the stationary phase due to the increased solvent-wall interaction in the narrow-bore tube. Furthermore, separation times have been too long for such applications. Nevertheless, some modifications such as a diminution of the revolutional radius of the centrifuge, while increasing rotating speeds have generated highly efficient systems with an effective analytical capability. In the analytical HSCCC system, operational speeds of up to 2000 rpm (57) and even 4000 rpm (58) can promote countercurrent flow of two solvent phases through a coiled tube of less than 1 mm i.d., resulting in efficient separations of microquantities of samples in a short period of time (59). In general, the observed resolution and speed are comparable to those of HPLC (60). By comparison with analytical HPLC, the major advantage of HSCCC is that it is always possible to reverse the elution mode, in the case of complex samples having a wide polarity range.

As analytical HSCCC chromatographs are now commercially available, new fields of applications are possible, and the method will certainly be more and more widely exploited. However, most of the separations already performed with analytical HSCCC apparatus have been carried out with user-built instruments. Typically, these instruments consist of a MLCPC equipped with a single coil of 50–70-m × 0.85-mm i.d. PTFE tubing (61, 62), with a total capacity of about 30–40 mL. In order to increase stationary-phase retention, higher rotational speeds (1500–2000 rpm) and higher centrifugal forces are required. These are achieved by a smaller revolutionary radius than in preparative HSCCC.

Two prototypes (HSCCC-4000) have been developed recently at the NIH (58, 63, 64). These prototypes are equipped with 0.3- or 0.5-mm i.d. columns (7 or 8-mL total coil capacity), with a 2.5-cm revolutional radius and are able to operate at a maximum speed of 4000 rpm.

In all of these instruments, flow-rates are directly related to the selected solvent system, and must be adapted in order to minimize the back pressure and leakage of the stationary phase. To date, analytical HSCCC instruments are available from two manufacturers (see also Table 7.4). In addition, the HSCCC-4000 apparatus may soon be commercialized by the following companies: Pharma-Tech Research Corp., Baltimore, MD; P.C. Inc., Potomac, MD; and Shimadzu, Kyoto, Japan (65).

7.3.1. Alkaloids

The solvent system n-hexane/ethyl acetate/methanol/water (3:7:5:5) was used by Lee et al. (60) to separate indole-3-acetic and three homologs: indole-3-acetamide, indole-3-butyric acid, and indole-3-acetonitrile. These plant growth hormones were base line separated within 90 min using the lower phase as the mobile phase. The MLCPC used for this application had a 5-cm revolutional radius, a total coil capacity of 38 mL and rotated at 2000 rpm.

Table 7.4. Selected Analytical Applications of HSCCC Techniques in the Separation of Natural Products

Substances Separated	Solvent System	Apparatus[a]	References
Alkaloids	nC_6H_{14}–EtOAc–MeOH–H_2O (1:1:1:1)	Pharma-Tech-2000	59
	nC_6H_{14}–EtOAc–MeOH–H_2O (3:7:5:5)	Pharma-Tech-2000	59
	nC_6H_{14}–EtOH–H_2O (6:5:5)	Experimental instrument	66
Anthraquinones	nC_6H_{14}–EtOAc–MeOH–H_2O (9:1:5:5)	Pharma-Tech-2000	57
Coumarins	$CHCl_3$–MeOH–H_2O (13:7:8)	"Micro" HSCCC	67–69
Flavonoids	$CHCl_3$–MeOH–H_2O (4:3:2)	Pharma-Tech-2000	70
	$CHCl_3$–MeOH–H_2O (4:3:2)	Experimental instrument	58
Indole	nC_6H_{14}–EtOAc–MeOH–H_2O (3:7:5:5)	Experimental instrument	60
	nC_6H_{14}–EtOAc–MeOH–H_2O (1:1:1:1)	Pharma-Tech-4000	65
Lignans	nC_6H_{14}–EtOH–H_2O (6:5:5)	Experimental instrument	72
	nC_6H_{14}–EtOAc–MeOH–H_2O (1:1:1:1)	"Micro" HSCCC	74
Macrolides	nC_6H_{14}–iPrOH–MeOH–H_2O (40:10:16:4)	"Micro" HSCCC	74
	nC_6H_{14}–EtOAc–MeOH–8% NH_3 (1:1:1:1)	Pharma-Tech-4000	65
Triterpenoids	nC_6H_{14}–EtOH–H_2O (6:5:1)	Experimental instrument	75

[a] Pharma–Tech CCC-2000: Analytical high-speed countercurrent chromatograph, about 70 m × 0.85 mm i.d. (∼ 40-mL coil volume), Pharma-Tech, Baltimore, MD. Pharma-Tech CCC-4000: Analytical high-speed countercurrent chromatograph, about 40 m × 0.3 mm i.d. (∼ 7-mL coil volume), Pharma-Tech, Baltimore, MD. "Micro" HSCCC: High-speed countercurrent chromatograph for analytical HSCCC, about 50 m × 0.85 mm i.d. (∼ 30-mL coil volume), P.C. Inc., Potomac, MD. Experimental instrument: 0.85-mm i.d. tubing; for further description refer to original paper.

213

The separation efficiency of the system was estimated at 1000–1300 theoretical plates.

A mixture of the first three indole auxins was also separated with a prototype of the HSCCC-4000 instrument. The 0.3-mm i.d. version of this apparatus, spinning at 4000 rpm, was used to perform the analysis. The lower phase of the solvent system n-hexane/ethyl acetate/methanol/water (1:1:1:1) was used as a mobile phase at a flow-rate of 0.1–0.2 mL/min. Under these conditions, base line separation was performed within 30 min; excellent theoretical plate numbers, ranging from 12000 to 5500 (65), were obtained.

The separation of an artificial mixture of vincamine (**29**), the major alkaloid of *Vinca minor* (Apocynaceae) and vincine (11-methoxyvincamine) (**30**), was performed by analytical HSCCC in hexane/ethanol/water (6:5:5) with a multilayer coil (0.85 mm i.d.). Comparison of this separation with results obtained from analytical RP-HPLC showed that both methods gave base line resolution, but it was possible to observe a small peak just preceding the vincine peak in analytical CCC, which was not resolved by RP-HPLC. Analysis of this sample by CCC-MS led to the postulate that this minor compound represented an isomer of vincine (66).

The alkaloids tetrandrine (**49**), fangchimoline (**50**), and cyclanoline (**51**) from dried roots of *Stephania tetrandra* (Menispermaceae), a Chinese medicinal plant with anti inflammatory properties, have been separated by analytical CCC. Tetrandrine and fangchinoline are bisbenzylisoquinoline alkaloids and cyclanoline is a water-soluble quaternary protoberberine-type alkaloid. While cyclanoline can be separated by conventional methods, the two other products elute together. An artificial mixture of these three alkaloids was resolved by a HSCCC-2000 chromatograph in two n-hexane/ethyl acetate/ methanol/ water solvent systems. When the lower phase of n-hexane/ethyl acetate/ methanol/water (1:1:1:1) was used as the mobile phase, the separation could be performed within 100 min. Utilization of n-hexane/ethyl acetate/methanol/water (3:7:5:5), pumping lower phase into the head end of the column led

49	tetrandrine	R = CH₃
50	fangchinoline	R = H

51

to the elution of cyclanoline and fangchinoline in 70 min. In order to elute tetrandrine, upper phase had to be pumped in the reverse direction (59).

7.3.2. Anthraquinones

Five hydroxy anthraquinones from the rhizome of *Rheum palmatum* (Polygonaceae) were separated by analytical CCC. Injection of an extract containing about 0.5% of anthraquinone derivatives into a Pharma-Tech CCC-2000 chromatograph with the solvent system *n*-hexane/ethyl acetate/methanol/water (9:1:5:5) (mobile phase: lower phase; 1 mL/min; 1800 rpm) successively afforded physicion (**52**), aloe-emodin (**53**), and rhein (**54**) within 40 min. Then the run was reversed by eluting with the upper phase, giving the less polar anthraquinones chrysophanol (**55**) and emodin (**56**), which otherwise would have remained in the stationary phase. The separation was achieved within approximately 70 min (57).

		R_1	R_2
52	Physicion	OCH_3	CH_3
53	Aloe-emodin	H	CH_2OH
54	Rhein	H	COOH
55	Chrysophanol	H	CH_3
56	Emodin	OH	CH_3

7.3.3. Coumarins

An artificial mixture (50 μg each) of three common plant coumarins (herniarin, scopoletin, and umbelliferone) and one flavanone (hesperetin) was separated with a coil centrifuge connected to a photodiode array detector. Reduction of detector noise caused by nonretained stationary phase was achieved by adding an auxiliary solvent (methanol or isopropanol) to the coil effluent. The lower phase of chloroform/methanol/water (13:7:8) was used as the mobile phase (0.8 mL/min) and base line resolution of the four compounds was achieved in less than 30 min (67, 68). The same separation has also been monitored by ELSD (evaporative laser light-scattering detection (69).

7.3.4. Flavonoids

A crude flavonoid mixture, obtained from the ethanolic extract of the fruits of sea buckthorn (*Hippophae rhamnoides*, Elaeaginaceae) has been successfully separated with an analytical HSCCC instrument. The separation of a 3-mg mixture with chloroform/ethanol/water (4:3:2) was complete in 15 min when the mobile lower phase was pumped at a flow-rate of 5 mL/min (rotation speed: 1800 rpm) (70). An increased separation time was observed with lower flow-rates (1 or 3 mL/min). Under these conditions, excellent retention of the stationary phase was observed. Isorhamnetin was detected as the major flavonoid of this extract. As a matter of fact, the five major components of the sea buckthorn fruit extract were separated within 8 min, a time scale wholly comparable with analytical HPLC separations. The instrument used in this analysis was a prototype CPC with 2.5-cm revolutional radius, 0.85 mm i.d., and a coil capacity reduced to 8 mL. This experimental instrument was operating at 3500 rpm, and the same solvent system as the one employed above, at a flow-rate of 2 mL/min, was used to perform the separation (58).

7.3.5. Lignans

Schisandra rubriflora (Schisandraceae) is a traditional Chinese herbal medicinal plant, which has been used for the treatment of chronic hepatitis. Lignans in the kernel were found to be responsible for the *in vivo* biological activity (71). However, the isolation of lignans from an ethanolic extract of the kernels of this plant presented some problems because of the structural similarities of these compounds. Schisanhenol (**17**) and its acetate (**18**) were not separated by RP-HPLC with methanol/water (34) but were resolved by preparative HSCCC with the two-phase solvent system *n*-hexane/ethyl acetate/methanol/water (10:5:5:1) (34). Utilization of an analytical high-speed CPC equipped with a multilayer coil (0.85 mm i.d.), enabled base line resolution of these two compounds, and the separation of four structurally related lignans. The lower (mobile) phase of the two-phase solvent system *n*-hexane/ethanol/water (6:5:5) enabled elution of a 0.15-mg sample solution within 80 min (60). When a thermospray MS was used for on-line detection, the six lignans were easily detected by means of selective ion chromatograms and mass spectra (72, 73).

The dichloromethane extract of the rhizomes of mayapple (*Podophyllum peltatum*, Berberidaceae) contains cytotoxic lignans, with podophyllotoxin being the major compound. Analytical HSCCC of a 3-mg sample of this extract with the *n*-hexane/ethyl acetate/methanol/water (1:1:1:1) solvent system (lower phase as mobile phase) was carried out at 0.5 mL/min. Virtually pure podophyllotoxin was obtained, and another compound that had almost the same R_f on TLC was easily separated. The retention of the stationary phase was 50% (74).

7.3.6. Macrolides

The analysis and detection of bryostatin 1 (**57**), an antitumor macrolide isolated from the marine bryozoan *Bugula neritina*, is a very tedious and time-

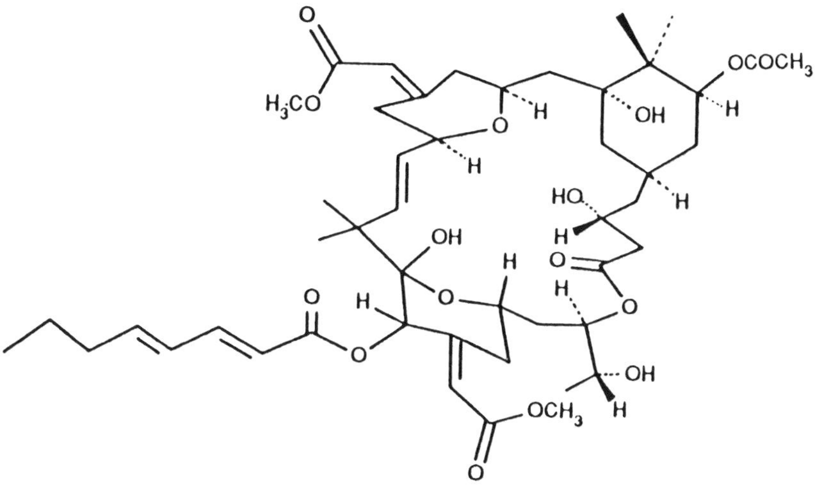

57

consuming operation. Indeed, RP-HPLC analysis of the crude extracts or fractions of *B. neritina* needs several sample clean-up steps before injection into a HPLC instrument. However, separation of the crude extract (10 mg) by analytical HSCCC using *n*-hexane/isopropanol/methanol/water (40:10:16:4), using the lower phase as the mobile phase, allowed the elution of bryostatin 1 and 2 of high enough purity for direct HPLC analysis (74).

The components from the class of mycinomycins, which are 16-membered macrolide antibiotics mainly used for animals, have been separated with the same analytical HSCCC prototype as the one used for the separation of indole auxins by Oka et al. (65). In this case, the lower phase of *n*-hexane/ ethyl acetate/methanol/8% aqueous ammonia (1:1:1:1) was pumped at 0.1–0.2 mL/min into the head end of the column and gave base line separation within 50 min. Compared with the separation of indole auxins, the results show lower theoretical plate numbers, which is probably due to a higher retention of the stationary phase (65).

7.3.7. Triterpenoids

Lee et al. (75) used an analytical CPC prototype (0.85 mm i.d., 38-mL total coil capacity, 5-cm rotational radius, 1500 rpm) to separate seven terpenoid carboxylic acids from a crude extract of *Boswellia carterii* (Burseraceae). With the lower phase of the solvent system *n*-hexane/ethanol/water (6:5:1) as the mobile phase (1 mL/min), the separation was achieved within 50 min. The HSCCC–UV chromatogram showed four distinct peaks. Thermospray HSCCC–MS was used to aid their characterization and, furthermore, allowed the identification of the minor components present in the extract.

7.4. CONCLUSIONS

Over the past few years the technique of HSCCC has established itself as a efficient method for the separation of natural products. Since separation is wholly accomplished with liquid phases, the phenomena of irreversible adsorption, sample loss, denaturation, and tailing, traditionally encountered in classical chromatography, are avoided. These considerations are particularly relevant to the chromatography of crude extracts and, furthermore, different selectivities (when compared with solid–liquid chromatography) for complex mixtures are possible. Both polar and apolar substances can be separated without any trouble. For the latter, nonaqueous solvent systems have also been introduced.

Applications have included many different classes of natural product and there seems to be no limitation as to the potential candidates for CCC. In contrast to DCCC, HSCCC is a rapid technique and can produce separations in less than 1 h. Analytical HSCCC instruments are now available, and although they will never replace HPLC, they represent a useful means for the separation of small quantities of sample and the testing of solvent systems for larger scale applications. So far, milligram-to-gram quantities of sample have been successfully treated in preparative applications but there is still a need to develop instruments that will be capable of separating 10 g, 100 g, or even kilogram mixtures. If this is possible, the technique will be extended to pilot-scale and industrial-scale applications, and the potential of HSCCC will be enormously increased. Finally, technical modifications of existing apparatus and refinement of construction are urgently required to make HSCCC more universally acceptable to users.

REFERENCES

1. T. Tanimura, J. J. Pisano, Y. Ito, and R. L. Bowman, *Science*, **169**, 57 (1990).
2. K. Hostettmann, M. Hostettmann, and A. Marston, *Nat. Prod. Rep.*, **1**, 471 (1984).
3. J. K. Snyder, K. Nakanishi, K. Hostettmann, and M. Hostettmann, *J. Liq. Chromatogr.*, **7**, 243 (1984).
4. Y. Ito, "Principles and Instrumentation of Countercurrent Chromatography," in N. B. Mandava and Y. Ito, Eds., *Countercurrent Chromatography: Theory and Practice*, Marcel-Dekker, New York, 1988, p. 79.

5. W. D. Conway, *Countercurrent Chromatography: Apparatus, Theory, and Applications*, VCH, New York (1989).

6. Y. Ito, *J. Biochem. Biophys. Methods*, **5**, 105 (1981).

7. I. Slacanin, A. Marston, and K. Hostettmann, *J. Chromatogr.*, **482**, 234 (1989).

8. Y. Ito, *CRC Crit. Rev. Anal. Chem.*, **17**, 65 (1986).

9. Y. Ito and F. E. Chou, *J. Chromatogr.*, **454**, 382 (1988).

10. Y. Ito, H. Oka, and J. L. Stemp, *J. Chromatogr.*, **475**, 219 (1989).

11. Y. Ito, *J. Chromatogr.*, **301**, 377 (1984).

12. W. Murayama, T. Kobayashi, Y. Kosuge, H. Yano, Y. Nunogaki, and K. Nunogaki, *J. Chromatogr.*, **239**, 643 (1982).

13. A. Berthod and D. W. Armstrong, *J. Liq. Chromatogr.*, **11**, 547 (1988).

14. K. Hostettmann and A. Marston, *Anal. Chim. Acta*, **236**, 63 (1990).

15. K. Hostettmann, M. Hostettmann, and A. Marston, *Preparative Chromatographic Techniques: Applications in Natural Product Isolation*, Springer Verlag, Berlin (1986).

16. A. Marston, I. Slacanin, and K. Hostettmann, *Phytochem. Anal.*, **1**, 3 (1990).

17. K. Hostettmann, *Planta Med.*, **39**, 1 (1980).

18. T. Okuda, T. Yoshida, T. Hatano, K. Yazaki, R. Kira, and Y. Ikeda, *J. Chromatogr.*, **362**, 375 (1986).

19. W. D. Conway and Y. Ito, *J. Liq. Chromatogr.*, **7**, 291 (1984).

20. G. M. Brill, J. B. McAlpine, and J. E. Hochlowski, *J. Liq. Chromatogr.*, **8**, 2259 (1985).

21. I. Slacanin, Mise au point et application de techniques chromatographiques pour analyses et isolement de produits naturels. Ph.D. Thesis, University of Lausanne Switzerland (1992).

22. A. Marston, I. Slacanin, and K. Hostettmann, *J. Liq. Chromatogr.*, **13**, 3615 (1990).

23. I. Slacanin, A. Marston, K. Hostettmann, N. Delabays, and C. Darbellay, *J. Chromatogr.*, **557**, 391 (1991).

24. M. Vanhaelen and R. Vanhaelen-Fastre, *J. Liq. Chromatogr.*, **11**, 2969 (1988).

25. P. Li, T.-Y. Zhang, and Y. Ito, *J. Chromatogr.*, **538**, 219 (1991).

26. T. Hatano, T. Fukuda, T. Miyase, T. Noro, and T. Okuda, *Chem. Pharm. Bull.*, **39**, 1238 (1991).

27. M. C. Recio, I. Slacanin, M. Hostettmann, A. Marston, and K. Hostettmann, *Bull. Liais. Groupe Polyphénols*, **15**, 215 (1990).

28. M. C. Recio-Iglesias, A. Marston, and K. Hostettmann, *Phytochemistry*, **31**, 1387 (1992).

29. T. Okuda, T. Yoshida, and T. Hatano, *J. Liq. Chromatogr.*, **11**, 2447 (1988).

30. T. Yoshida, T. Hatano, and T. Okuda, *J. Chromatogr.*, **467**, 139 (1989).

31. T. Okuda, T. Yoshida, T. Hatano, Y. Ikeda, T. Shingu, and T. Inoue, *Chem. Pharm. Bull.*, **34**, 4075 (1986).

32. T. Okuda, T. Hatano, T. Koneda, M. Yoshizaki, and T. Shingu, *Phytochemistry*, **26**, 2053 (1986).

33. J. K. Nitao, M. G. Nair, D. L. Thorogood, K. S. Johnson, and J. M. Scriber, *Phytochemistry*, **30**, 2193 (1991).

34. Y.-W. Lee, Q.-C. Fang, Y. Ito, and C. E. Cook, *J. Nat. Prod.*, **52**, 706 (1989).

35. O. Potterat, M. Saadou, and K. Hostettmann, *Phytochemistry*, **30**, 889 (1991).

36. F. Fullas, Y. H. Choi, A. D. Kinghorn, and N. Bunyapruphatsara, *Planta Med.*, **56**, 332 (1990).

37. B. Diallo, R. Vanhaelen-Fastre, and M. Vanhaelen, *J. Chromatogr.*, **558**, 446 (1991).

38. O. Potterat, K. Hostettmann, H. Stoeckli-Evans, and M. Saadou, *Helv. Chim. Acta*, **75**, 833 (1992).

39. J.-Y. Zhou, Q.-C. Fang, and Y.-W. Lee, *Phytochem. Anal.*, **1**, 74 (1990).

40. B. Diallo and M. Vanhaelen, *J. Liq. Chromatogr.*, **11**, 227 (1988).

41. J. E. Farthing and M. J. O'Neill, *J. Liq. Chromatogr.*, **13**, 941 (1990).

42. S. Nitz, M. H. Spraul, and F. Drawert, *J. Agric. Food Chem.*, **38**, 1445 (1990).

43. K. L. Rinehart, *J. Nat. Prod.*, **53**, 771 (1990).

44. H. H. Sun, S. Sakemi, N. Burres, and P. McCarthy, *J. Org. Chem.*, **55**, 4964 (1990).

45. F. J. Schmitz, F. S. DeGuzman, M. B. Hossain, and D. van der Helm, *J. Org. Chem.*, **56**, 804 (1991).

46. D. E. Schaufelberger and G. R. Pettit, *J. Liq. Chromatogr.*, **12**, 1909 (1989).

47. N. Fusetani, T. Sugawara, S. Matsunaga, and H. Hirota, *J. Am. Chem. Soc.*, **113**, 7811 (1991).

48. S. Sakemi and H. H. Sun, *J. Org. Chem.*, **56**, 4304 (1991).

49. S. Kohmoto, O. J. McConnell, A. Wright, F. Koehn, W. Thompson, M. Lui, and K. M. Snader, *J. Nat. Prod.*, **50**, 336 (1987).

50. S. Sakemi, L. E. Tottori, and H. H. Sun, *J. Nat. Prod.*, **53**, 995 (1990).

51. D. G. Martin, C. Biles, and R. E. Peltonen, *Am. Lab.*, **18**, 21 (1986).

52. J. E. Hochlowski, M. M. Mullally, G. M. Brill, D. N. Whittern, A. M. Buko, P. Hill, and J. B. McAlpine, *J. Antibiot.*, **44**, 1318 (1991).

53. D. G. Martin, R. E. Peltonen, and J. M. Nielsen, *J. Antibiot.*, **39**, 721 (1986).

54. R. R. Rasmussen, M. E. Nuss, M. H. Scherr, S. L. Mueller, J. B. McAlpine, and L. A. Mitscher, *J. Antibiot.*, **39**, 1515 (1986).

55. R. R. Rasmussen, M. H. Scherr, D. N. Whittern, A. M. Buko, and J. B. McAlpine, *J. Antibiot.*, **40**, 1383 (1987).

56. K. Harada, I. Kimura, A. Yoshikawa, M. Suzuki, H. Kakazawa, S. Hattori, K. Komori, and Y. Ito, *J. Liq. Chromatogr.*, **13**, 2373 (1990).

57. T.-Y. Zhang, L. K. Pannel, Q.-L. Pu, D.-G. Cai, and Y. Ito, *J. Chromatogr.*, **442**, 455 (1988).

58. H. Oka, F. Oka, and Y. Ito, *J. Chromatogr.*, **479**, 53 (1989).

59. T.-Y. Zhang, L. K. Pannel, D.-G. Cai, and Y. Ito, *J. Liq. Chromatogr.*, **11**, 1661 (1988).

60. Y.-W. Lee, C. E. Cook, Q.-C. Fang, and Y. Ito, *J. Chromatogr.*, **477**, 434 (1989).

61. Y. Ito and Y. W. Lee, *J. Chromatogr.*, **391**, 290 (1987).

62. R. J. Romanach and J. A. de Haseth, *J. Liq. Chromatogr.*, **11**, 91 (1988).

63. H. Oka, Y. Ikai, N. Kawamura, M. Yamada, K. I. Harada, M. Suzuki, F. E. Chou, Y. W. Lee, and Y. Ito, *J. Liq. Chromatogr.*, **13**, 2309 (1990).

64. H. Oka, Y. Ikai, N. Kawamura, M. Yamada, J. Hayakawa, K. I. Harada, K. Nagase, H. Murata, M. Suzuki, and Y. Ito, *J. High Resol. Chromatogr.*, **14**, 306 (1991).

65. H. Oka, Y. Ikai, N. Kawamura, J. Hayakawa, K. I. Harada, H. Murata, M. Suzuki, and Y. Ito, *Anal. Chem.*, **63**, 2861 (1991).

66. Y.-W. Lee, R. D. Voyksner, Q.-C. Fang, C. E. Cook, and Y. Ito, *J. Liq. Chromatogr.*, **11**, 153 (1988).

67. D. E. Schaufelberger, *J. Liq. Chromatogr.*, **12**, 2263 (1989).

68. D. E. Schaufelberger, *Planta Med.*, **55**, 584 (1989).

69. D. E. Schaufelberger, T. G. McCloud, and J. A. Beutler, *J. Chromatogr.*, **538**, 87 (1991).

70. T.-Y. Zhang, R. Xiao, Z.-Y. Xiao, L. K. Pannel, and Y. Ito, *J. Chromatogr.*, **445**, 199 (1988).

71. H. J. Wang and Y. Y. Chen, *Acta Pharm. Sinica*, **20**, 832 (1985).

72. Y.-W. Lee, R. D. Voyksner, T.-W. Pack, and C. E. Cook, *Anal. Chem.*, **62**, 244 (1990).

73. Y.-W. Lee and R. D. Voyksner, *Nature (London)*, **338**, 91 (1989).

74. D. E. Schaufelberger, *J. Chromatogr.*, **538**, 45 (1991).

75. Y.-W. Lee, T.-W. Pack, R. D. Voyksner, Q.-C. Fang, and Y. Ito, *J. Liq. Chromatogr.*, **13**, 2389 (1990).

76. A, Marston, C. Borel, and K. Hostettmann, *J. Chromatogr.*, **450**, 91 (1988).

77. T.-Y. Zhang, X. Hua, R. Xiao, and S. Kong, *J. Liq. Chromatogr.*, **11**, 233 (1988).

78. Y. Ito, H. Oka, and Y.-W. Lee, *J. Chromatogr.*, **498**, 169 (1990).

79. F. A. Tomas-Barberan and K. Hostettmann, *Planta Med.*, **54**, 266 (1988).

80. I. Slacanin, A. Marston, K. Hostettmann, D. Guédon, and P. Abbe, *Phytochem. Anal.*, **2**, 137 (1991).

81. G. R. Pettit, D. E. Schaufelberger, R. A. Nieman, C. Dufresne, and J. A. Saenz-Renauld, *J. Nat. Prod.*, **53**, 1406 (1990).

82. K. Chifundera, I. Messana, C. Galeffi, and Y. de Vincente, *Tetrahedron*, **47**, 4369 (1991).

83. G. R. Pettit and D. E. Schaufelberger, *J. Nat. Prod.*, **51**, 1104 (1988).

84. A. Marston and K. Hostettmann, *Planta Med.*, **54**, 558 (1988).

85. T. Hatano, K. Yazaki, A. Okonogi, and T. Okuda, *Chem. Pharm. Bull.*, **39**, 1689 (1991).

86. B. C. Van Wagenen, J. Huddleston, and J. H. Cardellina, *J. Nat. Prod.*, **51**, 136(1988).

87. N. Acton, D. L. Klayman, I. J. Rollman, and J. F. Novotny, *J. Chromatogr.*, **355**, 448 (1986).

88. D. Vargas, X. A. Dominguez, K. Acuna-Askar, M. Gutierrez, and K. Hostettmann, *Phytochemistry*, **27**, 1532 (1988).

89. M. Jaziri, B. Diallo, and M. Vanhaelen, *J. Chromatogr.*, **538**, 227 (1991).

90. T. Abbott, R. Peterson, J. McAlpine, L. Tjarks, and M. Bagby, *J. Liq. Chromatogr.*, **12**, 2281 (1989).

91. Y.-M. Yang, H. A. Lloyd, L. K. Pannel, H. M. Fales, R. D. McFarlane, C. J. McNeal, and Y. Ito, *Biomed. Environ. Mass Spectrom.*, **13**, 439 (1986).

92. D.-G. Cai, M.-J. Gu, J.-D. Zhang, G.-P. Zhu, T.-Y. Zhang, No Li, and Y. Ito, *J. Liq. Chromatogr.*, **13**, 2399 (1990).

93. B. Kanyinda, B. Diallo, R. Vanhaelen-Fastre, and M. Vanhaelen, *Planta Med.*, **55**, 394 (1989).

94. K. P. Manfredi, J. W. Blunt, J. H. Cardellina, J. B. McMahon, L. L. Pannell, G. M. Cragg, and M. R. Boyd, *J. Med. Chem.*, **34**, 3402 (1991).

95. J. Quetin-Leclercq, L. Angenot, L. Dupont, and N. G. Bisset, *Phytochemistry*, **27**, 4002 (1988).

96. J. Quetin-Leclercq and L. Angenot, *Phytochemistry*, **27**, 1923 (1988).

97. B. Diallo and M. Vanhaelen, *Phytochemistry*, **26**, 1491 (1987).

98. B. Kanyinda, B. Diallo, R. Vanhaelen-Fastre, and M. Vanhaelen, *Planta Med.*, **55**, 394 (1989).

99. R. C. Bruening, F. Derguini, and K. Nakanishi, *J. Chromatogr.*, **357**, 340 (1986).

100. L. A. Décosterd, Isolement et détermination de structure de constituants biologiquement acitifs de deux espèces de la famille des Guttifères: *Hypericum revolutum* VAHL et *Hypericum calycinum* L. Ph.D. Thesis, University of Lausanne, Switzerland (1990).

101. N. Fischer, B. Weinreich, S. Nitz, and F. Drawert, *J. Chromatogr.*, **583**, 193 (1991).

102. I. A. Sutherland, D. Heywood-Waddington, and Y. Ito, *J. Chromatogr.*, **384**, 197 (1987).

103. R. H. Chen, J. E. Hochlowski, J. B. McAlpine, and R. R. Rasmussen, *J. Liq. Chromatogr.*, **11**, 191 (1988).

104. R. L. Dillman and J. H. Cardellina, *J. Nat. Prod.*, **54**, 1056 (1991a).

105. R. L. Dillman and J. H. Cardellina, *J. Nat. Prod.*, **54**, 1159 (1991b).

106. D. B. Stierle and D. J. Faulkner, *J. Nat. Prod.*, **54**, 1134 (1991).

107. D. E. Schaufelberger, G. N. Chmurny, J. A. Beutler, M. P. Koleck, A. B. Alvarado, B. W. Schaufelberger, and G. M. Muschik, *J. Org. Chem.*, **56**, 2895 (1991).

108. G. P. Gunawardana, S. Kohmoto, and N. S. Burres, *Tetrahedron Lett.*, **30**, 4359 (1989).

109. M. F. Raub, J. H. Cardellina, M. I. Choudhary, C.-Z. Ni, J. Clardy, and M. C. Alley, *J. Am. Chem. Soc.*, **113**, 3178 (1991).

110. H. H. Sun and S. Sakemi, *J. Org. Chem.*, **56**, 4307 (1991).

111. S. Sakemi, H. H. Sun, C. W. Jefford, and G. Bernardinelli, *Tetrahedron Lett.*, **30**, 2517 (1989).

112. S. Sakemi, T. Higa, U. Anthoni, and C. Christophersen, *Tetrahedron*, **43**, 263 (1987).

113. S. Sakemi, T. Ichiba, S. Kohmoto, G. Saucy, and T. Higa, *J. Am. Chem. Soc.*, **110**, 4851 (1988).

114. S. Kohmoto, O. J. McDonnell, and A. Wright, *Experientia*, **44**, 85 (1988).

115. R. C. Bruening, E. M. Oltz, J. Furukawa, K. Nakanishi, and K. Kustin, *J. Nat. Prod.*, **49**, 193 (1986).

116. L. J. Putman and L. G. Butler, *J. Chromatogr.*, **318**, 85 (1985).

CHAPTER

8

HIGH-SPEED COUNTERCURRENT CHROMATOGRAPHY ON MEDICINAL HERBS

TIAN-YOU ZHANG

Beijing Institute of New Technology Application, Xizhimen, Beijing 100035, China

8.1. INTRODUCTION

Medicinal herbs are an important source of natural products for medicine, which include various chemical components ranging from fat-soluble to water-soluble compounds. The isolation of the biologically active components is the starting point of further research in chemistry and pharmacology as well as in the utilization of these compounds.

Traditional Chinese medicine is an extremely rich summation of the experience acquired by the Chinese people in thousands of years of struggle against disease. It is an important part of the brilliant Chinese ancient culture and has played a tremendous role in safeguarding the health of the people and the vitality of the Nation. In order to take traditional Chinese medicine to a higher level and utilize it on a larger scale, it is both necessary and interesting to find the bioactive components in the traditional drugs by modern scientific experimentation and to use them as leading compounds for new drug design. In fact, Chinese scientists have done much work in this regard and have developed many new drugs, such as anisodamine and the antimalarial agent Qinghaosu (artemisine).

The classic methods for extraction of medicinal herbs were combined with PPC (paper partition chromatography), thin-layer chromatography (TLC), and CLC. The development and application of such modern chromatographic techniques as gas chromatography (GC), high-performance liquid chromatography (HPLC), thin-layer (TL) scanning and electrophoresis as well as new types of high-efficiency chromatographic supports have significantly raised the technical level of isolation and have shortened the period of research projects. But all of these techniques still have the following limitations:

High-Speed Countercurrent Chromatography, Edited by Yoichiro Ito and Walter D. Conway.
Chemical Analysis Series, Vol. 132.
ISBN 0-471-63749-1 © 1996 John Wiley & Sons, Inc.

1. Solid chromatographic supports are used, inevitably leading to irreversible adsorption and contamination of fractions, and may even cause denaturation and deactivation, and so on.

2. The present chromatographic apparatus are mainly used for micro-analysis and determination of the contents of simples, but they are limited to applicable fields and provide a rather narrow useful range of solid supports.

3. The important need in the study of phytochemistry for isolation and micro- and semipreparative purification could not be met with classical column chromatography or preparative TCL, and so on, because they are manual techniques and show poor reproducibility. Semipreparative and preparative HPLC are also greatly limited in their use not only because they are expensive, but also because they consume a great amount of solvent and have strict requirements for the pretreatment of samples.

4. In contrast, the above chromatographic techniques prove to be more efficient for isolation of fat-soluble components with lower polarity than high molecular weight water-soluble components with medium polarity.

Countercurrent distribution has long been recognized as an effective means for purification of a wide variety of bioactive molecules. The major drawbacks have been the long separation time required and inconvenience in operating the countercurrent distribution train.

Ito's countercurrent chromatography (CCC), with the planet contrifuge system represents one of the most convenient methods for fractionating a variety of natural products and medicinal herbs. It has been rapidly developed and steadily improved during the past decade. It has been borne out that this technique has certain preparative capabilities based on the principle of liquid–liquid partition without using any solid supports. Its unique properties promise a fine future of exploitation and application (1–5).

8.2. THE PRESENT STATUS OF HIGH-SPEED COUNTERCURRENT CHROMATOGRAPHY ON ISOLATION OF MEDICINAL HERBS

Based on the achievement and experiences developed in the 1960s and 1970s by Ito, one type of horizontal flow-through coil planet centrifuge (CPC) was produced by Zhang at Beijing Institute of New Technology Application, Beijing, China in 1980. During the period of using this machine to separate and purify antibiotics, and organic phosphorus pesticides, Zhang and his co-worker Cai experienced initial success in separating flavones and alkaloids (6). After Ito's invention of HSCCC, Zhang worked with Ito at the National Institutes of Health, Bethesda, MD, to develop several kinds of HSCCC machines and applied them especially in the isolation of medicinal herbs. With the use of analytical HSCCC, made by Pharma-Tech Research Corp., Baltimore, MD, Zhang et al. successfully separated several chemical components, such as hydroxyanthraquinone (7), alkaloids (8), and flavonoids (9) from samples of various medicinal herbs. They have also isolated some alkaloids (10) and flavonoids (11) by using the Ito multilayer coil separator and extractor, made by P.C. Inc., Potomac, MD. In the meantime, Lee et al. (12–16) also successfully applied this method to separate nature products.

In 1989, a model GS-10A HSCCC was produced by Zhang's group in Beijing, China. With close cooperation from Cai and other Chinese scientists who worked with Zhang we used this apparatus for the separation of active components from crude extracts of medicinal herbs (17–19). These included separations of rhein, emodin, and aloe-emodin; a mixture of physcion and chrysophanol, as well as an unknown component from the total anthraquinone fraction of Rhuba; isorhamnetin, kaempferol, quercetin, and rutin from EtOAC extract of *Hippophae rhamnoides*; aristolochic acid-A and a flavonoid from a crude extract of *Aristochia debilis Siebet Zucc*; digoxin and a complicated component from digitoxin (biochemical reagent); phytolacca esculenta saponin A, B, C, D, and water-soluble saponins H and J from the total saponin fraction of *Phytolacca esculenta Van Houtte*. Some alkaloids were separated from six Chinese medicinal herbs by employing a simple solvent system. Good results from these studies will be described in Section 6.

Recently, more and more applications of HSCCC on medicinal herbs have been published (20–23). There is no doubt that HSCCC can become a useful technique and a method complementary to other chromatographic methods for the analysis and separation of different kinds of natural drugs. Great contributions will be made to the development and utilization of the great treasure of medicinal herbs that have a long history in treating diseases and are a rich source of active compounds.

8.3. SEPARATIONS OF SUBSTANCES PREPARED FROM MEDICINAL HERBS BY THE HORIZONTAL FLOW-THROUGH COIL PLANET CENTRIFUGE

8.3.1. Apparatus and Experimental Procedures

The CPC used in this study is driven by a rotary frame through a set of nylon gears around the stationary pipe mounted on the central axis of the centrifuge. The rotary frame, consisting of a pair of aluminum rotary arms rigidly bridged with links, holds a pair of column holders symmetrically at a distance of 14 cm from the central axis of the apparatus (center of revolution) and simultaneously rotate about their own axis (center of rotation). Both rotation and revolution of the holders are at the same angular velocity in the same direction. The revolutional speed can be continuously adjusted from 0 to 500 rpm with a control unit that has a three-digit rpm display.

The column unit was prepared by winding a piece of polytetrafluoroethylene (PTFE) tubing onto a metal pipe of 1.25 cm o.d. Two different tubing sizes were used, 2.6 mm i.d. with a 0.5-mm wall thickness for large-scale separation and 0.8 or 1.0 mm i.d. with 0.3- or 0.5-mm wall thickness for small-scale separation. Eight columns were symmetrically arranged around each holder and interconnected in a series with the suitable PTFE tubing connectors. A counterweight was added on the lighter side in order to obtain the necessary balance between the two different columns. The rotational radius for each column unit is 3.2 cm, which gives a β value (ratio of the radius of rotation to the radius of revolution) of 0.23.

Except for the main centrifuge and its control unit, the whole test system utilizes conventional liquid chromatographic equipment, including a mobile phase pump, sample injector, UV-detector, and fraction collector.

The normal operation procedure of CPC is as follows: The phase of the two-phase solvent system chosen as the stationary phase is first pumped into the column until the entire column space is filled and all air has been excluded. With the centrifuge running at the desired speed, the other phase chosen as the mobile phase is then pumped through the system. The sample can be dissolved in either phase and locally introduced at the sample port with an injector. The sample can be injected at the front of the mobile phase, but a short and suitable delay was often introduced to ensure perfect separation and symmetrical elution of components with a large partition coefficient into the mobile phase, which is then detected as the first peak near the solvent front.

8.3.2.   Separations of a Group of Similar Substances

Rutinum is a common herb used in Chinese medicine. When it is extracted from *Flos sophorae*, quercetin $(C_{15}H_{10}O_7)$ is always accompanied by the main component rutin $(C_{27}H_{30}O_{16})$. The preliminary application of this CPC has achieved complete separation of these two substances with excellent reproducibility.

The other application was to extract the principle alkaloids from a crude extract of *Herba scandentis*, which had an appearance of a thick brown oil. By using the large-scale preparative column, 500 mg of this sample was separated in each operation. One advantage of this apparatus is that there is no need for the crude sample to be cleaned carefully before application.

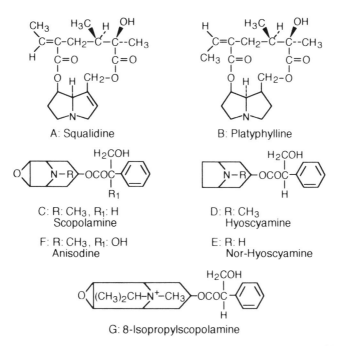

Figure 8.1. The chemical structures of seven similar compounds extracted from medicinal herbs.

In order to show the capability of this machine, a group of similar substances prepared from herbs including (a) squalidine, (b) platyphylline, (c) scopolamine, (d) hyoscyamine, (e) norhyoscyamine, (f) anisodine, and (g) 8-isopropylscopolamine were selected as the samples. Their chemical structures are shown in Fig. 8.1.

Figure 8.2 shows the chromatogram of the mixture of A and B obtained byCPC separation. These two substances were completely separated and each fraction was subsequently analyzed satisfactorily by TLC, as shown at the bottom of Fig. 8.2.

Figure 8.3 shows the result of separation of the sample mixture of 5 mg each of C, D, and E, which are dissolved in 0.2 mL of chloroform . In this experiment, the solvent system was chloroform/0.07 M phosphate buffer (pH 6.5)(1:1). The other conditions were the same as those of the above experiment.

Figure 8.4 shows the result of separation of the sample mixture of 5 mg each of C, F, and G, dissolved in 0.2 mL of chloroform. The solvent system of chloroform/0.07 M phosphate buffer (pH 6.3)(1:1) was used. Other conditions were the same as those of the above experiment.

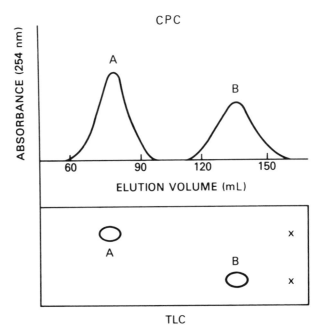

Figure 8.2. Top: Countercurrent chromatogram of squalidine and platyphylline obtained by the present CPC method. Solvent system consists of chloroform/0.2 M phosphate buffer (pH 6.2) (1:1); mobile phase is the lower nonaqueous phase; sample consists of squalidine (A, 5 mg) and platyphylline (B, 5 mg) dissolved in 0.2 mL of chloroform; column is a large-scale preparative column with four units; revolutional speed = 320 rpm; flow-rate = 60 mL/h; detection = 280 nm. Bottom: TLC analysis of the above fractions obtained with the CPC method.

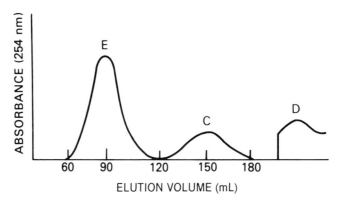

Figure 8.3. Countercurrent chromatogram of scopolamine (C), hyoscyamine (D), and nor-hyoscyamine (E) with the present CPC method. Solvent system, chloroform/0.07 M phosphate buffer (pH 6.5) (1:1); mobile phase, lower nonaqueous phase; sample, mixture of 5 mg each of the above three compounds dissolved in 0.2 mL of chloroform; other experimental conditions, see Fig. 8.2.

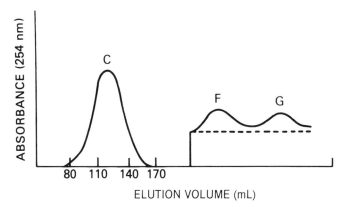

Figure 8.4. Countercurrent chromatogram of scopolamine (C), anisodine (F), and 8-isopropyl-scopolamine (G) with the present CPC method. Solvent system consists of chloroform/0.07 M phosphate buffer (pH 6.3) (1:1); mobile phase is the lower nonaqueous phase; sample mixture of 5 mg each of the above three compounds dissolved in 0.2 mL of chloroform; other experimental conditions, see Fig. 8.2.

8.3.3. Studies on the Preparative Capability of CPC and HPLC in the Separation of Polar Compounds

8.3.3.1. Sample Preparation

In these studies, above-ground parts of *Oxytropis ochrocephala Bunge* were collected in Qinhai Province of China and identified at the Institute of Plateau Biology, Academia Sinica. Milled raw plant tissue was extracted with hot ethanol under reflux over a water bath. The solvent was evaporated under vacuum, and the residue was dissolved in a 5% acetic acid aqueous solution. The supernatant was extracted with ether to remove chlorophylls and lipophilic compounds and then made alkaline with ammonia to pH 9.0 for extraction of the alkaloids with dichloromethane. The upper aqueous phase was then lyophilized to dryness and used as the sample (24).

8.3.3.2. CPC Separation

For the CPC fractionation, a two-phase solvent system composed of chloro-form/ethyl acetate/methanol/water (2:4:1:4) was selected. The lower non-aqueous phase was used as the mobile phase at a flow-rate of 1 mL/min, while the apparatus was run at 280 rpm. The sample solution was prepared by dissolving 60 mg of the crude extract in the upper phase for each separation. Effluent from the outlet of the column of four units was monitored at 254 nm.

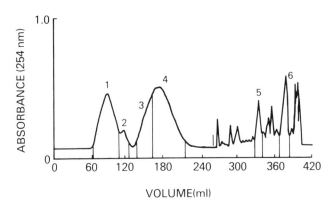

Figure 8.5. The CPC separation of ethanol extract of *O. ochrocephala B.* Solvent system consists of chloroform–ethyl acetate–methanol–water (2:4:1:4); mobile phase is the lower nonaqueous phase; sample is 60 mg of the extract; column 4 units with 120-mL capacity; revolutional speed = 280 rpm; detection = 254 nm.

After 270 mL of the mobile phase was eluted, the centrifuge was stopped, and the column contents were slowly purged with nitrogen gas to collect the peak fractions retained in the column.

Figure 8.5 shows a typical chromatogram of this separation (24). Multiple peaks were eluted in increasing order of polarity and six fractions corresponding to the main peaks, labeled 1–6, were manually collected and subjected to HPLC analysis.

8.3.3.3. HPLC Separation

The HPLC separations were done using the HP-1090M Model (Hewlett-Packard, Waldbronn, F.R.G.) with three different columns: μBondapak C_{18} (300 × 3.9 mm i.d.) 10 μm, Herpsil o.d. (100 × 2.1 mm i.d.) 5 μm, and LiChrosorb RP-18 (200 × 10 mm i.d.) 10 μm. A gradient-elution mode using methanol and water was selected. The analytical columns were eluted at a flow-rate of 1 mL/min and the semipreparative column at 4 mL/min. The effluents were monitored at 254 nm in all cases.

The HPLC separations were done with a reversed-phase column because polar compounds are most conveniently separated by this method.

Figure 8.6 illustrates an analytical chromatogram of the crude *O. ochrocephala B.* extract obtained by a gradient elution of methanol in water using a μBondapack C_{18} column (24). Because of the reversed-phase mode, peaks were eluted in a decreasing order of polarity in contrast to the CPC separation above. Among many compounds present in the crude extracts, only two

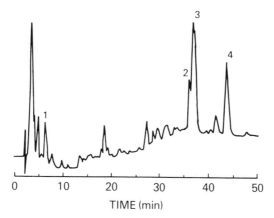

Figure 8.6. High-performance liquid chromatography analysis of ethanol of *O. ochrocephala B.*
Experimental conditions: column is μBondapak C_{18} (300 × 3.9 mm i.d.), 10 μm; mobile phase is
water/methanol, initial composition is 100:0, changed between 0 and 10 min to 90:10, between 10
and 15 min to 80:20, between 15 and 30 min to 50:50, between 30 and 50 min to 30:70;
flow-rate = 1 mL/min; detection = 254 nm with HP-1040 M photodiode array detector.

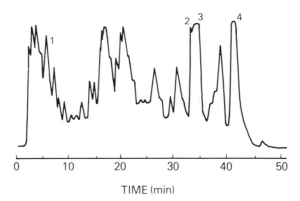

Figure 8.7. Semipreparative HPLC separation of ethanol extract of *O. ochrocephala B.* Experimental conditions: column is LiChrosorb RP-18 (200 × 10 mm i.d.), 10 μm; flow-rate = 4 mL/
min; other conditions, as in Fig. 8.5.

components corresponding to peaks 3 and 4 were obtained in relatively high
concentrations.

Figure 8.7 shows a semipreparative separation of the same sample similarly
obtained in a preparative mode using a 10-mm i.d. column operated under
overloaded conditions with a 3-mg sample size. Four peak fractions indicated
in the chromatogram were manually collected, concentrated by evaporation
in vacuum, and subjected to the analytical HPLC analysis.

8.3.3.4. *Comparison of Separation Efficiencies of CPC and HPLC*

As described above, peak fractions obtained from the preparative separations (Figs. 8.5 and 8.7) were analyzed with HPLC under the same conditions applied to the separation shown in Fig. 8.6. Qualitative analysis revealed that the compounds found in CPC fractions 1, 3, and 5 correspond to those found in HPLC fractions 4, 3, and 1, respectively. It can be seen directly from Figs. 8.5–8.7 that HPLC has a higher column efficiency and shorter separation time, but a lower sample loading capacity. The loading capacity of the semipreparative column is still limited to a milligram range where further increase of the sample size would cause substantial loss in peak resolution.

In contrast, CPC has a lower column efficiency, requires a longer separation time, but provides a much greater sample loading capacity. It is suitable for the separation of polar compounds such as those corresponding to peaks 1 and 3 in Fig. 8.5. In fact, the separation time of CPC in the case of Fig. 8.5 is five times that of HPLC in the case of Fig. 8.7, while the sample size in Fig. 8.5 is 20 times that applied in HPLC in Fig. 8.7. The elution profile of the desired peaks in Fig. 8.5 suggests that the sample size may be further increased without a detrimental loss in peak resolution.

The purities of the peak fractions obtained from CPC and HPLC were quantitatively determined by analytical HPLC to compare the efficiency in preparative separations between these two methods. The results are summarized in Fig. 8.8 (A–F) where chromatograms on the left were obtained from CPC fractions 1–6, and those on the right were from semipreparative HPLC fractions 4–1. It can be seen that the purities of the compound separated by CPC and HPLC are quite similar in Fig. 8.8A, but CPC yields substantially higher purity in Fig. 8.8C. Fig. 8.8E further indicates that the more polar compound in CPC fraction 5 shows a much higher purity than the same compound in semipreparative HPLC fraction 1. The most polar compound in CPC fraction 6 contains some impurities, but it was not well separated by the semipreparative HPLC, as suggested from the elution profile in the chromatogram in Fig. 8.7.

——▶

Figure 8.8. The HPLC analyses of peak fractions obtained from CPC (I) and semipreparative HPLC (II). Fractions (fr.) 1–6 in I (left column) are corresponding to peak fractions 1–6 in Fig. 8.5 and fractions 4–1 in II (right column) to peak fractions 4–1 in Fig. 8.7. Matched pairs of chromatograms in A, C, D, and E indicate analysis of the same components separated by the two different methods. Experimental conditions: I, column is Herpsil ODS (100×2.1 mm i.d.), 5 μm; mobile phase is water/methanol, (A–D) initial composition 54:46, changed between 0 and 10 min to 32:68, (E, F) 95:5; otherwise same as described in Fig. 8.6. II, mobile phase is water/methanol (E) initial compostion 100:0, changed between 0 and 10 min to 90:10, (C, D) 60:40, (A) initial composition 60:40; other conditions are same as in Fig. 8.7.

HPLC ANALYSES OF PREPARATIVE FRACTIONS OF
O. OCHROCEPHALA BUNGE ETHANOL EXTRACT

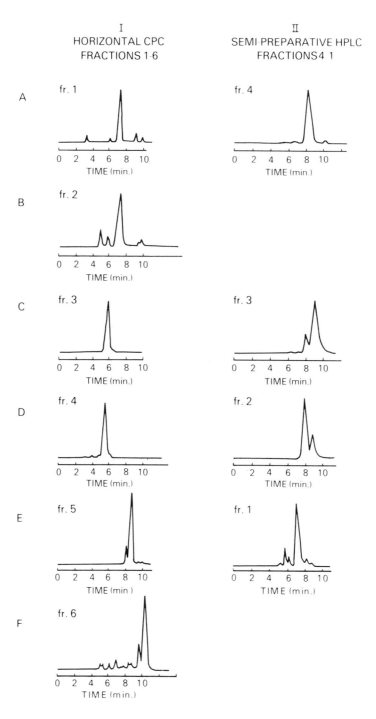

I
HORIZONTAL CPC
FRACTIONS 1-6

II
SEMI-PREPARATIVE HPLC
FRACTIONS 4-1

Overall results of these comparative studies indicate that CPC requires a longer separation time, but it has the important advantages of yielding higher purity fractions and providing a much greater sample loading capacity. This technique is especially suitable for separating polar compounds from natural products including medicinal herbs.

8.4.   SEPARATION OF ALKALOIDS, HYDROXYANTHRAQUINONES, AND FLAVONOIDS BY ANALYTICAL HSCCC

8.4.1.   Apparatus and Experimental Procedures

The apparatus used in these experiments was a Pharma-Tech Model CCC-2000 analytical countercurrent chromatograph (made by Pharma-Tech Research Corp., Baltimore, MD). It is equipped with a column holder at a 2.5-in revolution radius. A multilayer coil prepared from about a 70 m length of heavy wall 0.85-mm i.d. PTFE tubing is coaxially mounted on the holder. The β values range from 0.5 at the internal terminal to 0.75 at the external terminal. The total capacity of the column is 43 mL, which includes 3 mL, in the flow-tubes. The maximum revolutional speed of this centrifuge is 2000 rpm. This commericial CCC model is equipped with an LDC/Milton Roy metering pump, a speed controller (with a digital rpm display), and pressure gauge.

In each experiment, the coiled column was first filled with the stationary phase with the pump, followed by injection of the sample solution through the sample part. Then the apparatus was rotated at a speed of 1800 rpm while the mobile phase was pumped into the head end of the column at a proper flow-rate. In some cases, the reversal elution mode, from the tail toward the head of the column, was necessary. In the normal elution mode, effluent from the tail of the column was continuously monitored with an LKB Uvicord S (LKB Instruments, Bromma, Stockholm, Sweden) and collected into test tubes with an LKB fraction collector. An aliquot of each fraction was dilluted with methanol and the absorbance was determined with a spectrophotometer.

A two-phase solvent system was thoroughly equilibrated in a separatory funnel at room temperature, and the two phases were separated shortly before use.

8.4.2.   Separation of Alkaloids Extracted from *Stephania tetrandra S. Moore*

Dried roots of *Stephania tetrandra S. Moore* (Menispermaceas) (8) or Fenfangji in China, is one of the famous traditional Chinese drugs used for rheumatism and arthritis. The total alkaloid content or the active compounds in the

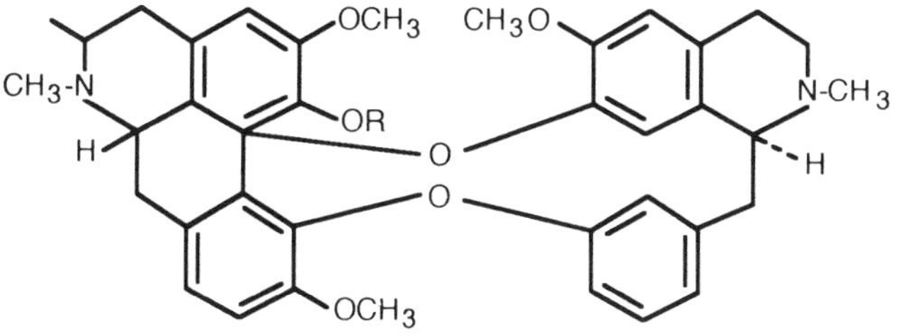

	R
tetrandrine (I)	CH_3
fangchinoline (II)	H

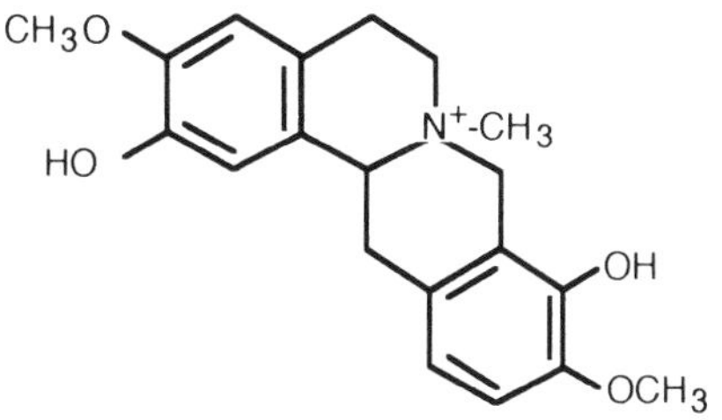

cyclanoline (III)

Figure 8.9. The chemical structures of tetrandrine (I), fangchinoline (II), and cyclanoline (III).

natural products is 2.3%. Three major alkaloids were identified as tetrandrine (I, 1%), fangchinoline (II, 0.5%), and cyclanoline (III, 0.2%). Compounds I and II were inseparable by conventional methods resulting in a mixed product, while III was well separated from the other two compounds. As illustrated in Fig. 8.9, compounds I and II are both bisbenzylisoquinoline alkaloids, whereas III is a water-soluble quaternary protoberberine-type alkaloid.

The sample solution was prepared as follows: A mixture of I and II was added to purified III to obtain a 10:5:2 weight ratio of the three compounds to simulate their composition in the natural drug. A 3-mg quantity of this sample mixture was dissolved in 0.5 mL of the upper stationary phase of the selected solvent system. Solvent systems composed of *n*-hexane/ethyl acetate/methanol/water at two different volume ratios of 3:7:5:5 in the first experiment and 1:1:1:1 in the second one were used. In both cases the lower phase was used as the mobile phase at a flow-rate of 60 mL/h in the normal elution mode.

Figure 8.10 shows the chromatogram obtained from the first experiment. In the normal elution, peaks 1 and 2 were completely resolved and collected in 70 min. This was followed by the reversed elution mode without interrupting the centrifuge run to collect the third peak in an additional 30 min. As shown in the chromatogram, a very small amount of impurity present between peaks 1 and 2 was also resolved. In this separation, the retention of the stationary phase was 50%, and the maximum pressure at the outlet of the pump measured 70 psi.

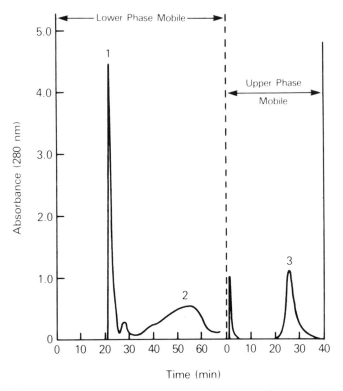

Figure 8.10. Chromatogram of the sample mixture of tetrandrine/fangchinoline/cyclanoline (10:5:2). Solvent system *n*-hexane–ethyl acetate–methanol–water (3:7:5:5).

The chromatogram obtained from the second experiment is shown in Fig. 8.11. It demonstrates an alternative approach where the solvent composition was adjusted to modify the partition coefficients of the compounds to shorten the separation time without the use of a reversed elution mode. The retention of the stationary phase was substantially increased to 72.5%, while the maximum pressure increased to 110 psi, probably due to the increased interfacial tension between the two phases. Because the partition coefficient $K(m/s)$ of purified III in this solvent system measured as high as 120, it was eluted as the first peak immediately after the solvent front, as shown in Fig. 8.11. The partition coefficients of I and II can be estimated from the chromatogram by using the following equation:

$$K(m/s) = (C - R_f)/(R - R_f) \qquad (8.1)$$

where $K(m/s)$ is the partition coefficient expressed as solute concentration in

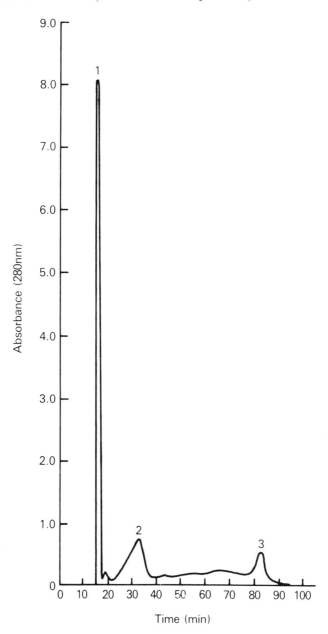

Figure 8.11. Chromatogram of the sample mixture of tetrandrine/fangchinoline/cyclanoline (10:5:2). Solvent system is *n*-hexane–ethyl acetate–methanol–water (1:1:1:1).

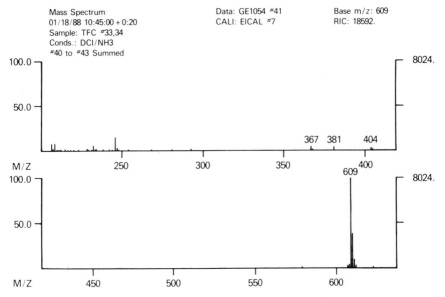

Figure 8.12. The mass spectrum of the fraction from peak 2 in Fig. 8.11.

the mobile phase divided by that in the stationary phase; C is the total column capacity; R_f the retention volume of the solvent front; and R, the retention volume of the solute peak. From the chromatogram shown in Fig. 8.11, the partition coefficients $K(m/s)$ of I and II were computed as 0.38 and 1.37, respectively.

To identify the compounds in peaks 2 and 3 in Fig. 8.11, a Finnigan MAT mass spectrometer was used to analyze the peak fractions. The mass spectrum, in Fig. 8.12, obtained from the fraction of peak 2 shows the molecular weight of this material is 609. It indicated that II must be the major compound in peak 2. The molecular weight of the compound from the fraction of peak 3 in Fig. 8.13 was determined to be 623, which indicates that I is the major compound in peak 3 in Fig. 8.11.

8.4.3. Separation of Hydroxyanthraquinone Derivatives Extracted from Rhubarb

Rhubarb is one of the most well-known drugs produced in China as a purgative and stomachic. A mixture of hydroxyanthraquinones was extracted from the rhizome of *Rheum palmatum L.* This extract contains about 0.5% free hydroxyanthraquinones (7). The chemical structures of five major compounds,

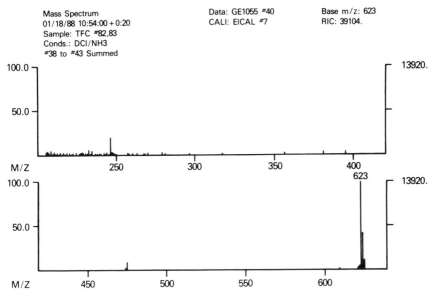

Figure 8.13. The mass spectrum of the fraction from peak 3 in Fig. 8.11.

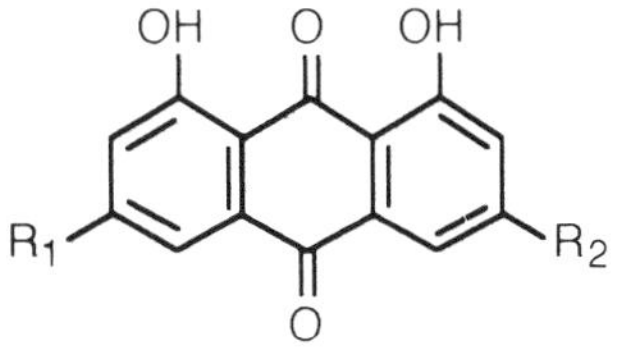

	R$_1$	R$_2$
chrysophanol (I)	H	CH$_3$
emodin (II)	OH	CH$_3$
rhein (III)	H	COOH
physcion (IV)	OCH$_3$	CH$_3$
aloe-emodin (V)	H	CH$_2$OH

Figure 8.14. The chemical structures of chrysophanol (I), emodin (II), rhein (III), physcion (IV), and aloe-emodin (V).

chrysophanol (I), emodin (II), rhein (III), physcion (IV), and aloe-emodin (V), are given in Fig. 8.14.

In the present experiment, a two-phase solvent system composed of *n*-hexane/ethyl acetate/methanol water (9:1:5:5) was used. The sample solution was made by dissolving 1 mg of crude power in 1 mL of the upper phase of the selected solvent system. The lower mobile phase was pumped into the inlet of the column at a flow-rate of 60 mL/h. The effluent from the outlet of the column was continuously monitored at 278 nm and 1-mL fractions were collected in test tubes. After 40 min of elution, when peaks 1, 2, and 3 had appeared, the run was reversed by eluting with the upper phase through the external terminal of the multilayer coil column at the same flow-rate of 60 mL/h. During the reversed running, a narrow capillary tube was applied at the outlet of the column to avoid surging of the effluent due to a reversed pressure gradient formed in the column. This reversed elution was continued until the retained two peaks A and B were eluted, which took place in about 30 min. An aliquot of each fraction obtained from the normal and reversed runs was diluted with a known volume of methanol and the absorbance determined at 280 nm with a Zeiss PM6 spectrophotometer.

Figure 8.15 shows a chromatogram of these five major compounds from the crude extract. The combined use of two elution modes can shorten the separation time and yield a high solute concentration in fractions. The fractions of all five peaks were analyzed with a Finnigan MAT mass spectrometer. The results indicated that peak A, B, 1, 2, and 3 corresponded to I, II, IV, V, and III, respectively.

The above results clearly indicate that a 1-mg complex mixture can be separated and purified in about 70 min using less than 100 mL of solvent for each phase. This experiment can be repeated several times a day. The use of conventional selective solvent extraction, CLC or TLC alone, has failed to separate and purify these compounds completely.

8.4.4. Rapid Separation of Flavonoids Extracted from Sea Buckthorn (*Hippophae rhamnoides*)

The sample of the crude flavonoid mixture used in the present experiment was prepared from the dried fruits of Sea buckthorn (*H. rhamnoides*) by ethanol extraction (9). A two-phase solvent system of chloroform/methanol/water (4:3:2) was selected. The lower nonaqueous phase of this system was used as the mobile phase in the head-to-tail elution mode. For evaluating the analytical capability of this method and equipment, a small amount of sample, 3 mg of the mixture, was loaded in each separation. A series of experiments was performed to study the effect of flow-rates from 60 to 300 mL/h on partition efficiency.

Figure 8.16 shows the chromatogram obtained at a flow-rate of 60 mL/h. Five main peaks were completely resolved in about 90 min. The high partition efficiency of the present method was evident from a minor peak present between the first and second major peaks, which could not be detected in the semipreparative separation (see Fig. 8.21 of Section 8.5). The fraction collected from peak 2 was analyzed by a Finnigan MAT mass sepectrometer. The result shown in Fig. 8.17 clearly indicates that the isorhamnetin present in peak 2 was in a highly purified state.

A set of chromatograms obtained at higher flow-rates of 120, 200 and 300 mL/h are shown in Fig. 8.18. Separation times of these experiments were

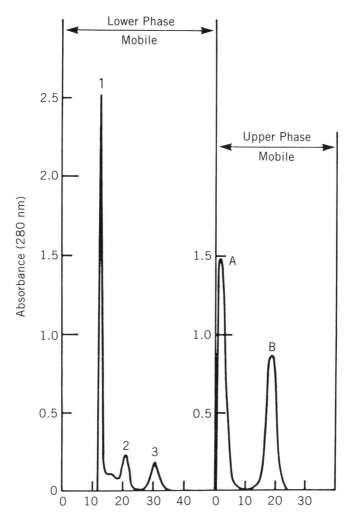

Figure 8.15. Chromatogram of hydroxyanthraquinone derivatives from a crude extract of rhizome of *Rheum palmatum L.* Solvent system is *n*-hexane/ethyl acetate/methanol/water (9:1:5:5).

40, 25, and 15 min, respectively. The retention values of the stationary phase decreased from 77.5% at the flow-rate of 120 mL/h to 68% at the maximum flow-rate of 300 mL/h, while the maximum pressure was increased from 250 to 300 psi.

By increasing the flow-rate of the mobile phase in these analytical HSCCC separations, the separation time for the crude sample mixture was shortened to within 15 min, which is quite comparable with that of analytical HPLC. The

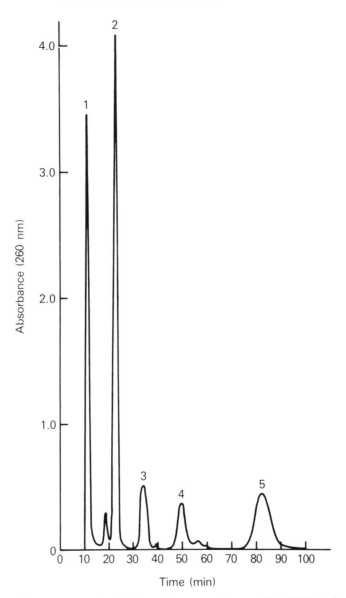

Figure 8.16. Chromatogram obtained from a 3-mg extract by analytical CCC at a flow-rate of 60 mL/h.

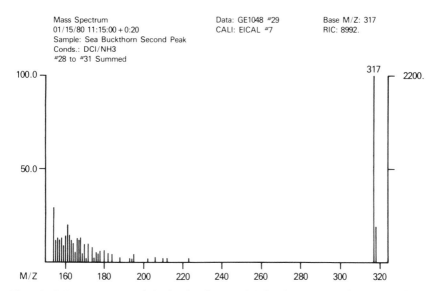

Figure 8.17. Mass spectrum of the fraction from peak 2 in Fig. 8.15. Sample consists of Sea buckthorn second peak. Condition: Finnigan 4500 mass spectrometer scanned from 150 to 500 amu in 0.5 s with sample applied to the "Direct Exposure Probe" using CI/ammonia ionsitation conditions.

sample size in each separation was 3 mg (0.5 mL), much greater than that applied in other analytical separation techniques. This provides an advantage in that the effluent of the analytical HSCCC can be fractionated into test tubes to recover the purified material by evaporating the solvent.

8.5. SEPARATIONS OF FLAVONOIDS AND ALKALOIDS BY MULTILAYER COIL SEPARATOR AND EXTRACTOR

8.5.1. Apparatus and Experimental Procedures

The apparatus used in these experiments was a commericial model of MLCPC called the Ito multilayer coil separator and extractor (P.C. Inc., Potomac, MD) (10, 11). The column holder is positioned at a distance of 10 cm from the central axis of the centrifuge. The separation column was prepared by winding a long piece of PTFE tubing, 1.6 mm i.d. and 0.3-mm wall thickness, directly onto the holder hub of 10 cm diameter making multiple coiled layers. The β value of this equipment ranges from 0.5 at the internal terminal to 0.8 at the external terminal. The total capacity of the multilayer coil measures about 280 mL.

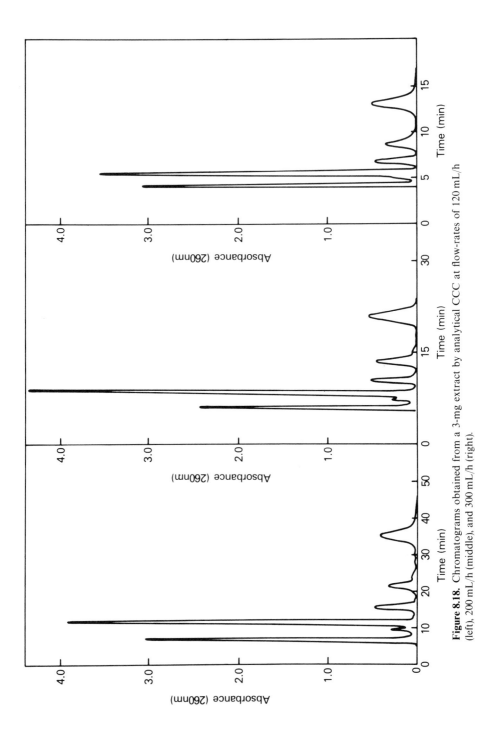

Figure 8.18. Chromatograms obtained from a 3-mg extract by analytical CCC at flow-rates of 120 mL/h (left), 200 mL/h (middle), and 300 mL/h (right).

The separations were performed with a standard procedure used of the analytical HSCCC described above. The optimum revolution speed of this apparatus is 800 rpm.

8.5.2. Separation of *Daphne genkwa* Flavonoids

The sample used in this separation was a mixture of (a) 3'-hydroxygenkwanin, (b) apigenin, and (c) luteolin extracted from *Daphne genkwa Siebet Zucc*, a traditional Chinese medicinal (TCM) herb (10). The chemical structures of these three components are colsely related to each other, as illustrated in Fig. 8.19. A two-phase solvent system composed of chloroform/methanol/water (4:3:2) (v/v/v) was selected and the lower phase was used as the mobile phase.

Partition coefficients of these three samples in the above solvent system are as the following: $K_a(L/U) = 9.65$, $K_b(L/U) = 2.70$, $K_c(L/U) = 0.84$, where $K(L/U)$ is the sample concentration in the lower phase divided by that in the upper phase. The sample solution was prepared by dissolving 10 mg each of the above three compounds in 9 mL of the solvent mixture.

Figure 8.20 shows the chromatogram obtained by this experiment. Three peaks were resolved completely and eluted in 2 h. This result is substantially

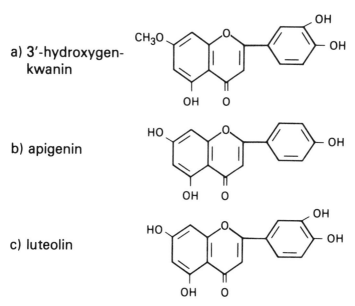

a) 3'-hydroxygen-kwanin

b) apigenin

c) luteolin

Figure 8.19. The chemical structures of (a) 3'-hydroxygenkwanin (b) apigenin, and (c) luteolin.

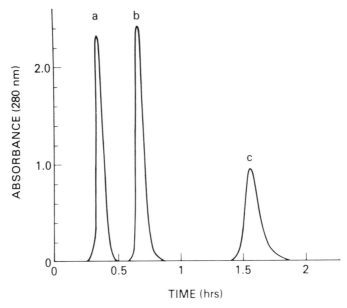

Figure 8.20. Chromatogram of separation of the synthetic sample mixture of (a) 3′-hydroxy-genkwanin, (b) apigenin, and (c) luteolin.

better than that from conventional column chromatography in terms of peak resolution, fraction concentration, and separation time.

Partition efficiency of the multilayer coil column can be calculated from the chromatogram according to the conventional GC formula

$$N = (4R/W)^2 \qquad (8.2)$$

where N is the partition efficiency expressed in terms of theoretical plate (TP), R is the retention time of the peak maximum, and W is the peak width expressed in the same unit as R. Partition efficiencies of peaks b and c calculated from Eq. (8.2) are 704 and 576, respectively.

Resolution of two peaks in a chromatogram can be computed by

$$R_s = 2\Delta R/(W_1 + W_2) \qquad (8.3)$$

where R_s is the resolution between peaks 1 and 2, ΔR is the difference between retention time of peaks 1 and 2, and W_1, W_2 are the peak widths of peaks 1 and 2. According to Eq. (8.3), the resolution between peaks a and b (R_{sab}) is 3.6, and the resolution between peaks b and c (R_{sbc}) is 4.5.

By using Eq. (8.1), partition coefficients of these three compounds were estimated from the chromatogram as $K_a = 9.67$, $K_b = 2.67$, and $K_c = 0.81$. These calculated values are very close to those obtained by the test tube partition measurement of each compound.

8.5.3.   Separation of Flavonoids from Crude Extracts of Sea Buckthorn (*Hippophae rhamnoides*)

A crude flavonoid sample used for this separation was a yellow powder prepared by ethanol extraction from the dried fruit of Sea buckthorn (*H. rhamnoides*) (11). About 100 mg of this sample was used for each separation. A two-phase solvent system composed of chloroform/methanol/water (4:3:2) (v/v/v) was selected in this experiment. The sample is insoluble in water, and the sample solution was prepared by dissolving it in the solvent mixture at a concentration of 2.2%.

Figure 8.21 shows the chromatogram obtained with the Ito multilayer separator and extractor. All components were completely resolved from each other as symmetrical peaks and eluted from the column in 2.5 h. Partition efficiencies computed from Eq. (8.2) range from 800 TP (2nd peak) to 530 TP

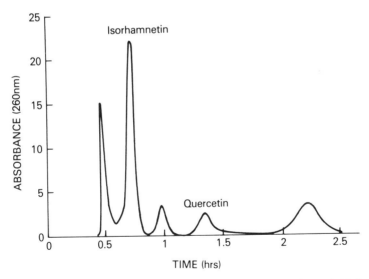

Figure 8.21. Countercurrent chromatogram of flavonoids in crude extract from dried fruits of Sea buckthorn by a MLCPC. Sample size = 100 mg; solvent system consists of chloroform/methanol/ water (4:3:2); mobile phase is the lower phase; flow-rate = 200 mL/h; revolution = 800 rpm; fraction volume = 6 mL.

(5th peak). By calculating the partition coefficients of each peak according to Eq. (8.1) and comparing the obtained values with those of the pure compounds, quercetin and isorhamnetin peaks were identified as labeled in the chromatogram.

Compared to the separation of the same sample with a horizontal flow-through CPC, the present application yielded a sample loading five-fold larger, much higher peak resolution, and was completed in a considerably shorter period of time.

8.5.4. Separation of Alkaloids from *Anisodus tangulicus* (*Maxin*) Pasch

Crude ethanol extract from *Anisodus tangulicus* (*Maxin*) *Pasch* (10) is known to contain a series of alkaloids, such as hyoscyamine, scopolamine, anisodamine, anisodine, and cuscohygrine.

In this experiment, two-phase solvent system composed of chloroform/ 0.07 M sodium phosphate (pH 6.4) (1:1) was used. The sample solution of this crude ethanol extract was prepared by mixing 0.4 mL of the extract with 6 mL of the solvent mixture. The separations were performed in two different elution modes. In the first experiment, the upper phase was used as the stationary phase, and the lower phase was passed through the column at a flow-rate of 200 mL/h.

Figure 8.22 shows a chromatogram obtained by this elution. Three peaks labeled I, II, and III in the left chromatogram were eluted by the mobile phase. After 300 mL of the eluate was collected, the centrifuge was stopped, and the column contents were collected by connecting the column inlet to a pressured nitrogen line. This produced peaks A, B, and C shown in the right chromatogram in Fig. 8.22. The partition coefficients of these compounds corresponding to peaks I, II, and III were calculated from Eq. (8.1) as $K_I(L/U) = 15.08$, $K_{II}(L/U) = 2.79$, and $K_{III}(L/U) = 2.04$.

Figure 8.23 shows the chromatogram obtained from the second experiment in which the lower phase was the stationary phase, and the upper phase was fed in the reversed elution mode but was otherwise under identical experimental conditions. In this case, the partition coefficients of the three peaks labeled a, b, and c were similarly calculated from Eq. (8.1) yielding $K_a(U/L) = 10$, $K_b(U/L) = 2.78$, and $K_c(U/L) = 1.80$.

Fractions containing each compound from the above two experiments showed distinct color so that they were identifiable by the naked eye. The colors observed in peaks I, II, III, A, B, and C in Fig. 8.22 were found to correspond to those in peaks 1, 2, 3, a, b, and c, respectively, in Fig. 8.23.

Partition coefficients of scopolamine and hyoscyamine in the present two-phase solvent system were also determined with purified samples by test tube partition measurement, which gave $K_s(L/U) = 2.72$ and $K_h(L/U) = 1.96$.

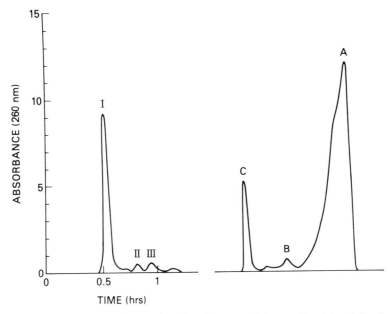

Figure 8.22. Chromatogram of separation of crude extract of *A. tangulicus* (*Maxin*) *Pasch* with chloroform–0.07 *M* sodium phosphate (pH 6.4) (1:1), lower phase mobile.

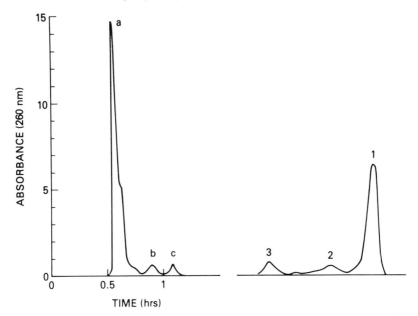

Figure 8.23. Chromatogram of separation of crude extract of *A. tangulicus* (*Maxin*) *Pasch* with chloroform–0.07 *M* sodium phosphate (pH 6.4) (1:1), upper phase mobile.

These partition coefficients values are closely correlated to those of peaks II and III shown in Fig. 8.22 computed from Eq. (8.1).

8.6.   SEMIPREPARATIVE SEPARATION OF ALKALOIDS BY GS-10A HSCCC

8.6.1.   Apparatus and Experimental Procedures

These experiments were performed with the GS-10A HSCCC, a MLCPC for HSCCC, constructed at the Beijing Institute of New Technology Application, Beijing, China. The apparatus holds a pair of column holders symmetrically on the rotary frame at a distance of 8 cm from the central axis of the centrifuge. The multilayer coil separation column was prepared by winding a 130-m length of 1.5-mm i.d. PTFE tubing directly onto the holder hub making 14 coiled layers between a pair of flanges spaced 7.4 cm apart. The total capacity of the column measured 230 mL. The β values range from 0.5 at the internal terminal to 0.8 at the external terminal. An adjustable counterweight was mounted on the opposite side of the column holder for balancing the centrifuge system. The revolutional speed of this machine was regulated with a speed controller in a range between 0 and 1000 rpm, while 800 rpm was applied as an optimum speed for the conventional two-phase solvent systems.

The experimental procedures are the same as those for the Ito multilayer coil separator and extractor described above.

8.6.2.   Separation of Alkaloids from *Sophora flavescens Ait* and *Datura mete L.*

The root of *S. flavescens Ait* is used for the treatment of various diseases in traditional Chinese medicine as an antifebrile, a diuretic, an antihelminthic, and an antidote. It contains two main alkaloids, matrine (I) and oxymatrine (II). The crude sample solution of this herb was obtained by solvent extraction.

The flowers of *D. mete L.* are said to be effective for relieving asthma, a cold cure, and alleviating pain. The main components of the crude sample prepared by solvent extraction are scopolamine (III) and hyoscyamine (IV).

Figure 8.24 shows a chromatogram of crude alkaloids extracted from *S. flavescens Ait* obtained by HSCCC. Separation was performed with a two-phase solvent system composed of chloroform/0.07 M phosphate–0.04 M citrate buffer (pH 6.4), (1:1) (17). The lower nonaqueous phase was used as the mobile phase. As indicated in the diagram, peaks 1, 2, and 3 were eluted during the centrifugal run, and peaks 4 and 5 were collected from the column contents after stopping the centrifuge.

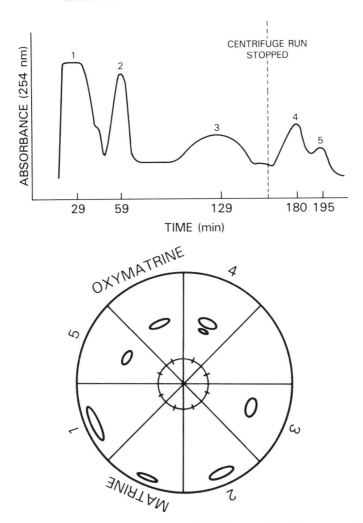

Figure 8.24. Top: HSCCC separation of crude alkaloids from *Sophora flavescens Ait*. Bottom: PPC analysis of HSCCC fractions.

By means of the TLC and PPC analysis peaks 1 and 5 were identified as I and II, peak 4 contained a mixture of II and a minor impurity of unknown nature. Peaks 2 and 3 were trace components that remain to be identified.

Multiple sample injections were performed to obtain much greater amounts of purified compunds. About 50 mg of I and a small amount of peaks 2 and 3 were obtained from four successive sample injections at 1.5-h intervals. The results demonstrated that multiple sample injection may be a very useful technique for isolation of active compounds from medicinal herbs.

Figure 8.25 shows a similar chromatogram of crude alkaloids extracted from *D. mete L.* and obtained by HSCCC. The experimental procedures and the selection of two-phase solvent systems were the same as those in the above experiment, except that the pH of the buffer solution was 6.5. Peaks 1–4 were eluted during the centrifugal run, peak 5 was collected from the column contents. By means of a TLC analysis, peaks 3 and 5 were identified as III and IV, respectively, while peak 2 was a trace alkaloid and peak 4 was a trace non-alkaloid. Peak 1, eluted near the solvent front, consisted of a mixture of alkaloids.

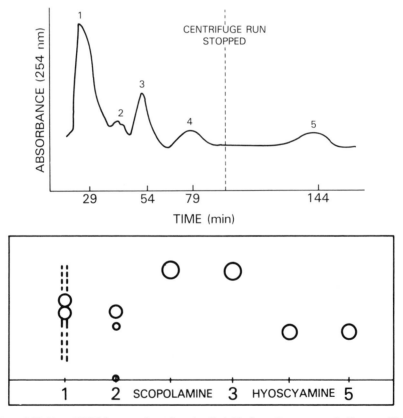

Figure 8.25. Top: HSCCC separation of crude alkaloids from *Datura mete L.* Bottom: TLC analysis of HSCCC fractions from *Datura mete L.*

8.6.3. Separation of Alkaloids from *Cephalotaxus fortunei Hook F.*

Three anticancer active compounds, harringtonine (V), isoharringtonine (VI), and homoharringtonine (VII), were isolated from *C. fortunei Hook. F.*

Figure 8.26 shows that these components possess similar chemical structures: VII and V differ only by a bond $-CH_2$ group, while VI and VII are isomers, which could be hardly separable by adsorption chromatography.

The sample of crude alkaloids used for this experiment was obtained from the leaves and branches of *C. fortunei Hook F.* by solvent extraction of the buffered solution pH 6.7. A two-phase solvent system composed of chloroform/0.07 M sodium phosphate – 0.04 M citrate buffer solution (pH 5.0) (1:1) was selected. The sample solution was prepared by dissolving 20–30 mg of crude alkaloids in the mobile phase. The nonaqueous mobile phase eluted peaks 1, 2, 3, 4, and 5 at a flow-rate of 2 mL/min. Peak 6 was obtained from the solvent retained in the column by stopping rotation and displacing the stationary phase by continued pumping of the mobile phase. Figure 8.27 gives the chromatogram of this separation.

It has been shown by CO-TLC and CO-PPC analysis that peaks 2, 4, and 6 were identical with the Compounds V, VI and VII, respectively. The

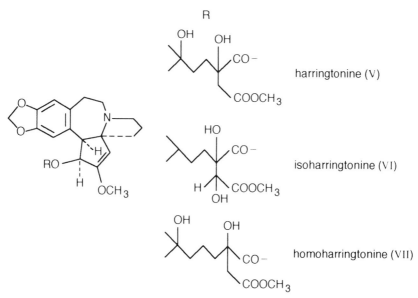

Figure 8.26. The chemical structures of harringtonine (V), isoharringtonine (VI), and homoharringtonine (VII).

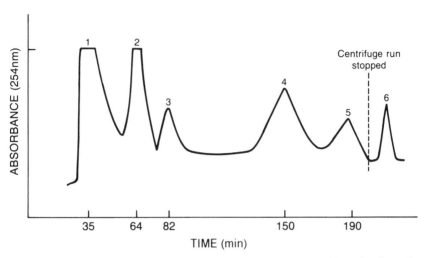

Figure 8.27. The HSCCC separation of harringtonine, isoharringtonine, and homoharringtonine from a crude extract of *Cephalotaxus fortunei Hook. f.*

molecular weights of these three compounds were determined by means of mass spectrometry as 531 V, 545 VI and 531 VII.

In order to collect much larger amounts of the important medicinal Compound VI in a purified state, the sample size of each injection was increased further. When a sample quantity of 300 mg was applied, 70 mg of purfied Compound VI was obtained.

8.6.4.   Separation of Alkaloids from *Senecio fuberi Hemsl*

A mixture of the active compounds squalidine VIII, platyphylline IX, and neoplatyphylline X, can be obtained from the medicinal herb, *S. fuberi Hemsl.*

Figure 8.28 shows the chemical structures of these three compounds. The Compounds IX and X were confirmed as cis–trans isomers, which could hardly be separated by conventional chromatographic methods or the horizontal flow-through CPC. In this study two sample mixtures were prepared. First, two kinds of alkaloid mixtures A and B were prepared by solvent pH gradient extraction of the plant material. From these two mixtures, the sample solution 1 was made by dissolving a mixture of 5 mg of alkaloid mixture A with 10 mg of alkaloid mixture B in 1 mL of mobile phase. Sample solution 2 was made by adding 10–12 mg of B to 1 mL of sample solution 1. A two-phase solvent system composed of chloroform/$0.07\ M$ sodium phosphate–$0.04\ M$ citrate buffer (pH 6.2–6.45) (1:1) was selected and the lower nonaqueous phase

Squalidine (VIII)

R_1	R_2	
H	CH_3	Platyphylline (IX)
CH_3	H	Neoplatyphylline (X)

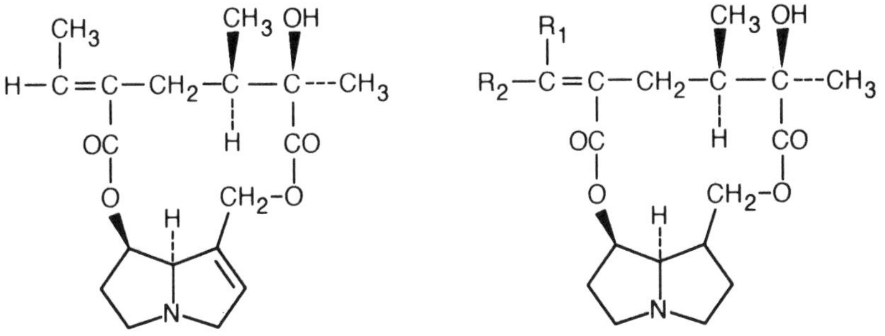

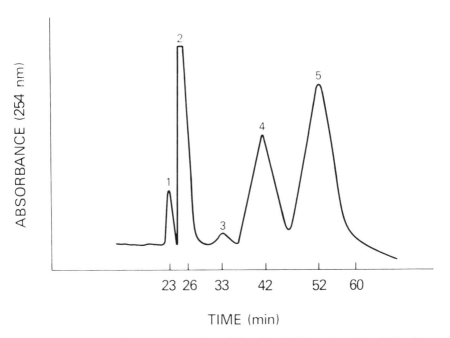

Figure 8.28. The chemical structure of squalidine (VIII), platyphylline (IX), and neoplatyphylline (X).

Figure 8.29. The HSCCC chromatogram of squalidine, platyphylline, and neoplatyphylline from *Senecio fuberi Hemsl.*

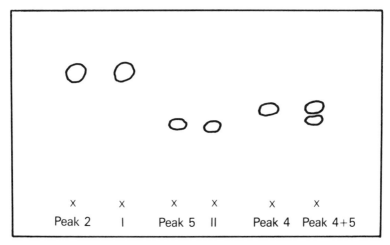

I: SQUALIDINE   II: PLATYPHYLLINE

Figure 8.30. TLC analysis of HSCCC fractions from *Senecio fuberi Hemsl.*

was used as the mobile phase in the head-to-tail elution mode at a flow-rate of 2 mL/min.

Figure 8.29 shows the chromatogram of sample solution 1. The alkaloid components were resolved by HSCCC into five peaks in about 1 h. The TLC analysis of the peak fractions revealed that the contents of peaks 2 and 5 were authentic squalidine and platyphylline, respectively (Fig. 8.30). Peaks 1 and 3 were trace constituents that have not yet been assayed. The R_f value of the fraction of peak 4 was 0.55. The chromatogram obtained from sample solution 2 showed that the ratio between peaks 4 and 5 was about 2:3 measured by the peak areas with traces of peaks 1, 2, and 3. In order to determine the chemical structure of the compound in peak 4, about 30 mg of the pure alkaloid was obtained from the six consecutive runs. Finally, it was identified as neo-platypylline on the basis of melting point (mp), specific rotation (α), infrared (IR), proton nuclear magnetic resonance (^{1}H NMR), and mass spectrometry (MS) data.

By raising the pH value of the buffer from 6.21 to 6.45, the retention times of peaks 4 and 5 were shortened from 70 to 42 min and 92 to 52 min, respectively, and the retention volumes of these peaks were correspondingly reduced from 140 to 84 mL and 184 to 104 mL, respectively. The detailed experimental conditions are listed in Table 8.1. At all three pH values, peaks 4 and 5 were completely resolved.

Table 8.1. Effects of Flow-Rate of the Mobile Phase on Various Parameters[a]

Flow-Rate (mL/h)	Retention of Stationary Phase (%)	Maximum Pressure (psi)	Separation Time (min)
60	86	170	90
120	77.5	250	40
200	72.8	270	25
300	68	330	15

[a]Revolution speed is 1800 rpm.

8.6.5. Separation of Vincamine and Vincine

Vincamine (1), a major alkaloid in *Vinca minor*, is used for the treatment of cerebral vascular diseases. The mother liquor obtained from recrystallization of vincamine always contains vincine (2) in a ratio of almost 1:1. These two compounds differ only by a methoxy group in their chemical structure, as shown in Fig. 8.31. Conventional purification methods, such as recrystallization, preparative TLC, and open-column chromatography, failed to separate these two compounds.

In our laboratory, the HSCCC separation of vincamine and vincine (19) was performed with a two-phase solvent system of hexane/ethanol/water (6:5:5). The lower phase was used as the mobile phase. The sample solution was prepared by dissolving 40 mg of the simple mixture in 140 mL of lower phase

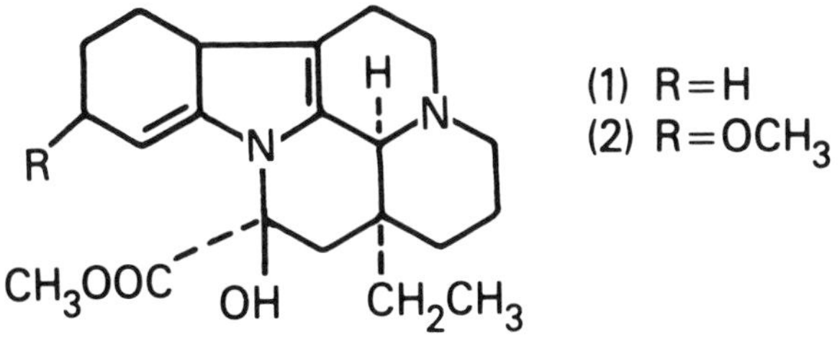

(1) R=H
(2) R=OCH$_3$

Chemical structures of vincamine (1) and vincine (2).

Figure 8.31. The chemical structures of vincamine (1) and vincine (2).

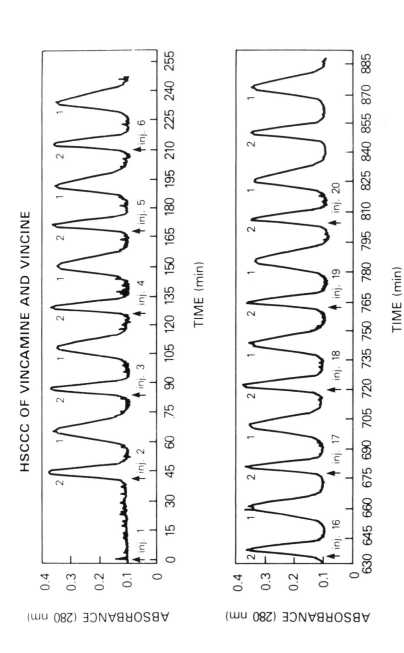

Figure 8.32. The HSCCC chromatogram of separation of vincamine (1) and vincine (2) with multiple injections.

Sample: Vincamine 1 + Vincine 2; Solvent System: Hexane:ethanol:water (6:5:5); Mobile Phase: Lower phase; UV: 280 nm; Flow Rate: 2.6 mL/min; Chart Speed: 10 cm/h.

solvent. A 6.0-mL aliquot of sample solution was injected in each experiment. For preparative purification, multiple injections of the same sample solution were introduced at 42-min intervals.

A shown in Fig. 8.32, the mixture of vincamine and vincine was completely resolved. It was found that the GS-10A HSCCC system was capable of performing the separation by applying more than 20 consecutive injections without changing the stationary phase in the coiled column. As shown in Fig. 8.32, peak resolution was unaltered throughout the entire experiment. After 20 consecutive injections of 1.7 mg of sample mixture, 16.5 mg of vincamine and 14 mg of vincine were obtained. In addition, the resolution obtained by this method remained unchanged when the machine was shut down overnight and restarted on the next day with the retained stationary phase in the coiled column.

Figure 8.33 shows the results of HPLC analysis of the mixture (A) and the HSCCC fractions of vincamine (B) and vincine (C). The purity of collected fractions were examined by using an analytical Zorbax-ODS column (4.6 mm × 25 cm). The chemical structure of each compound was also confirmed by mp, MS, and NMR.

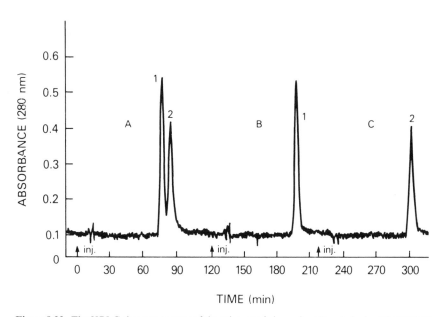

Figure 8.33. The HPLC chromatograms of the mixture of vincamine (1) and vincine (2), HSCCC purified vincamine (1) and HSCCC purified vincine (2).

8.7. CONCLUSION

High-speed CCC, which was invented by Ito, represents the newest achievement in chromatography with its character of rapid separation, semipreparation and high efficiency. In addition to the three kinds of equipment, including the horizontal flow-through CPC, analytical HSCCC, and semipreparative HSCCC, the new large-scale preparative CCC named the cross-axis CPC has been developed. The capability of this apparatus was demonstrated in the separation of a crude Sea buckthorn extract on a 100-mg scale (25). The main advantages in HSCCC mentioned above are as follows.

Since no solid supports are used, various complications such as irreversible adsorption and contamination of samples can be eliminated . All components in the sample solution injected into the column will be recovered. The crude sample can be injected directly into the column, which simplifies sample preparation.

Numerous two-phase solvent systems with a broad spectrum in polarity can be applied in this method. Either the aqueous or organic phase can be used as the mobile phase.

With its strong resolving power, HSCCC can shorten the separation time to 1–2h. The quantity of purified compounds may range from 1 to 100 mg. Consecutive injections can be applied for semipreparative separations.

As described in the above sections, the present method is especially suitable for separation of alkaloids from medicinal herbs using a simple solvent system of chloroform/buffer. The flavonoids and hydroxyanthraquinones can be easily separated with high partition efficiency.

High-speed CCC technology will become widely used in the research and production of natural drugs, since the instrument is inexpensive, convenient to operate, solvent saving, and has semipreparative capacity.

If the method is combined with other efficient chromatographic methods, such as HPLC, the efficacy of separation and purification is enormously improved. In the rapidly growing field of biotechnology, purification of high molecular weight and water soluble components is especially needed, but currently only few means of separation are available. Therefore, HSCCC should be extensively utilized in this field in the near future.

ACKNOWLEDGMENTS

I wish to thank Dr. Yoichiro Ito for his help in the development of HSCCC equipment and its application. I also wish to express my thanks to Dr. D. G. Cai for his long-term cooperation and his contribution to this chapter and to Dr. Q. C. Fang for his helpful suggestions. I am grateful to Mrs. Carol Kosh for her fine work in preparation of the manuscript.

REFERENCES

1. Y. Ito, *Trends in Analytical Chemistry*, **5** (6), 142 (1986).

2. Y. Ito, *CRC Crit. Rev. Anal. Chem.*, **17**, 65 (1986).

3. Y. Ito, *Nature (London)*, **326**, 419 (1987).

4. Y. Ito, *Countercurrent Chromatography: Theory and Practice*, Chapter 3, Marcel-Dekker, New York, 1988, pp. 79–442.

5. Y. Ito, *J. Chromatogr.*, **538**, 3 (1991).

6. T. Y. Zhang, *J. Chromatogr.*, **315**, 287 (1984).

7. T. Y. Zhang, L. K. Pannell, Q. L. Pu, D. G. Cai, and Y. Ito, *J. Chromatogr.*, **442**, 455 (1988).

8. T. Y. Zhang, L. K. Pannell, D. G. Cai, and Y. Ito, *J. Liq. Chromatogr.*, **11**(8), 1661 (1988).

9. T. Y. Zhang, R. Xiao, Z. Y. Xiao, L. K. Pannell, and Y. Ito, *J. Chromatogr.*, **445**, 199 (1988).

10. T. Y. Zhang, D. G. Cai, and Y. Ito, *J. Chromatogr.*, **435**, 159 (1988).

11. T. Y. Zhang, X. Hua, R. Xiao, and S. Kong, *J. Liq. Chromatogr.* **11**(1), 233 (1988).

12. Y. W. Lee, R. D. Voyksner, Q. C. Fang, C. E. Cook, and Y. Ito, *J. Liq. Chromatogr.*, **11**(1), 153 (1988).

13. Y. W. Lee, Q. C. Fang, and C. E. Cook, *J. Liq. Chromatogr.*, **11**(1), 75 (1988).

14. Y. W. Lee, Q. C. Fang, C. E. Cook, and Y. Ito, *J. Nat. Prod.* **52** (4), 706 (1989).

15. Y. W. Lee, Q. C. Fang, C. E. Cook, and Y. Ito, *J. Chromatogr.*, **477**, 434 (1989).

16. Y. W. Lee, R. D. Voyksner, C. E. Cook, Q. C. Fang, T. W. Pack, and Y. Ito, *Anal. Chem.* **62**, 244 (1990).

17. D. G. Cai, M. J. Gu, J. D. Zhang, G. P. Zhu, T. Y. Zhang, and Y. Ito, *J. Liq. Chromatogr.*, **13** (12), 2399 (1990).

18. D. G. Cai, M. J. Gu, G. P. Zhu, J. D. Zhang, T. Y. Zhang, and Y. Ito, *J. Liq. Chromatogr.*, **15** (15,16), 2873 (1992).

19. J. Y. Zhoz, Q. C. Fang, and Y. W. Lee, *Pytochem. Anal.*, **1** (1), 1 (1990).

20. Y. W. Lee, T. W. Pack, R. D. Voyksner, Q. C. Fang, and Y. Ito, *J. Liq. Chromatogr.*, **13** (12), 2389 (1990).

21. K. Harada, I. Kimura, A. Yoshikawa, M. Suzuki, H. Nakazawa, S. Hattori, K. Komori, and Y. Ito, *J. Liq. Chromatogr.*, **13** (12), 2373 (1990).

22. A. Weisz, A. J. Langowski, M. B. Meyers, M. A. Thieken, and Y. Ito, *J. Chromatogr.*, **538**, 157 (1991).

23. H. J. Cahnmann, E. Gonçalves, Y. Ito, H. M. Fales, and E. A. Sokoloski, *J. Chromatogr.*, **538**, 165 (1991).

24. P. Li, T. Y. Zhang, X. Hua, and Y. Ito, *J. Chromatogr.*, **538**, 219 (1991).

25. T. Y. Zhang, Y. W. Lee, Q. C. Fang, R. Xiao, and Y. Ito, *J. Chromatogr.*, **454**, 185 (1988).

.

CHAPTER

9

ISOLATION OF MARINE NATURAL PRODUCTS BY HIGH-SPEED COUNTERCURRENT CHROMATOGRAPHY

NANCY L. FREGEAU* AND KENNETH L. RINEHART

Roger Adams Laboratory, University of Illinois, Urbana, Illinois 61801

9.1. INTRODUCTION

Natural products have been extensively studied, especially as a major source of new pharmaceuticals—both as actual drugs and as templates from which analogs with better efficacy or reduced toxicity can be developed. Until now most of this research focused on terrestrial species, especially plants and microorganisms, but other sources, including marine, are now being explored more extensively (1). Marine organisms produce some types of compounds, such as brominated tyrosine derivatives, that are not usually obtained from plants and microbes.

Several different considerations have encouraged marine natural products research. New diseases, such as acquired immunodeficiency syndrome (AIDS), and old diseases for which there are currently inadequate treatments (e.g., some types of cancer) have motivated the search for new drugs. Many researchers now use new bioassay-guided approaches to find compounds that may become useful pharmaceuticals (2). Also, new spectroscopic techniques allow the identification of ever smaller amounts of material. Thus, a bioactive compound that is present in a marine organism in only trace amounts can be isolated and characterized.

The potential impact of marine natural products can be illustrated by three clinically tested compounds. In the 1950s Bergmann and co-workers (3–5) found that the sponge *Cryptotethia crypta* contained large amounts of novel nucleosides, including spongothymidine, spongouridine, and spongosine (**1**–**3**).

*Taken in part from the Ph.D. Thesis of N. L. Fregeau, University of Illinois, Urbana, 1992.

High-Speed Countercurrent Chromatography, Edited by Yoichiro Ito and Walter D. Conway. Chemical Analysis Series, Vol. 132.
ISBN 0-471-63749-1 © 1996 John Wiley & Sons, Inc.

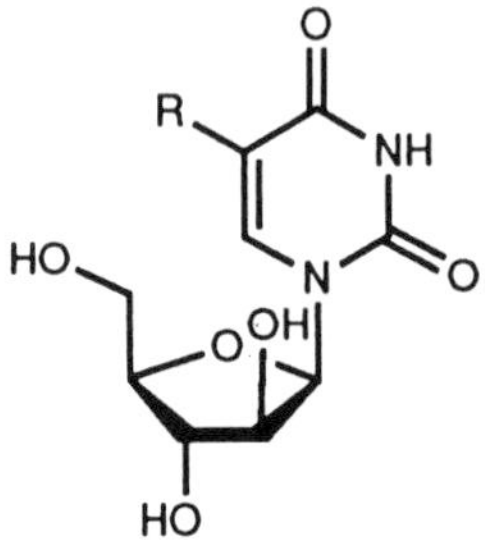

1 R = CH$_3$

2 R = H

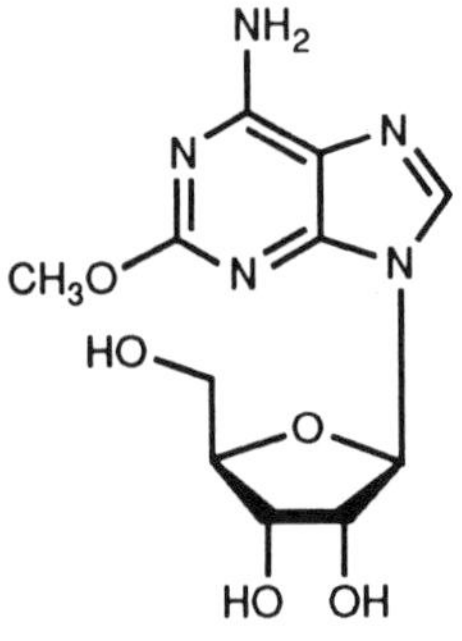

3

The first two, 3-β-D-arabinofuranosyl derivatives of thymidine and uracil, respectively, led directly to the development of the antiviral and anticancer drugs ara-A and ara-C and, eventually, to second and third generation "unnatural" nucleoside drugs, such as zidovudine (AZT) and acyclovir (2).

Another important series of marine natural products consists of the didemnins, first discovered in the early 1980s in our group (see also Section 9.4.5) (6–8). These cyclic depsipeptides, containing several unusual units, including hydroxyisovalerylpropionic acid and (3S, 4R, 5S)-isostatine, all possess antitumor, antiviral, and immunosuppressive activities. Among the most potent is

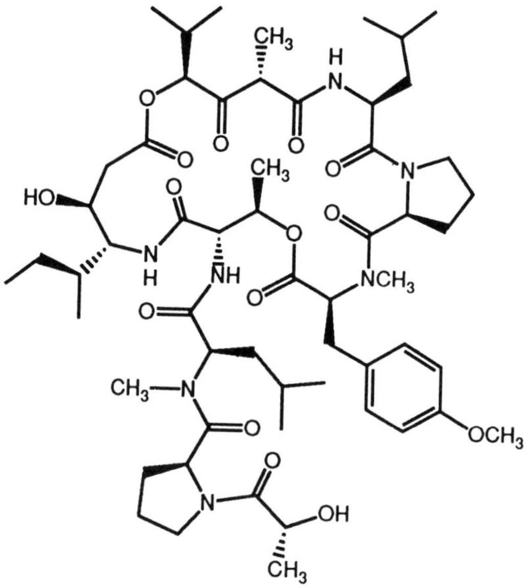

4

didemnin B (**4**), which is the first marine natural product in clinical testing and is now in Phase II clinical trials (7).

A third important class of marine natural products, the bryostatins, was first reported by Pettit et al. in 1982 (9). The more than 15 compounds in this series (see also Section 9.4.1) are macrocyclic lactones that possess potentially clinically useful antitumor activity and also bind to and activate protein kinase C. However, unlike the phorbol esters with which they compete for binding sites, they are not tumor promoters (10).

9.2. ALTERNATIVE SEPARATION TECHNIQUES

Marine natural products are often found in yields of 1 part per million (ppm) or less based on the wet weight of the organism. Moreover, they are often isolated from very complex mixtures of compounds. Purification of small amounts of material from extremely complex matrices, such as crude extracts of marine organisms, generally requires a wide variety of different separation methods. In addition, little is usually known about the target compound and its reactivity during the early stages of the isolation, so mild methods are best.

One such procedure involves partitioning the sample between two immiscible solvents. This simple technique allows complete recovery of the material, but is very limited in its ability to differentiate compounds. A third, and sometimes fourth, solvent can be added to change the separation or a series of different partitionings can be used, but the resolving power of this method is very low. Thus it is used most often at the beginning of an isolation to effect a rough division of the crude extract by solubility. One partitioning scheme used with various modifications by a number of researchers is that developed by Kupchan et al. (11).

Probably the most common separation technique used for marine natural products has been liquid–solid chromatography (LSC), which is generally applicable and has much greater resolving power than solvent partitioning. Silica chromatography is the most popular, but many different normal and reversed-phase packing materials are now available. Silica has the disadvantage of irreversibly adsorbing some compounds and has been known to catalyze the decomposition of others. Reversed-phase (RP) and bonded-phase supports, especially octadecylsilane (C-18), provide different adsorption characteristics and resolving abilities from the normal phase (NP), and are less reactive and less prone to irreversible adsorption.

Gravity column chromatography is being supplanted by flash chromatography, which allows rapid separation (12). Blunt et al. (13) also developed an RP flash technique for separating crude marine extracts, which are used in place of solvent partitioning. High-performance liquid chromatography (HPLC) is used mostly near the end of the isolation.

In solid phase extraction (SPE) a short column is used to absorb the desired material while the impurities are washed out. A more powerful eluting solvent is then applied to extract the compound of interest from the column. This method is particularly useful for removing salts and buffers from a sample. Prepacked columns (e.g., Waters Sep-Pak cartridges) and resins such as XAD-2 can be useful.

Compounds can also be differentiated by characteristics other than adsorption and solubility. Gel filtration, employing, for example, Sephadex LH-20, separates compounds partly on the basis of molecular weight. Charged compounds can be separated by ion exchange resins.

9.3.   COUNTERCURRENT CHROMATOGRAPHY

Countercurrent chromatography (CCC) is a separation technique that is finding increased usage among marine natural product chemists, who usually employ more than one type of purification technique. As others have extensively reviewed CCC (see earlier chapters in this volume), only a brief description will be given here (14–18). Countercurrent chromatography has many advantages compared to LSC. It employs a different type of separation than conventional chromatography and as such is a good complement to it. A technical comparison of CCC with HPLC has been provided by Foucault (14).

Countercurrent chromatography does not provide as good resolution as HPLC, generally less than or equal to 1000 theoretical plates. Nonetheless, since CCC has a higher ratio of stationary phase/total volume, it needs fewer theoretical plates to effect the same resolution as HPLC (14). In part because of its lower resolving capabilities, CCC is best used as a preparative tool (19).

Among marine natural product chemists, the multilayer coil planet centrifuge (MLCPC) (20), manufactured by P. C. Inc., Potomac, MD, appears to be the most prevalent, although other HSCCC instruments have also been used.

9.4.   EXAMPLES OF MARINE NATURAL PRODUCT ISOLATIONS USING HIGH-SPEED COUNTERCURRENT CHROMATOGRAPHY

High-speed CCC has been used for the isolation of many different marine natural products, as illustrated below. Some reports of HSCCC use may be missing due to the difficulty in finding such papers, because most workers do not report isolation methods in their abstracts. The examples given here have been organized by the class of compound isolated. The abbreviations UP (upper phase), LP (lower phase), and MP (mobile phase) are used for simplicity in describing the solvent systems.

9.4.1. Macrolides

Several of the bryostatins, which were discussed briefly above, have been purified by HSCCC. Petti et al. (21) isolated two new components, bryostatins 14 (**5**) and 15 (**6**), during the purification of bryostatins 1 (**7**) and 2 (**8**) from two 1000-kg samples of the bryozoan *Bugula neritina*. From one of these samples, a 37.51-g fraction enriched in **7** was obtained, which was subjected to MLCPC in 46 batches, using the solvent system hexane/EtOAc/MeOH/H$_2$O (45:15:10:3; MP = UP). Silica flash chromatography and recrystallization gave 1.5 g of **7**. The mother liquor was chromatographed on a silica HPLC column to yield 8.6 mg of **6**. Another fraction, enriched in **8** (14.11 g), was further purified in an analogous fashion, using hexane/EtOAc/MeOH/H$_2$O (40:25:12:5; MP = UP) for a total of 14 HSCCC runs. A similar isolation procedure was used for the second 1000-kg sample of bryozoan. In this case, hexane/EtOAc/MeOH/H$_2$O (3:7:5:5; MP = UP) was used to give, after further HPLC purification, 102 mg of **5**. The new compounds differ from previously known bryostatins by having a free hydroxyl (**5**) or a new esterifying group (**6**) on C20. Both of these compounds also show significant activity against P388 lymphocytic leukemia *in vitro* with ED$_{50}$ values of 0.33 and 1.4 μg/mL, respectively.

Schaufelberger et al. (22) also investigated the bioactive compounds from *B. neritina*. They reisolated **7** and **8** along with bryostatin 3 (**9**) and its previously unknown 26-ketone analog (**10**). Schaufelberger noted that **9** and **10** are more unstable than other bryostatins and thus require quick isolation, using nondestructive methods, such as HSCCC, particularly for small quantities of sample (23). Multilayer CPC was used for the final purification of **10**, effecting a separation that was not attainable by the previous RP-HPLC step. The solvent system was hexane/EtOAc/MeOH/H$_2$O (14:6:10:7; MP = UP). This isolation of **9** and **10** led to revision of the structure of bryostatin 3 to that shown here (**9**). This paper suggested that **10** may be an artifact produced by the isolation procedure. The protein kinase C binding was assayed by measuring the displacement of [^{3}H]phorbol-12,13-dibutyrate. Bryostatins 1–3 (**7**–**9**) showed comparable activity (IC$_{50}$ values 0.8, 2.5, and 6.6 ng/mL, respectively) while **10** was approximately an order of magnitude less active (29 ng/mL) than **7**–**9**.

Schaufelberger (19) also reported on the use of analytical HSCCC for the isolation of bryostatins. He used the solvent system hexane/2-PrOH/20% aq. MeOH (4:1:2; MP = LP) on a 10-mg sample of the crude extract of *B. neritina*.

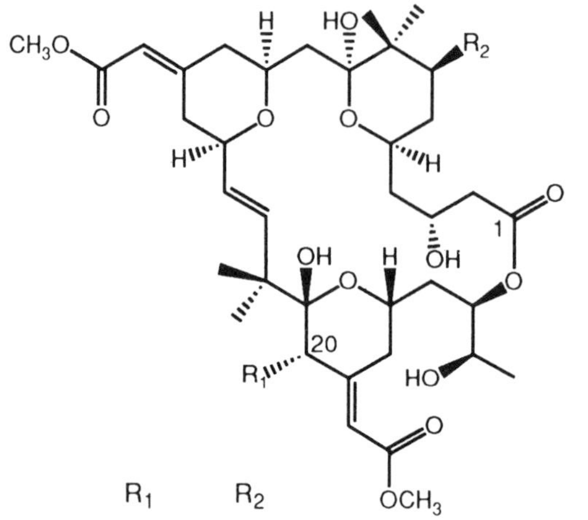

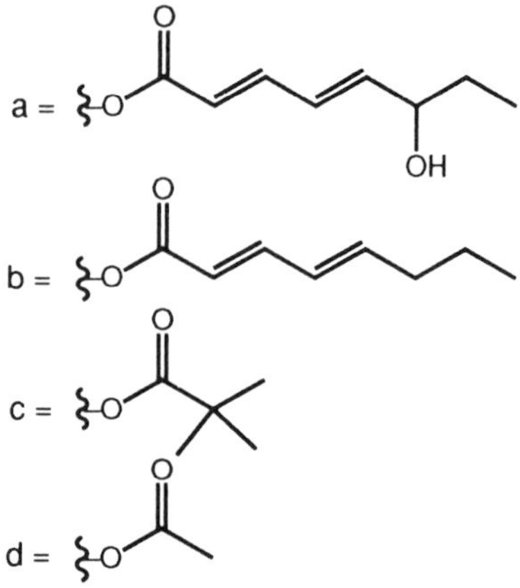

	R_1	R_2
5 :	OH	c
6 :	a	d
7 :	b	d
8 :	b	OH

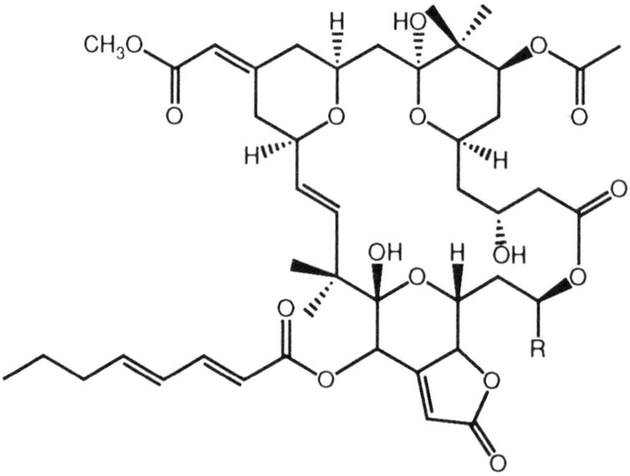

9 : R =

10 : R =

Fractions could then be analyzed by HPLC for the presence of **7**, eliminating several purification steps that previously had been required for this analysis.

Bryostatins are not the only macrolides found in *B. neritina*. Pettit et al. (24) also isolated the closely related neristatin 1 (**11**) while looking for other bryostatins for structure-activity analysis. Few details of the isolation were given other than the general methods employed. However, it appears that **11** was isolated from the same 1000-kg sample of bryozoan from which **5** was obtained (21). Thus, the same or a similar solvent system may have been employed. The 8.0-mg sample of **11** was found to decompose rapidly in several solvents, but was finally recrystallized from acetone–hexane, yielding crystals from which an X-ray structure was determined. Neristatin 1 was found to have a binding affinity for protein kinase C one and two orders of magnitude less than bryostatin 4 and phorbol 12,13-dibutyrate, respectively. Also, **11** showed only weak activity against P388 lymphocytic leukemia *in vitro* (IC$_{50}$ 10 μg/mL).

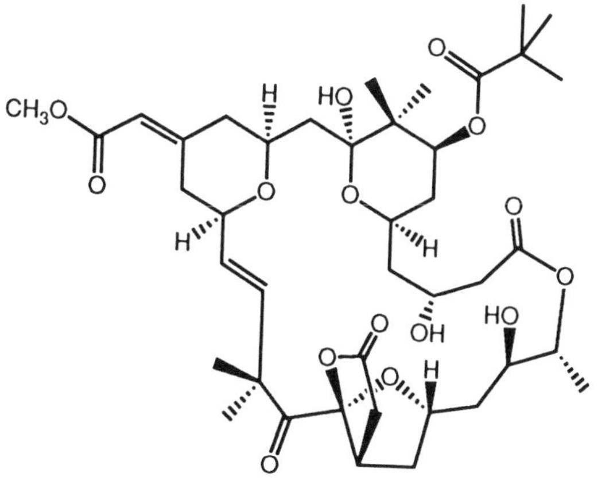

11

Faulkner and co-workers (25) examined the extracts of the Spanish dancer nudibranch *Hexabranchus sanguineus* and two *Halichondria* sp. sponges from Kwajalein Atoll, Marshall Islands and Palau, Western Caroline Islands. They isolated eight related macrolides (**12–19**), two of which were previously identified as kabiramide C (**12**) and halichondramide (**13**). Three of these compounds (**17–19**) appear to be oxidation artifacts of **13**. High-speed CCC was used to purify the crude extract of the *Halichondria* sp. from Kwajalein Atoll, employing the solvent system hexane/EtOAc/MeOH/H$_2$O (3:7:5:5; MP = UP). This yielded pure **17** and two fractions that were further purified by RP-HPLC to give **13**, **14**, **16**, **18**, and **19**. The purification of **15** from *H.*

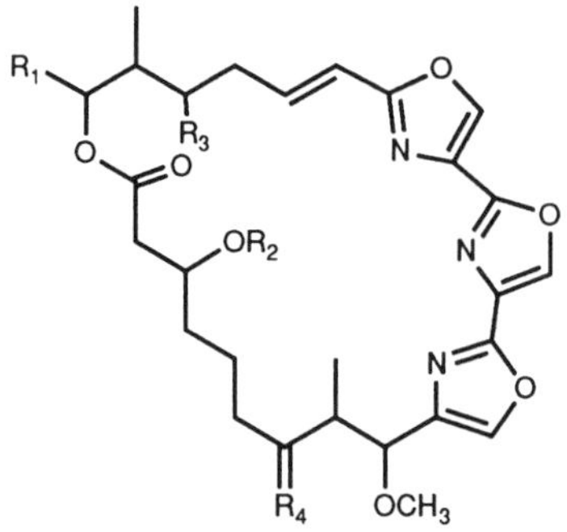

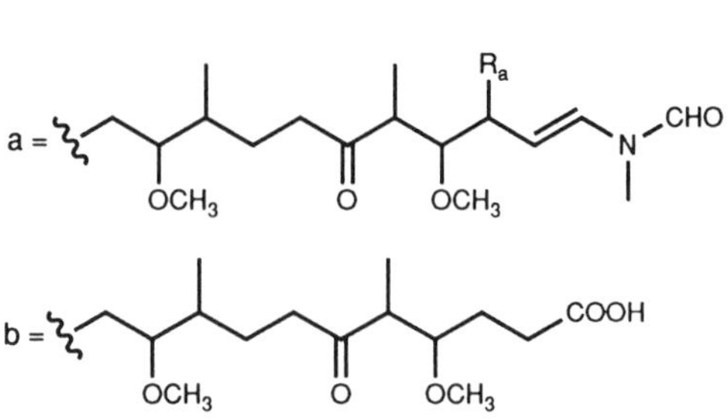

	R₁	R₂	R₃	R₄	Rₐ
12 :	a	CONH₂	OH	OH,H	CH₃
14 :	a	H	H	O	H
15 :	a	H	H	OH, H	H

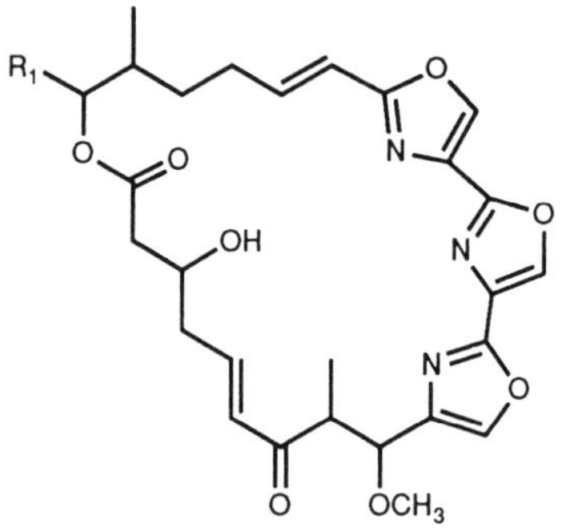

13 : R₁ = a, Rₐ = H
17 : R₁ = b

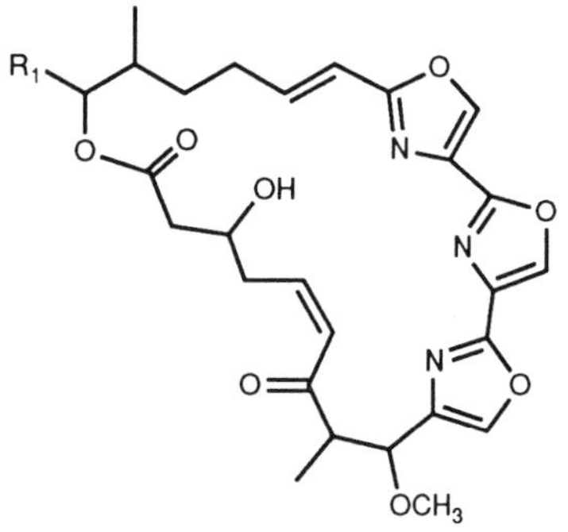

16 : R₁ = a, Rₐ = H

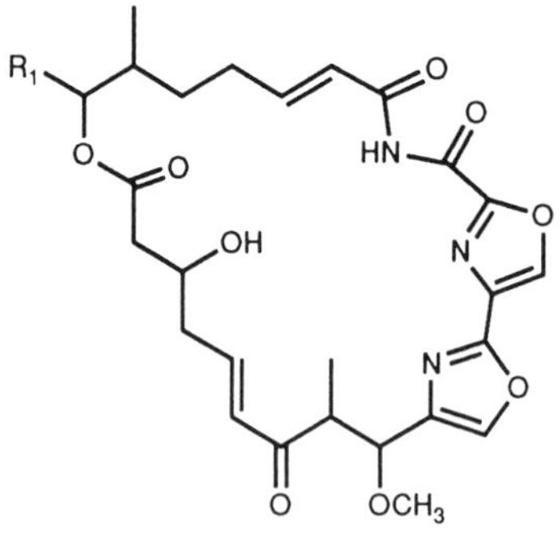

18 : R₁ = a, Rₐ = H

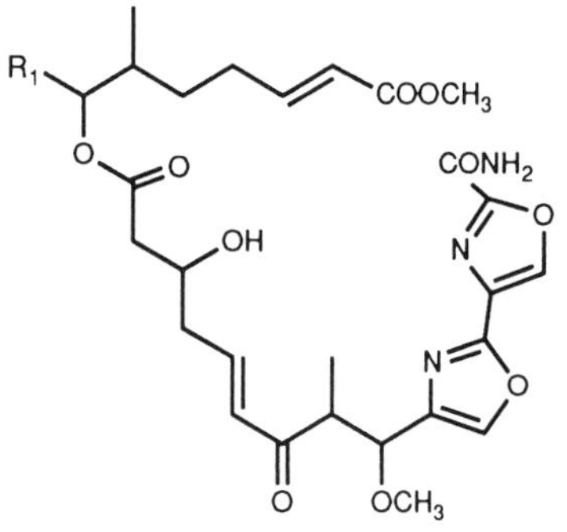

19 : R₁ = a, Rₐ = H

sanguineus did not involve HSCCC. These compounds showed antifungal activity against *Candida albicans* (*C. albicans*), especially **13**, which also showed activity against *Trichophyton mentagrophytes*. They also inhibited the cell division of fertilized sea urchin eggs.

9.4.2. Polyethers

Polyethers have been isolated from a number of marine sources. They are often highly toxic compounds that originate from dinoflagellates, but are accumulated by the other organisms through feeding. Because many of the cases of shellfish and other seafood poisoning result from contamination of the shellfish by these toxins, they have been extensively studied. These compounds are usually present in vanishingly small quantities (ppb range or less) so mild purification methods such as HSCCC with the least amount of sample loss are desirable.

Recently, okadaic acid (**20**) and dinophysistoxin-1 (**21**) have been reisolated from a new source, a *Phakellia* sp. sponge, along with the previously unknown 14,15-dihydrodinophysistoxin-1 (**22**) (26). Multilayer CPC was used in two different ways for this purification. First, the solvent system heptane/MeCN/CH$_2$Cl$_2$ (50:35:15; MP = LP) was employed to effect a rough separation of the bioactive components from the majority of the inactive material. The second HSCCC solvent system [heptane/EtOAc/MeOH/H$_2$O (4:7:4:3); MP = UP] cleanly and reproducibly separated **20** from **21** and **22**. This second system was repeated on the fractions containing **20** as a further purification step. These

20 : R = H
21 : R = CH$_3$

22

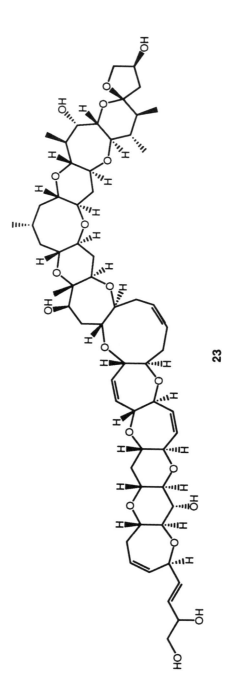

23

compounds all showed cytotoxicity against CV-1 monkey kidney cells, L1210 leukemia cells, and brine shrimp along with marginal antiviral activity against *Herpes simplex* virus type 1 (HSV-1) and vesicular stomatitis virus (VSV).

Another well-known polyether toxin is ciguatoxin (**23**), which has been implicated in about 20,000 cases of food poisoning a year (27). Although it has been studied for over 30 years, its structure was only recently solved by Yasumoto and co-workers (27) after the isolation of 0.35 mg from the viscera of about 4000 kg of the moray eel *Gymnothorax javanicus.* Scheuer's group has also studied this compound and they investigated the use of HSCCC for the purification of **23** (28). A number of different solvent systems were assayed by adding equivalent amounts of both phases to a crude sample. The layers were separated and both tested for toxicity in mice. The one that gave nearly equal toxicity in both layers was hexane/EtOAc/MeOH/H_2O (3:7:5:5). This solvent system was used for MLCPC (MP = LP).

9.4.3. Other Acetate-Derived Metabolites

Higa and co-workers (29) isolated onnamide A (**24**) from an Okinawan sponge, a *Theonella* sp. After chromatography on styrene–divinylbenzene resin and silica, the bioactive fractions were purified by HSCCC, using the solvent system $ClCH_2CH_2Cl/CHCl_3/MeOH/H_2O$ (2:3:10:6; MP = UP). This result-ed in 470 mg of pure **24**, which exhibited strong antiviral activity *in vitro* against HSV-1, VSV, and coronavirus A-59. This compound showed a marked resemblance to mycalamide A, from a New Zealand sponge (*Mycale* sp.), and pederin, from the terrestrial beetle *Paederus fuscipes.*

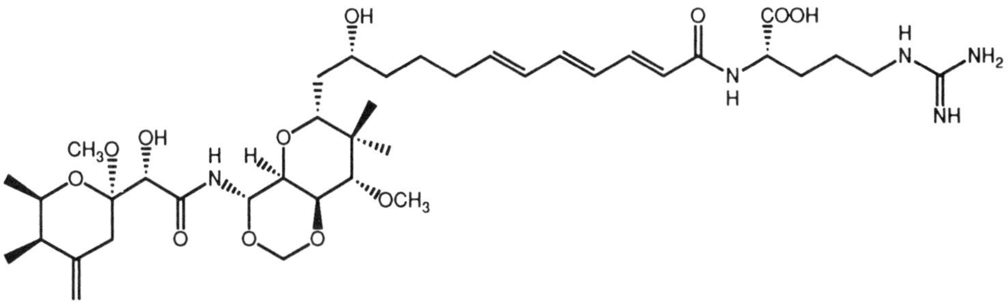

24

From another Okinawan sponge, *Plakortis lita,* Higa and co-workers (30) isolated a series of cyclic peroxides. The major bioactive compound was the previously known chondrillin (**25**), which was isolated with small amounts of four similar peroxides (**26–29**). These novel compounds were separated from **25** by HSCCC with the solvent system heptane/CH_2Cl_2/MeCN (5:1:4;

MP = UP). All of these compounds showed cytotoxicity against P388 leukemia cells, although **25** was significantly less active (IC$_{50}$ 5 μg/mL) than the C-6 epimers **26–29** (IC$_{50}$ 0.05–0.1 μg/mL).

	R$_1$	R$_2$
25 :	–(CH$_2$)$_{15}$CH$_3$	OCH$_3$
26 :	OCH$_3$	–(CH$_2$)$_{11}$CH$_3$
27 :	OCH$_3$	–(CH$_2$)$_9$
28 :	OCH$_3$	–(CH$_2$)$_7$
29 :	OCH$_3$	–(CH$_2$)$_9$

9.4.4. Terpenes and Steroids

Sun et al. (31) purified weinbersterol disulfates A (**30**) and B (**31**) from the sponge *Petrosia weinbergi*. After solvent partitioning of the crude extract, HSCCC was used with the solvent system CHCl$_3$/2-PrOH/MeOH/H$_2$O (9:2:12:10; MP = UP). Three bioactive fractions were further purified by RP-HPLC to give **30** and **31**. These compounds both showed antiviral activity against feline leukemia virus (EC$_{50}$ 4.0 and 5.2 μg/mL, respectively), while **30** also showed activity against the human immunodeficiency virus (EC$_{50}$ 1.0 μg/mL).

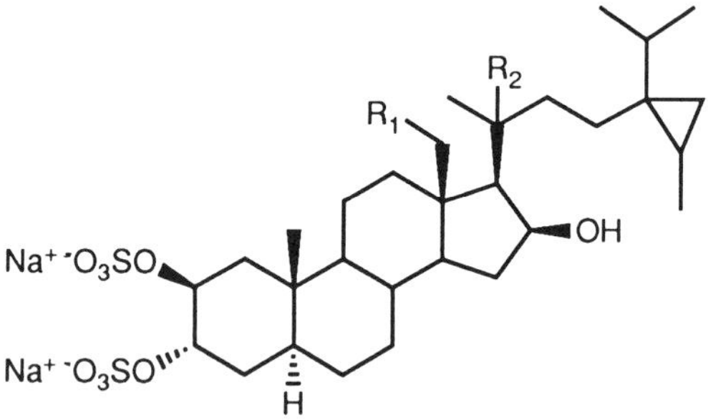

	R$_1$	R$_2$
30 :	H	OH
31 :	OH	H

Three bioactive tetracyclic furanoditerpenes have been isolated from a deep water Caribbean sponge, a *Spongia* sp. (32). Two of these compounds, spongiadiol (**32**) and epispongiadiol (**33**), had been isolated previously, while the third, isospongiadiol (**34**) was new. Repeated use of MLCPC with the solvent system heptane/EtOAc/MeOH/H_2O (4:7:4:3; MP = UP) gave pure **32** and **33** and another fraction, which upon recrystallization yielded pure **34**. The compounds **32**–**34** exhibited IC_{50} values of 0.5, 8, and 5 μg/mL against P388 cells and 0.25, 12.5, and 2 μg/mL against HSV-1, respectively.

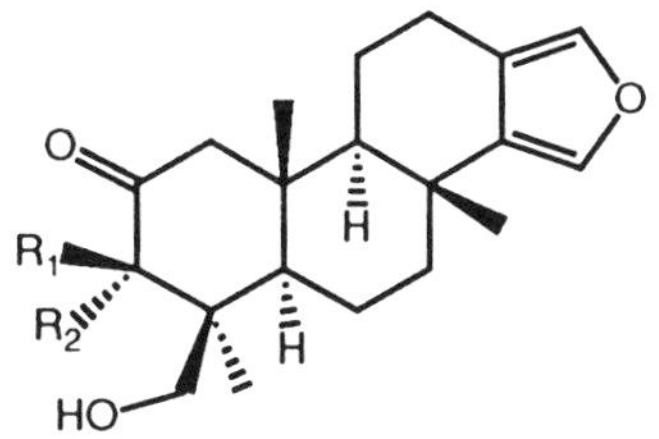

32 : R_1 = H, R_2 = OH
33 : R_1 = OH, R_2 = H

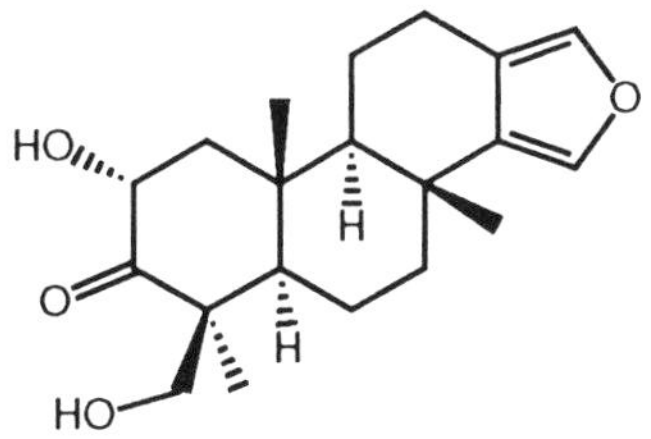

34

The sesquiterpene quinone methide puupehenone (**35**) has been reisolated from a new source, the deep water sponge *Strongylophora hartmani* van Soest (33), after previously having been found in several different sources by Scheuer and co-workers (34). Pure **35** was obtained from the crude extract by one chromatographic run on an MLCPC instrument with the solvent system heptane/CH_2Cl_2/MeCN (10:3:7; MP = UP). This compound showed activity against P388, A-549, HCT-8, and MCF-7 cancer cells, with IC_{50} values ranging from 0.1 to 10 μg/mL, and against *C. albicans*, with a minimum inhibitory concentration (MIC) of 3 μg/mL.

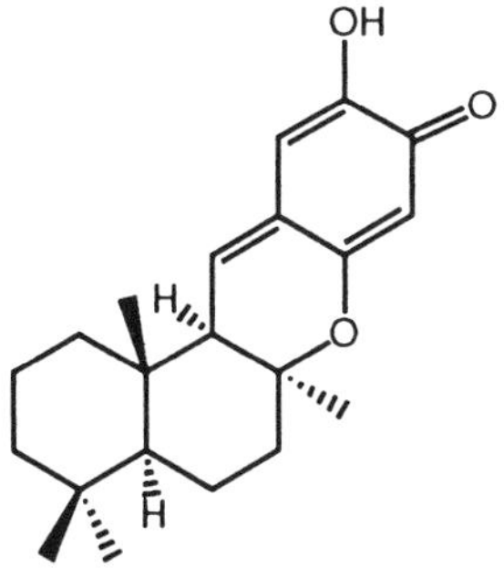

35

9.4.5. Peptides

As previously noted (Section 9.1), the didemnins are a series of cyclic depsipeptides with potent antitumor, antiviral, and immunosuppressive activities.
More polar fractions left from the isolation of didemnin B (**4**) have been
investigated for other, novel didemnins (26, 35). One such fraction from a silica
column was subjected to HSCCC, using the solvent system toluene/EtOAc/
MeOH/H_2O (6:7:7:4; MP = LP). This gave a series of four fractions that after
further purification yielded didemnins D (**36**), E (**37**), Y (**38**), and X (**39**),
respectively. Another side fraction, produced during the large-scale isolation
of **4** for clinical trials, was further fractionated by MLCPC with the solvent
system cyclohexane/EtOAc/toluene/MeOH/H_2O (2:7:2:4:4; MP = UP). The
second of five fractions thus produced was purified to give isodidemnin A_1 (**40**),
didemnin N (**41**), and nordidemnin N (**42**). Isodidemnin A_1 appeared to be
a stable conformer of didemnin A (**43**), which could be isolated separately from
43, but interconverted with it on silica.

Semisynthetic analogs of the didemnins have also been investigated (26).
In order to synthesize analogs with different side chains, **43** was used as
a starting material, since it has a D-MeLeu unit with a free amino group
attached to the ring. Didemnin A was purified from a sample that also
contained nordidemnin A (**44**), a compound that only differs from **43** by one
methyl group. High-speed CCC separated these two compounds with very
little overlap. By using the solvent system heptane/EtOAc/MeOH/H_2O
(4:7:4:3; MP = UP), 30 fractions were produced from a 413-mg sample.
Fractions 16–23 contained pure **43** (192 mg), while fractions 26–30 yielded
pure **44** (85 mg). The intervening fractions 24 and 25 contained a mixture of the
two (32 mg).

Several cyclic dipeptides (diketopiperazines) have been isolated from the
sponge *Tedania ignis* and an associated microorganism. Cardellina and co-
workers (36) isolated a *Micrococcus* sp. growing on this sponge and also
isolated several other compounds from both of these organisms (see Section 9.4.6). The marine bacterium was cultured and the resulting broth was
examined for secondary metabolites. A partially purified fraction was subjected to HSCCC, using the solvent system $CHCl_3$/MeOH/H_2O (25:34:20). No
indication of which phase was the mobile phase was given. This resulted in the
isolation of three diketopiperazines, *cyclo*-(Pro-Leu) (**45**), *cyclo*-(Pro-Val) (**46**),
and *cyclo*-(Pro-Ala) (**47**). These compounds were previously isolated from *T.
ignis* by Schmitz et al. (37), suggesting that **45**–**47** were actually produced by
the bacterium and accumulated by the sponge. The stereochemistry was not
reported, but **45**–**47** exhibited spectral characteristics identical to those of the
compounds isolated by Schmitz's group, which had found that they were
derived from L-amino acids.

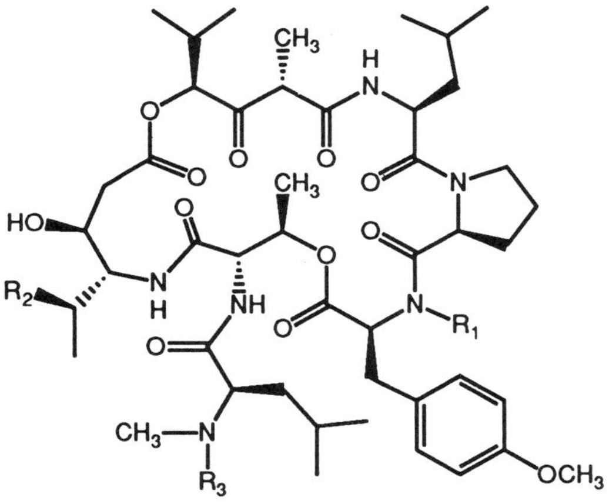

	R_1	R_2	R_3	n
4 :	CH_3	CH_2CH_3	a	—
36 :	CH_3	CH_2CH_3	b	2
37 :	CH_3	CH_2CH_3	b	3
38 :	CH_3	CH_2CH_3	c	3
39 :	CH_3	CH_2CH_3	c	4
41 :	H	CH_2CH_3	a	—
42 :	H	CH_3	a	—
43 :	CH_3	CH_2CH_3	H	—
44 :	CH_3	CH_3	H	—

279

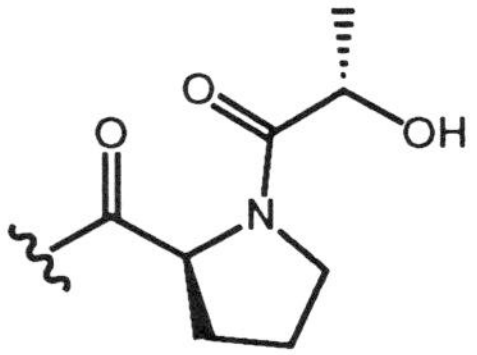

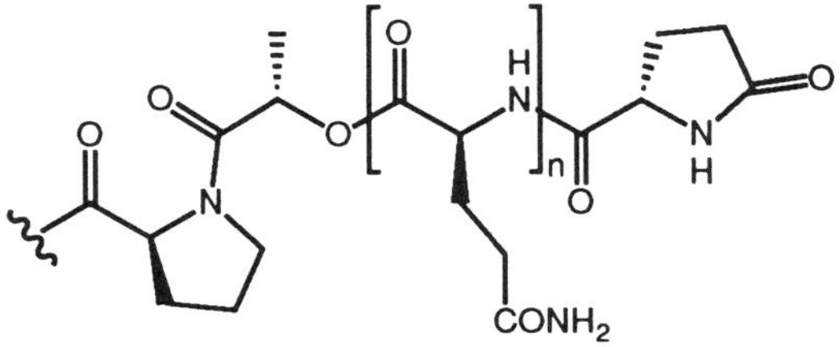

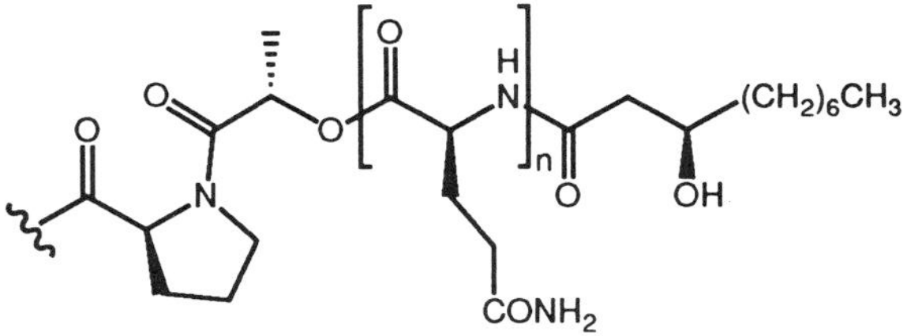

Cardellina's group (38) further investigated the metabolites of the sponge *Tedania ignis* and found two more diketopiperazines, *cyclo*-(L-Pro-L-Phe)(**48**) and *cyclo*-(L-Pro-L-thiaPro)(**49**). Other compounds isolated from this sponge by these authors are discussed in Section 9.4.6. High-speed CCC was employed as the final purification step after three gel permeation chromatography columns. The solvent system used was $CHCl_3/MeOH/H_2O$ (25:34:20). While no direct indication of which layer was used as the mobile phase, it was reported to be in the "ascending" mode, a droplet CCC term suggesting that the upper phase was the mobile phase. Cardellina and co-workers (39) previously isolated **48** from the fungus *Alternaria alternata*. The sulfur-containing **49** was a novel compound whose absolute stereochemistry was confirmed by synthesis. This compound showed no activity in brine shrimp cytotoxicity, phytotoxicity, plant growth regulatory, antimicrobial, or insecticidal assays.

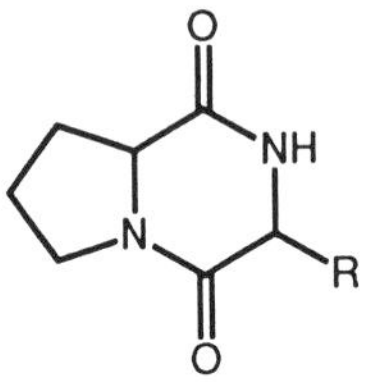

45: R = CH₂CH(CH₃)₂
46: R = CH(CH₃)₂
47: R = CH₃

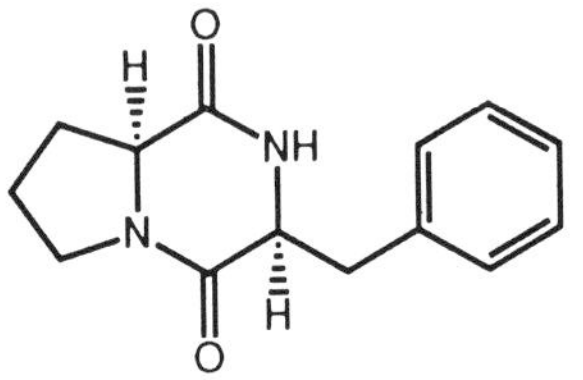

48

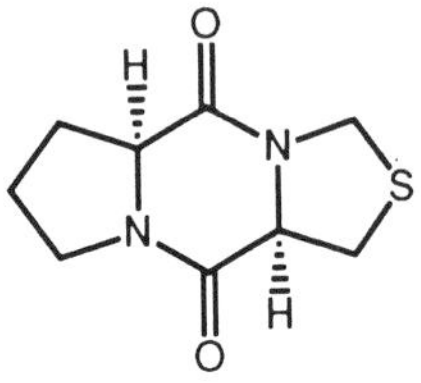

49

9.4.6. Nitrogen-Containing Heterocycles

Probably the largest and most diverse group of marine natural products studied so far consists of nitrogen-containing heterocycles. These compounds frequently exhibit biological activity and numerous examples of the use of HSCCC for their isolation can be found. Countercurrent chromatography may be of particular use for the purification of nitrogen-containing heterocycles since these compounds often tail badly on LSC. The discussion that follows has been divided according to the ring systems involved, but the subheadings are by no means mutually exclusive. Reduced or partially reduced systems are included with their aromatic counterparts.

9.4.6.1. *Benzothiazoles*

As noted in Section **9.4.5**, Cardellina and co-workers (40) isolated a bacterial symbiont from the sponge *Tedania ignis*. Further investigation of this *Micrococcus* sp. resulted in the isolation of four benzothiazoles (**50–53**). This communication noted that HSCCC was used for the final step in the isolation, but no further details were given. Although **50** and **51** had previously been isolated and all had been synthesized, this was the first report of the isolation of benzothiazoles from a marine source. It was suggested that a bacterial source of this type of compound might be useful as a chemical feedstock for the industrial production of aldose reductase inhibitors.

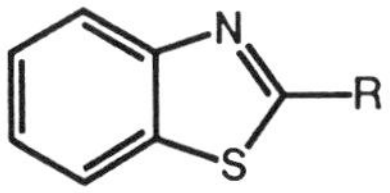

50 : R = SH   **53**
51 : R = CH_3
52 : R = OH

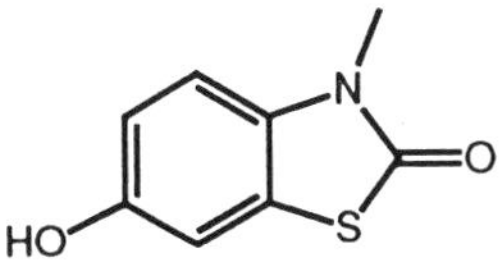

9.4.6.2. Quinolizidines

The first quinolizidines from a tunicate were isolated by Cardellina et al. (41), clavepictines A (**54**) and B (**55**) from the Bermuda tunicate *Clavelina picta*. High-speed CCC was used as the final purification step with the solvent system hexane/CH_2Cl_2/MeCN (10:3:7), but the mobile phase was not indicated. Clavepictine B produced crystals suitable for X-ray analysis, which established the relative, but not absolute, stereochemistry. These compounds exhibited cytotoxicity against a series of murine leukemia and human solid tumor cell lines (P388, A-549, U-251, and SN12K1), with IC_{50} values ranging from 1.8 to 8.5 μg/mL.

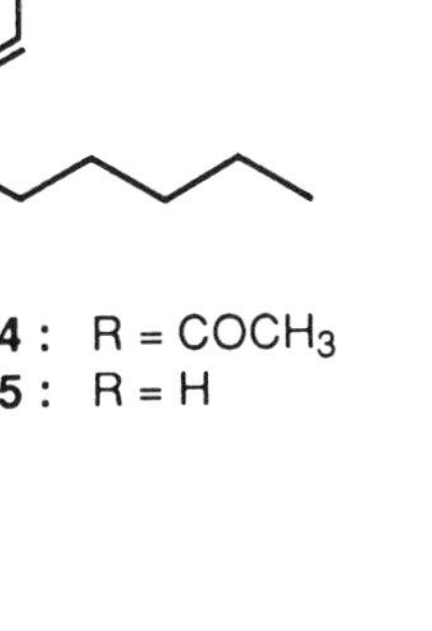

54 : R = COCH_3
55 : R = H

9.4.6.3. Pyridines

A series of pyridines substituted at the β position with long-chain methoxylamines have been isolated from a sponge collected in the Bahamas (42). Sun and co-workers (42) purified xestamines A (**56**), B (**57**), and C (**58**) from *Xestospongia wiedenmayeri*. Multilayer CPC with the solvent system heptane/CH_2Cl_2/MeCN (10:3:7; MP = UP) was used to separate **56** from the other two compounds, all of which were further purified by HPLC. Although **56–58** were structurally related to the cytotoxic niphatynes A and B, they showed no activity versus P388 cells.

Stierle and Faulkner (43) isolated other compounds in this series from the Bahamian sponge *Calyx podatypa*. Xestamines A and B were reisolated along with the new xestamines D (**59**) and E (**60**) and the *N*-methylpyridinium salts xestamines F–H (**61–63**). These workers used the same MLCPC solvent system as Sun and co-workers (42), but no mobile phase was indicated. This gave **56**, **57**, and **61** in pure form and a mixture of **59** and **60**, which could only be separated by GC. High-speed CCC was not used for the purification of **62** and **63**, which were also isolated as an inseparable mixture. The novel compounds, some as mixtures, were assayed for antimicrobial activity and cytotoxicity. In these assays, the pyridinium salts were about 100 times more active as antimicrobial agents, but about 100 times less active against brine shrimp.

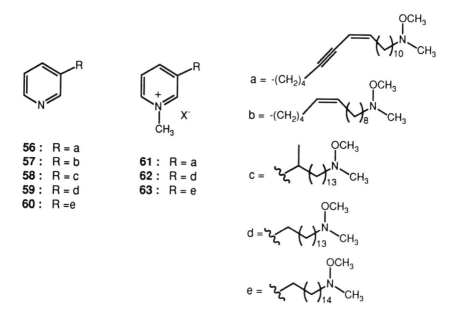

56 : R = a
57 : R = b
58 : R = c
59 : R = d
60 : R = e

61 : R = a
62 : R = d
63 : R = e

9.4.6.4. Indoles and β-Carbolines

A common type of nitrogen-containing heterocycles consists of indole deriva-tives. Kohmoto et al. (44) isolated 1,1-dimethyl-5,6-dihydroxyindolinium chlor-ide (64) from a deep water sponge, a *Dercitus* sp. This simple indolinium salt was isolated along with another nitrogen-containing heterocycle described below. The crude methanol extract was purified by HSCCC with the solvent system $CHCl_3/MeOH/H_2O$ (5:10:6) to give 64 after recrystallization. The authors did not indicate which phase was used as the stationary phase.

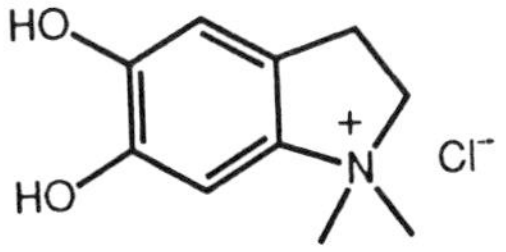

64

Sun and Sakemi (45) purified a brominated indole from the deep-water sponge *Discodermia polydiscus*. They used HSCCC with the solvent system $CHCl_3/MeOH/H_2O$ (5:10:6; MP = UP) to fractionate the crude extract. After Sephadex LH-20 chromatography and HPLC on an amino column, they isolated the novel compound discodermindole (65). This compound showed cytotoxicity against P388, A-549, and HT-29 cancer cell lines, with IC_{50} values of 1.8, 4.6, and 12 $\mu g/mL$, respectively. The absolute stereochemistry of 65 was not assigned.

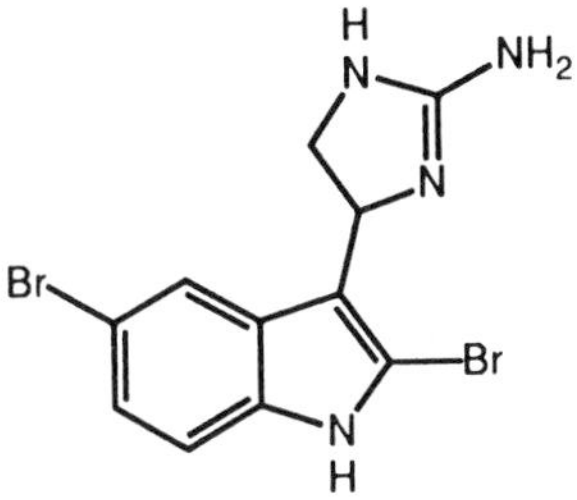

65

Dillman and Cardellina (46) isolated a variety of indole derivatives in their further investigation of the sponge *Tedania ignis* (see above for other com-pounds isolated from this sponge and an associated bacterium). High-speed CCC with the solvent system $CHCl_3/MeOH/H_2O$ (25:34:20; MP = LP) was used to purify 1-acetyl-β-carboline (66), 1-methylcarbazole (67), 3-(hy-droxyacetyl)-indole (68), and 6-(indole-3-yl)-5-methylhepta-(3E,5E)- dien-2-one (69). Four other compounds were also isolated, but without using

MLCPC. While **69** was a novel compound, **66–68** had previously been isolated. Compound **68** was from a *Micrococcus* sp. (bacterium), which was found in association with the sponge and the other two were from terrestrial sources. Since these compounds were isolated in very small amounts and several were similar, if not identical to, bacterial products, the authors suggested that they were produced by microbes associated with the sponge. Several of these compounds exhibited cytotoxicity against brine shrimp.

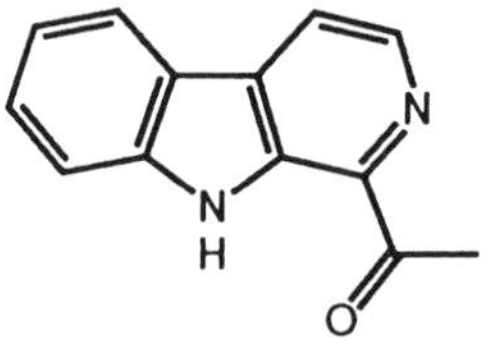

66 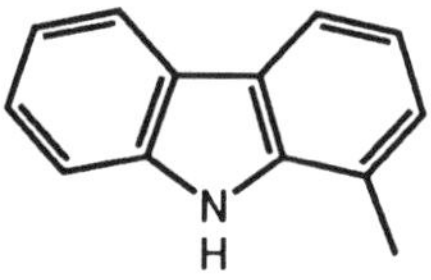   **67**

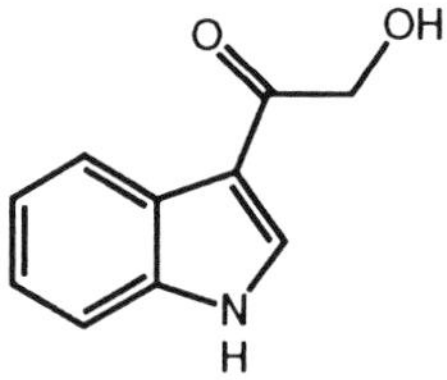

68 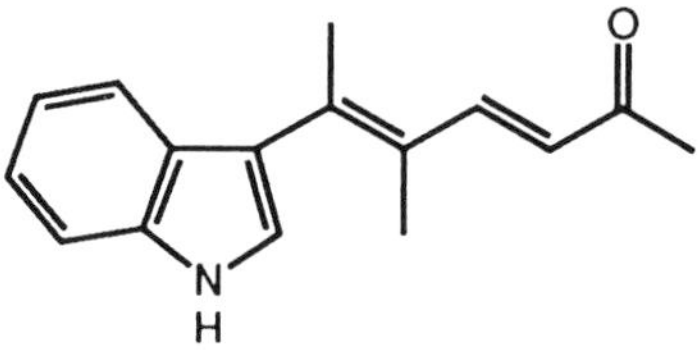   **69**

More complex 1-substituted β-carboline derivatives have been described by Higa and co-workers (47), from an Okinawan sponge, a *Haliclona* sp. Repeated MLCPC separations of the mother liquor resulting from recrystallization of the previously known manzamine A (**70**) yielded, after further purification, manzamines B (**71**) and C (**72**) as minor components. The solvent system employed here was heptane/EtOAc/MeOH/H_2O (4:7:4:3; MP = UP). The structures of **71** and **72** were solved by X-ray crystallography; both showed moderate activity against P388 cells *in vitro*, with IC_{50} values of 6 and 3 $\mu g/mL$ respectively.

Bobzin and Faulkner (48) isolated a series of tryptophan- and tyrosine-derived heterocycles from a sponge of the genus *Chelonaplysilla*, formerly classified as a *Dendrilla* sp. Chelonins A (**73**) and C (**74**) possessed 2,6-disubstituted morpholine rings, a structural subunit not observed from a natural source before this report, while chelonin B (**75**) and its bromo derivative (**76**) were ring-opened versions of this type of compound. After extraction,

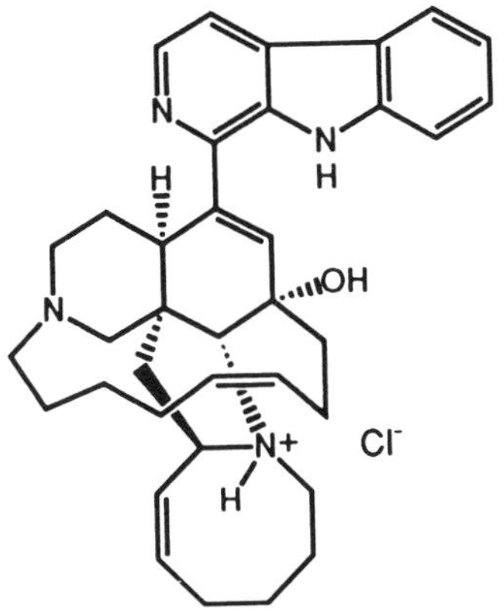

70

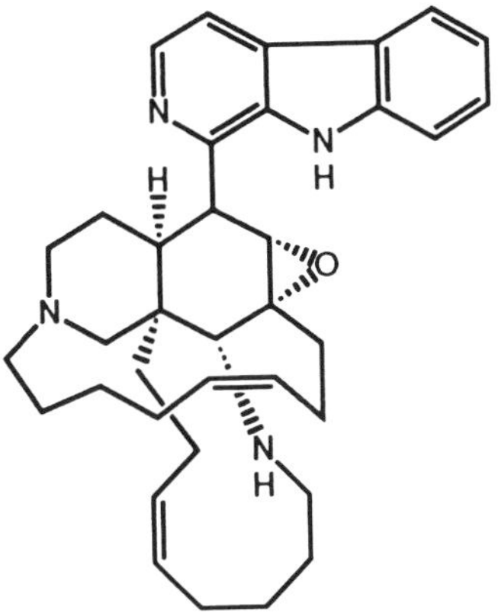

71

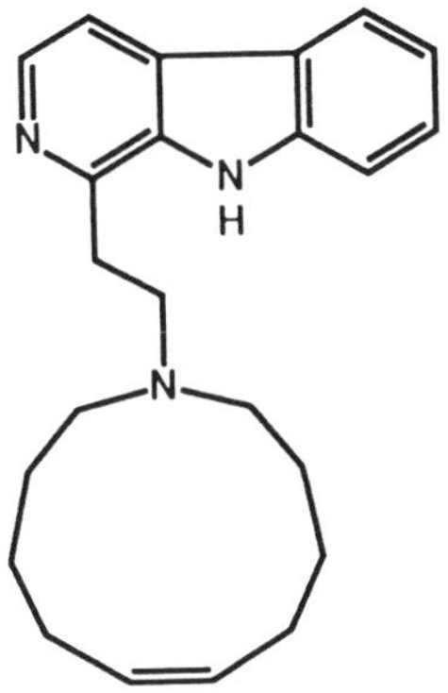

72

chromatography over a Sephadex LH-20 column and a silica flash column gave three bioactive fractions, A–C. The first was separated by HSCCC, using the solvent system hexane/EtOAc/MeOH/H$_2$O (3:7:5:5; MP = LP), which yielded pure **73**. This same solvent system was used on fraction B, yielding two new antimicrobial fractions, D and E. Fraction D was subjected to three more MLCPC separations, using first the same solvent system but with MP = UP, then the original solvent system (MP = LP), and finally EtOAc/95% EtOH/H$_2$O (2:1:2; MP = UP). The resulting bioactive fraction was finally purified by HPLC to give **75**. Fractions C and E were both purified by amino HPLC to give **73** and **74** from the former and **76** from the latter. All of these compounds except **74** exhibited antimicrobial activity against *Bacillus subtilis* at 100 μg/disk. Chelonin A (73) also inhibited 60% of the PMA-induced inflammation of the mouse ear at 50 μg/ear.

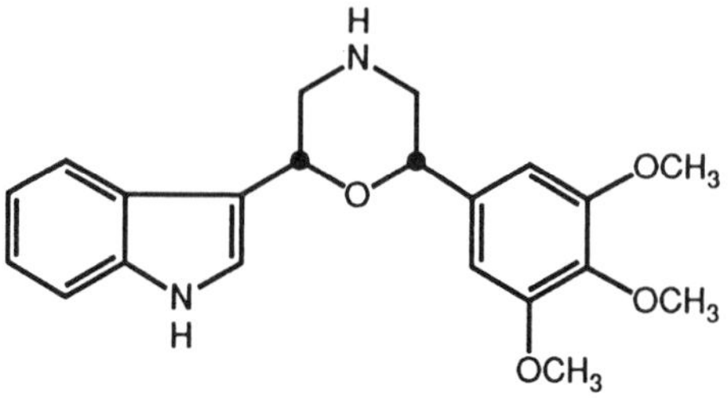

73

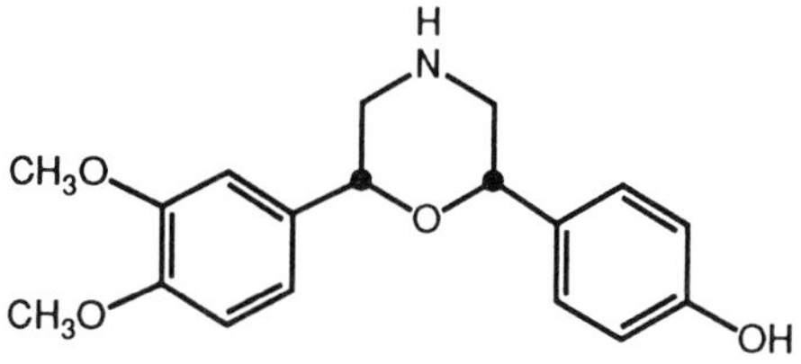

74

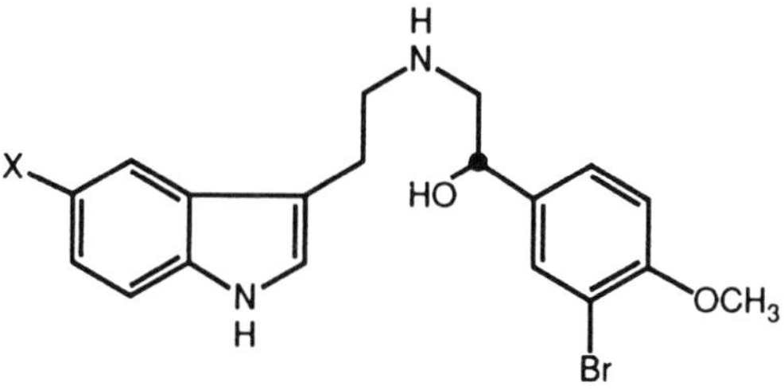

75 : X = H
76 : X = Br

Sakemi and Sun (49) isolated a series of bisindoles from the deep-water sponge *Spongosorites ruetzleri*. Nortopsentins A–C (77–79) were purified, along with the previously known (50) topsentin (80) and bromotopsentin (81). During the isolation, the authors noted that a fluorescent thin-layer chromatography (TLC) spot changed from colorless to reddish brown on exposure to light. This spot proved to be an overlapping mixture of 77–79, which could not be separated by conventional LSC (51). To avoid decomposition, the isolation was performed in dim light. Repeated HSCCC using the solvent systems heptane/EtOAc/MeOH/H$_2$O (4·7:4:3 and 5:7:4:3; MP = UP) resulted in the separation of **77** from a mixture of **78** and **79**, which could then be separated by preparative TLC. Compounds **80** and **81** were also isolated using MLCPC with an unspecified solvent system. These compounds were separable on C18, but their poor solubility limited the injection size for HPLC. With HSCCC, injections of up to 300 mg, sometimes even as a suspension, could be used, in contrast to a few milligrams per HPLC run (51). All five of these compounds showed cytotoxicity against P388 cells with IC$_{50}$ values of 7.6, 7.8, 1.7, 2.0, and 7.0 μg/ml for **77–81**, respectively. The new compounds also showed antifungal activity against *C. albicans* with MIC values of 3.1, 6.2, and 12.5 μg/mL for **77–79**, respectively, but did not exhibit any of the antiviral activity that **80** and **81** possessed.

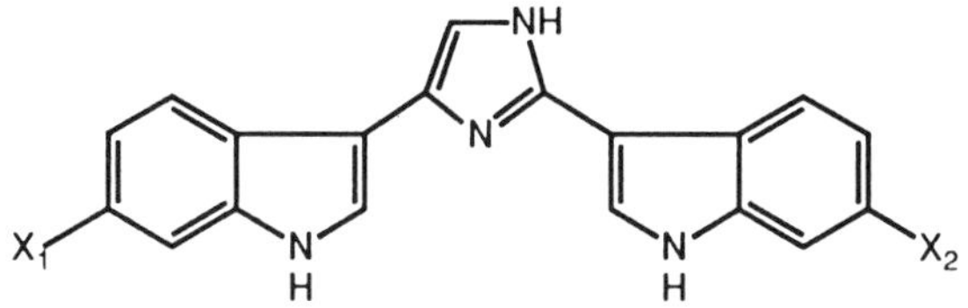

77 : X$_1$ = X$_2$ = Br
78 : X$_1$ = Br, X$_2$ = H
79 : X$_1$ = H, X$_2$ = Br

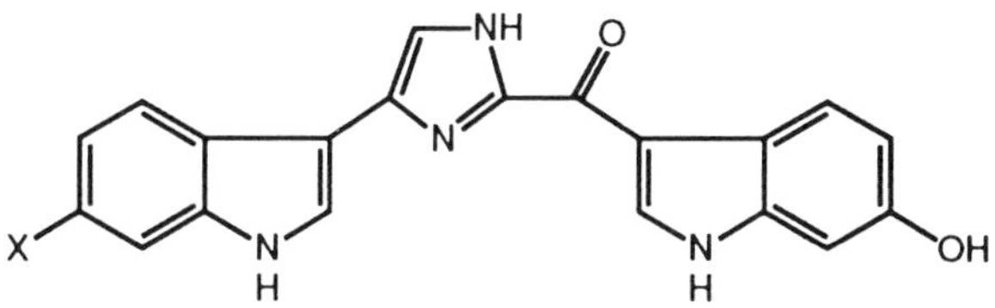

80 : X = H
81 : X = Br

9.4.6.5.   Pyrroles, Imidazoles

Sceptrin diacetate (**82**) has been repurified by HSCCC (52). Previously, our group used countercurrent distribution (CCD) a progenitor of HSCCC to isolate eight bromopyrroles from the sponges *Agelas conifera* and *Agelas* n.

sp., collected in the Caribbean (53). After initial extraction and trituration with methanol and acetone, the crude material was partitioned by CCD, using the solvent system $H_2O/1$-$BuOH/AcOH/EtOH$ (10:8:1:1). The resulting fractions were further purified by RPHPLC to give the known sceptrin diacetate (**82**) and seven other novel oroidin dimers (**83–89**). These compounds possess antiviral and antibacterial activity and are active in barnacle settlement and biochemical prophage induction assays. In order to obtain more **82** for further study, a sample enriched in this compound from the original isolation was subjected to HSCCC, using the solvent system $CH_2Cl_2/MeOH/H_2O$ (2:1:2) with the lower phase as the stationary phase, followed by HPLC. This work used a prototype analytical column built for us by W. D. Conway, State University at Buffalo, for use with our MLCPC (P. C. Inc.) in place of a preparative coil.

Schaufelberger and Pettit (54) used HSCCC to isolate the imidazolyl-pyrroloazepines, axinohydantoin (**90**), hymenialdisine (**91**) and debromohy-menialdisine (**92**) from sponges of the genera *Axinella* and *Hymeniacidon* (23, 55). These compounds exhibit low solubility in a number of solvents, thus making them difficult to isolate. They are most soluble in basic aqueous solutions, so an HSCCC solvent system using a phosphate buffer (pH 11.2) was developed. A 1:1 mixture of 1-BuOH and the buffer solution (0.01 M K_3PO_4 and 0.01 M K_2HPO_4) was used (MP = LP). This resulted in base line separation of the three compounds, which could be obtained in pure form by precipitation after adjusting the pH of the HSCCC fractions. The solvent system 1-BuOH/AcOH/H_2O (4:1:5; MP = LP) was also used to isolate **91** and **92** from *A. carteri* (55). The structure of the novel **90** was solved by X-ray crystallography, while the other two were known compounds. All three showed activity against PS leukemia *in vitro* with IC_{50} values of 18, 2.0, and 2.5 μg/mL for **90–92**, respectively.

9.4.6.6. *Condensed Polycyclic Aromatic Heterocycles*

A number of biologically active polycyclic aromatic nitrogen-containing compounds with condensed rings have been isolated from marine organisms. Schmitz et al. (56, 57) isolated meridine (**93**) and isomeridine (**94**) from the ascidian *Amphicarpa meridiana*. The former compound was purified both by a silica chromatography-based separation scheme and by HSCCC, while **94** was only obtained from the first isolation. The MLCPC solvent system used was $CHCl_3/MeOH/5\%$ aq. HCl (5:5:3; MP = LP). Isomeridine was probably not isolated from the HSCCC separation because it tautomerizes to **93** on standing in chloroform solution. An X-ray crystal structure was obtained of the trifluoroacetate salt of **93**. Meridine was found to inhibit topoisomerase II at 75 μM and to be cytotoxic towards P388 leukemia cells at

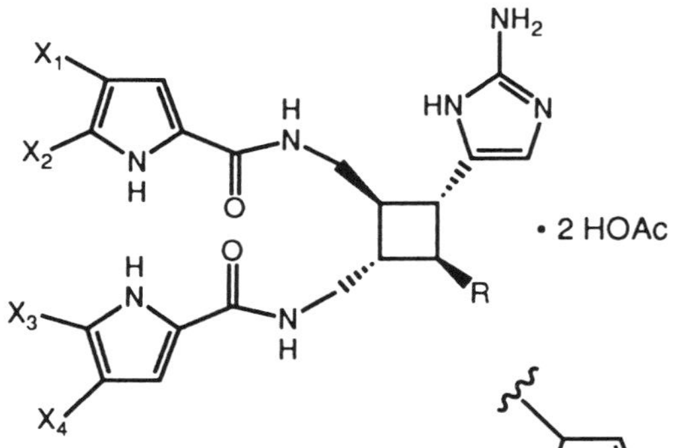

• 2 HOAc

	X_1	X_2
87:	H	H
88:	Br	H
89:	Br	Br

• 2 HOAc

A =

B =

	X_1	X_2	X_3	X_4	R
82:	Br	H	H	Br	A
83:	Br	H	H	H	A
84:	H	Br	Br	Br	A
85:	Br	H	H	Br	B
86:	Br	H	H	Br	B

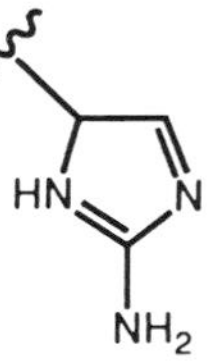

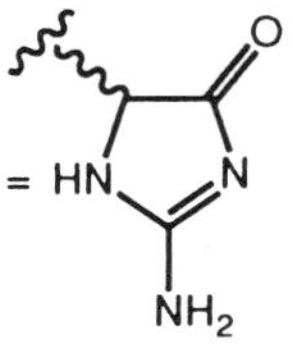

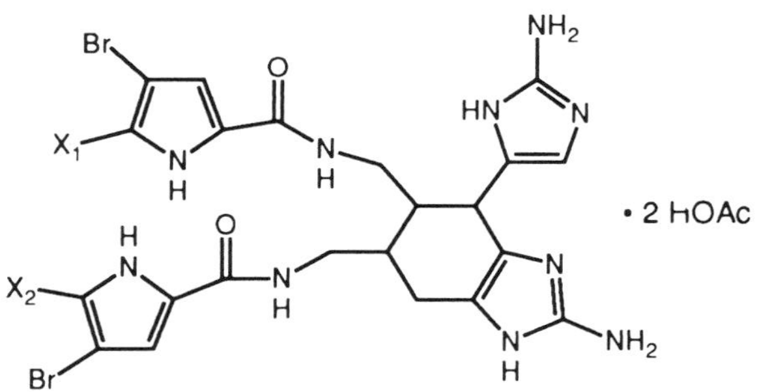

289

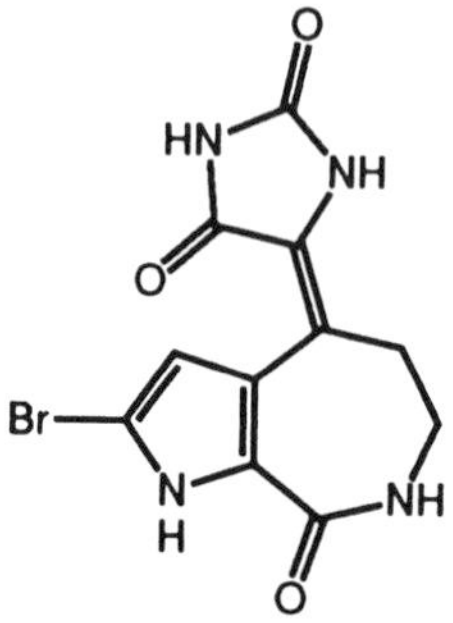

90

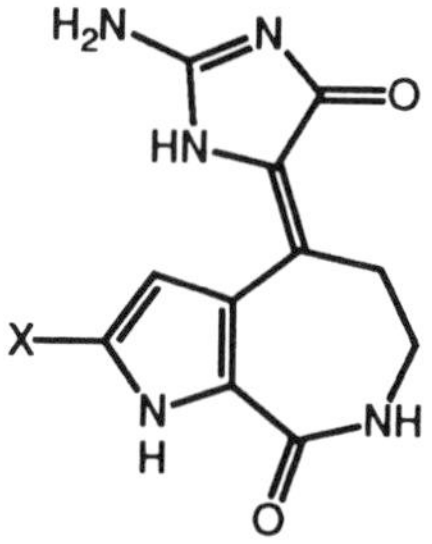

91 : X = Br
92 : X = H

0.3–$0.4\,\mu g/mL$. Because **94** tautomerized so readily, it was not assayed for biological activity.

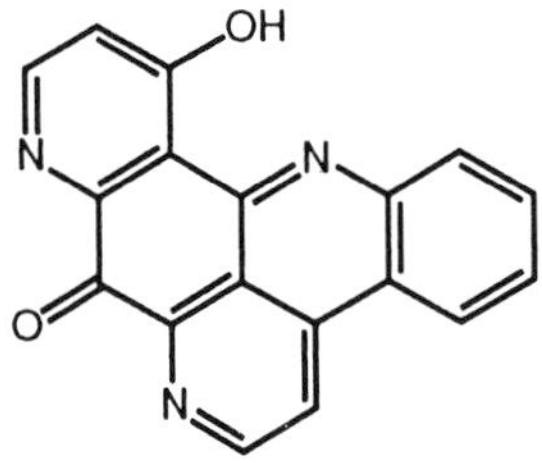

93 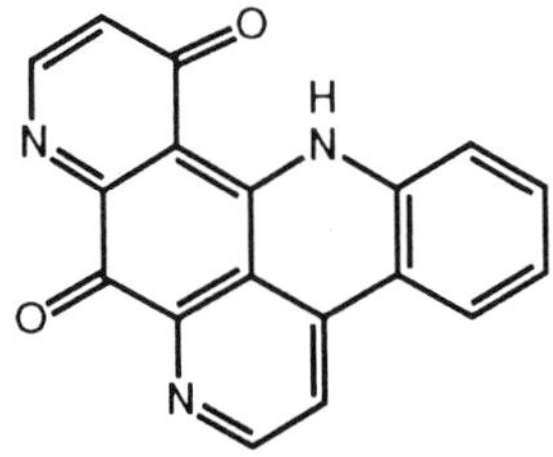   **94**

Another pentacyclic nitrogen-containing compound from a marine source is dercitin (**95**), the first compound in this class shown to contain a thiazole. Gunawardana et al. (58, 59) reported this violet pigment from a *Dercitus* sp., a deep-water sponge. While the authors indicated that they used HSCCC to optimize the yield of **95**, no solvent system was reported. This compound was isolated at the same time as the indolinium salt **64**, so possibly the same system was used (see above) (44). Dercitin exhibited cytotoxicity against P388, HCT-8, A-549, and T47D cell lines, with IC_{50} values of 0.05–$1.0\,\mu g/mL$. It also showed immunosuppressive activity in the two-way mixed lymphocyte reaction (MLR) assay (0% MLR at $0.01\,\mu g/mL$) and antiviral activity against HSV-1 and A-59 viruses at 5 and $1\,\mu g/mL$, respectively.

Further investigation of this *Dercitus* sp. sponge along with a *Stelletta* sp. sponge resulted in the reisolation of **95** along with a series of related compounds (59, 60). The isolation procedure for both sponges involved using HSCCC with the solvent system $MeOH/CH_2Cl_2/H_2O$ (5:5:3; no MP indicated) followed by amino HPLC. The first sponge produced dercitin (**95**), cyclodercitin (**96**), and two N-oxides of **95**, while the *Stelletta* sp. gave nordercitin (**97**), dercitamine (**98**) and dercitamide (**99**). Cyclodercitin was observed to be oxidized to the fully aromatic compound **100**. All of these compounds exhibited cytotoxicity against P388 cells, with IC_{50} values of 0.08, 1.9, 4.79, 26.7, 12.0, and 9.89 μM for **95–100**, respectively. Immunosuppressive activity was also reported for **97–99**. Originally the wrong regiochemistry for the thiazole group was proposed for **95–100** (58, 60). An X-ray crystal structure for stellettamine (**101**), recently isolated from the *Stelletta* sp. sponge, and further NMR studies on these compounds led to reassignment to the structures shown here (59).

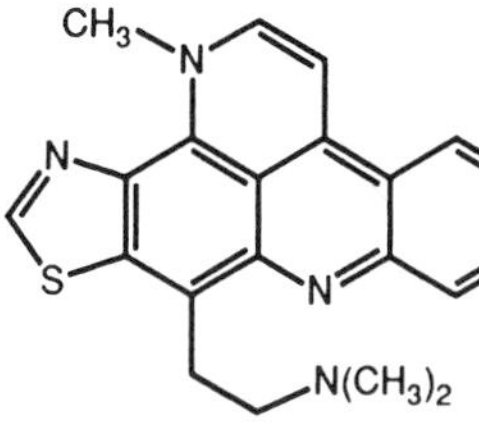

95 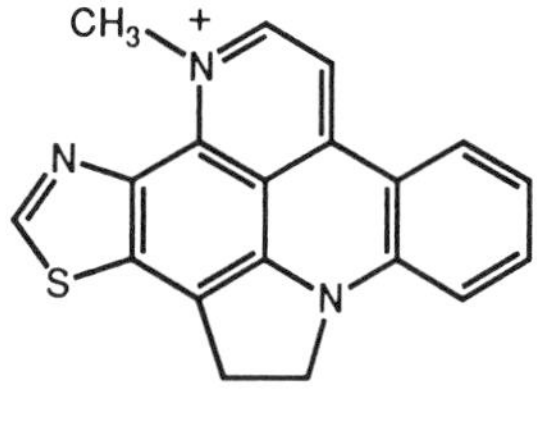   96 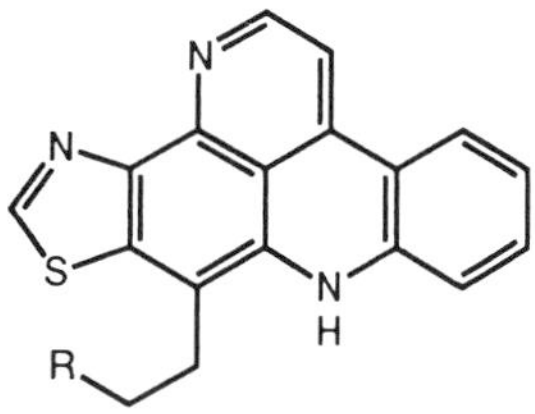   97 : R = N(CH₃)₂
                                                    98 : R = NHCH₃
                                                    99 : R = NHCOCH₂CH₃

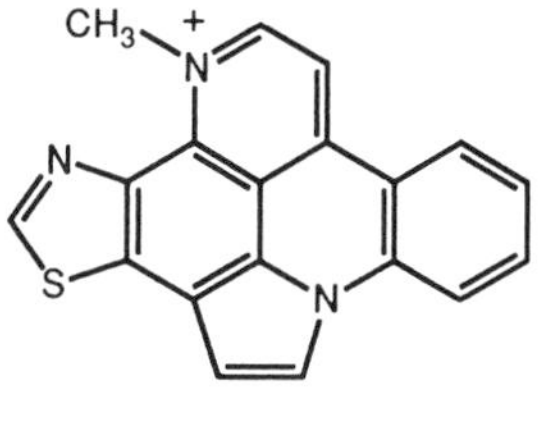

100 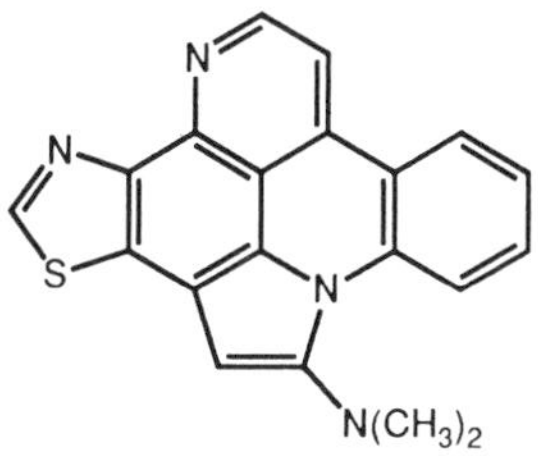   101

Another series of fused nitrogen-containing heteroaromatic compounds, from the deep-water sponge *Batzella* sp., has been reported by Sun and co-workers (61). Batzelline A (**102**) was purified by two HSCCC runs with the solvent systems heptane/EtOAc/MeOH/H₂O (4:7:4:3 and 2:7:6:3; both MP = UP). A third MLCPC separation using the system $CHCl_3$/diisopropylamine/MeOH/H₂O (7:1:6:4; MP = UP) on a fraction from the second MLCPC run gave batzellines B (**103**) and C (**104**). The structure of the major component **102** was solved by X-ray crystallography. These compounds may be biosynthetic precursors to the discorhabdins, another series of polycyclic heteroaromatic compounds (see below). Further examination of the extracts of the Caribbean deep water sponge *Batzella* sp. by Sun et al. (62) led to the isolation of four related pyrroloquinolines, isobatzellines A–D (**105–108**). After extraction and solvent partitioning, HSCCC was employed with the solvent system heptane/$CHCl_3$/MeOH/H_2O (2:7:6:3; MP = UP), giving **105–107**. The stationary phase from that MLCPC separation was further purified by a second HSCCC run with heptane/EtOAc/MeOH/H_2O (4:7:4:3; MP = UP) to give **108**. Isobatzelline A is somewhat unstable, oxidizing to **108** on an amino TLC plate. These compounds showed cytotoxicity against P388 cells with IC_{50} values of 0.42, 2.6, 12.6, and 20 µg/mL for **105–108**, respectively, and antifungal activity against *C. albicans* with MIC values of 3.1, 25, 50, and 25 µg/mL, respectively. Surprisingly, the batzellines (**102–104**) were inactive in these assays.

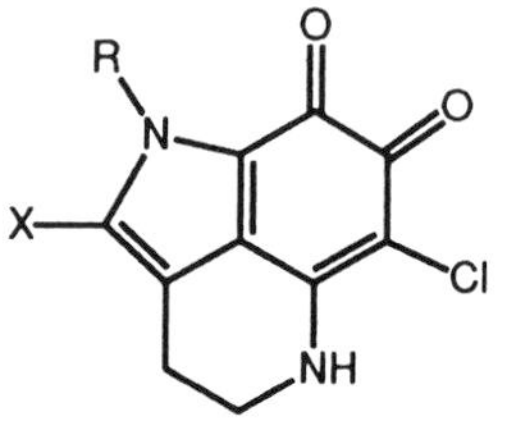

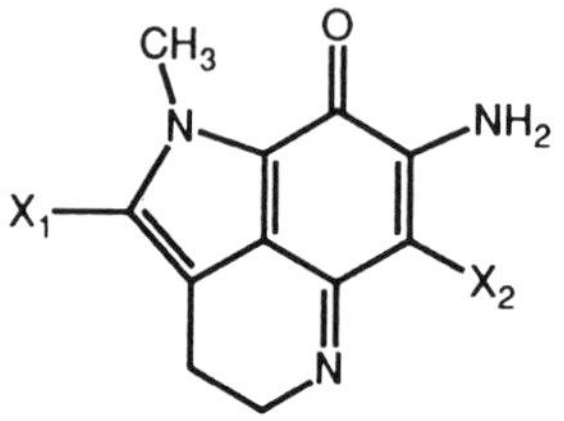

102 : R = CH₃, X = SCH₃
103 : R = H, X = SCH₃
104 : R = CH₃, X = H

105 : X₁ = SCH₃, X₂ = Cl
106 : X₁ = SCH₃, X₂ = H
107 : X₁ = H, X₂ = Cl

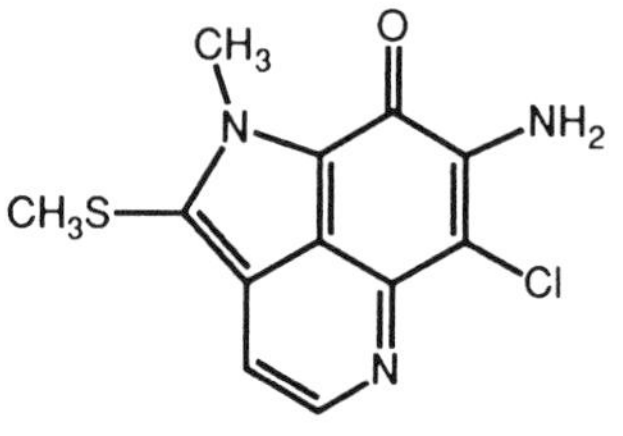

108

Yet another polycyclic heteroaromatic compound has been isolated by HSCCC. Extracts of both the New Zealand sponge *Latrunculia brevis* and an Okinawan sponge, a *Prianos* sp., gave discorhabdin D (**109**) along with the previously known discorhabdin A (**110**) (63). Two different separation procedures were used for the two sponges. While **109** and **110** were isolated by reversed-phase chromatography from *L. brevis*, HSCCC with the solvent system $CHCl_3/MeOH/H_2O$ (5:5:6; MP = UP) was used in their purification from the *Prianos* sp. Discorhabdins A–C demonstrated more potent cytotoxicity against P388 cells *in vitro* (IC_{50} values 0.01–0.03 µg/mL) than **109** (IC_{50} value 6 µg/mL), but only **110** had any activity (slight) against P388 *in vivo* (T/C 132% at 20 mg/kg). Discorhabdin D also exhibited antimicrobial activity (slight) against *B. subtilis*, *C. albicans*, and *Escherichia coli*, but not against *Pseudomonas aeruginosa* at 30 µg per disk.

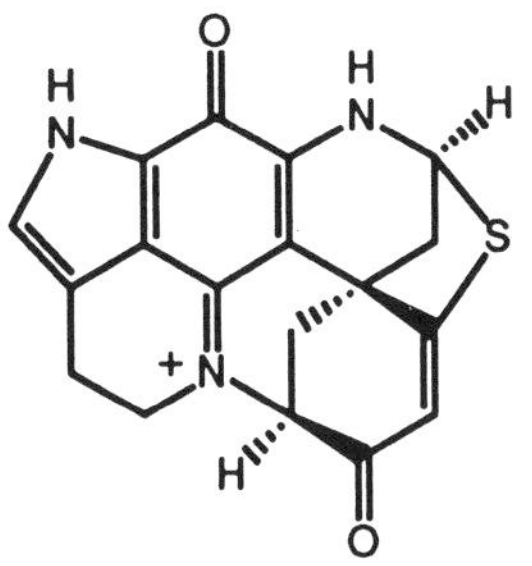

109 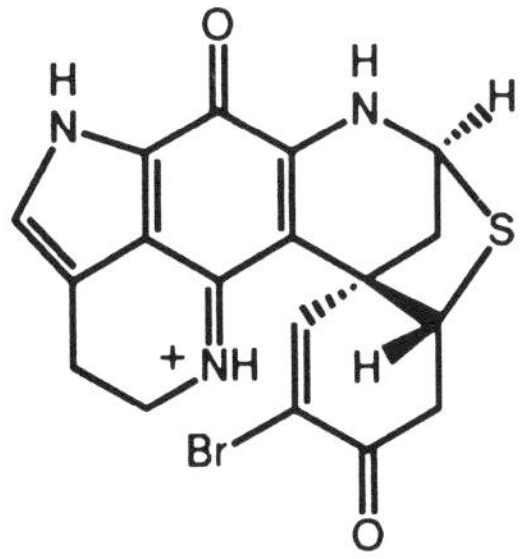   **110**

9.4.6.7. Cyclic Guanidines

Our group isolated a series of pentacyclic guanidines from the Western Mediterranean sponge *Crambe crambe* (64). Crambescidins 800 (**111**), 816 (**112**), 830 (**113**), and 844 (**114**) were obtained from the same sponge as crambines A and B (**115, 116**), which were previously isolated by droplet CCC (65). Crambescidins are most likely biosynthetically related to crambines, as well as (more closely) to ptilomycalin A, which was obtained from other marine sources (66). High-speed CCC was used in the isolation of **111–114**, although this purification procedure was not discussed in the published report. The solvent system heptane/EtOAc/MeOH/H_2O (4:7:4:3; MP = UP) separated **111–114** from the bulk of the sample, leaving only the crambescidin complex in the stationary phase. The resulting fraction could then be easily separated by HPLC. These compounds showed antiviral activity against HSV-1 at 1.25 µg/mL and cytotoxicity against L1210 at 0.1 µg/mL.

9.4.6.8. Isoquinolines

A series of potent antitumor nitrogen-containing heterocycles has been isolated from the Caribbean tunicate *Ecteinascidia turbinata* (26, 67–71). Ecteinascidins 743 (**117**), 729 (**118**), 745 (**119**), 759A (**120**), 759B (**121**), 770 (**122**),

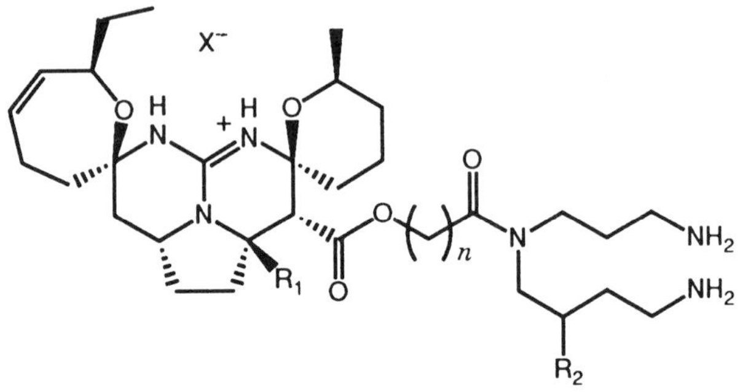

111 : R_1 = H, R_2 = OH, n = 15
112 : R_1 = R_2 = OH, n = 15
113 : R_1 = R_2 = OH, n = 16
114 : R_1 = R_2 = OH, n = 17

756 (**123**), 722 (**124**), and 736 (**125**) are di- and trimeric tetrahydroisoquinolines related to the saframycin-safracin class of antitumor antibiotics. The structure of **120** has not yet been solved. All of these compounds show cytotoxicity against CV-1 and L1210 cell lines. Ecteinascidins 729, 743, and 722 exhibit potent activity in vivo against several different types of tumors.

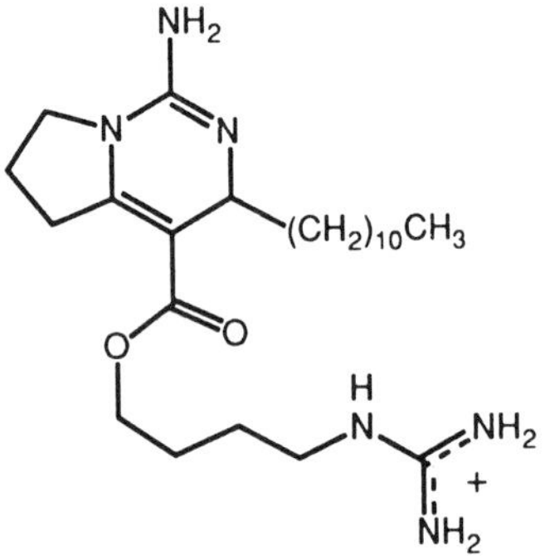

115

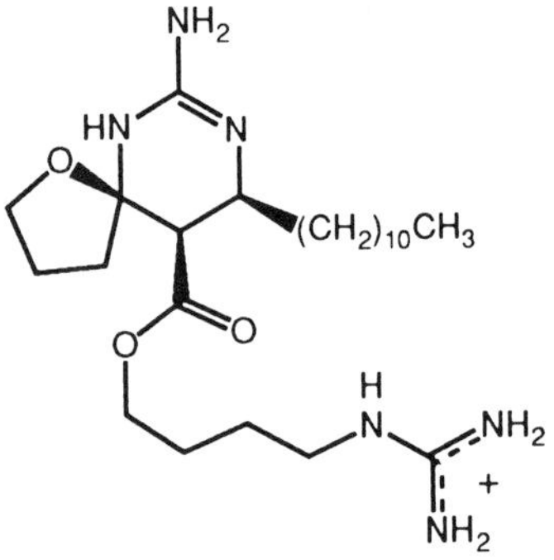

116

These compounds tend to decompose during the isolation, so mild separation techniques such as HSCCC are important for good recovery of material. Of the several different workers involved with these compounds in our labs, each has developed a different purification procedure, but all have used HSCCC. Holt (68), the first to isolate some of the ecteinascidins, used MLCPC with the solvent system cyclohexane/$CHCl_3$/MeOH/H_2O (1:2:4:2; MP = LP) to separate the crude extract into two bioactive and six inactive fractions. The first cytotoxic fraction yielded **117** and **119** after MPLC and HPLC, while the second gave **118**. This same solvent system was used in a second isolation, followed by a second HSCCC run with the solvent system hexane/ EtOAc/MeOH/H_2O (1:1:1:1; MP = UP) to give a fraction that contained **117** and **119–121** as shown by LC/FABMS (68). Application of the first solvent system also led to purification of **117** and **120–122** (68).

One of the authors of this chapter has developed several other HSCCC systems for the purification of these compounds (69). Hexane/EtOAc/ MeOH/H_2O (4:7:4:7; MP = UP) not only separates the ecteinascidins from inactive material, but this solvent system also separates **117** from **118**. A second solvent system (CCl_4/$CHCl_3$/MeOH/H_2O, 5:5:8:2; MP = LP) was found to separate **117** from other ecteinascidins, although it did not give good stationary-phase retention. In an attempt to stabilize the ecteinascidins during the isolation process, potassium cyanide was added to the crude extracts of *E. turbinata*. This converted **117** and **118** to their cyano derivatives, ecteinascidins 770 (**122**) and 756 (**123**), respectively. While **122** had previously been isolated as a natural product, **123** is a new semisynthetic compound. The solvent system CCl_4/$CHCl_3$/MeOH/H_2O (7:3:8:2; MP = LP) was used in the purification of these cyano compounds.

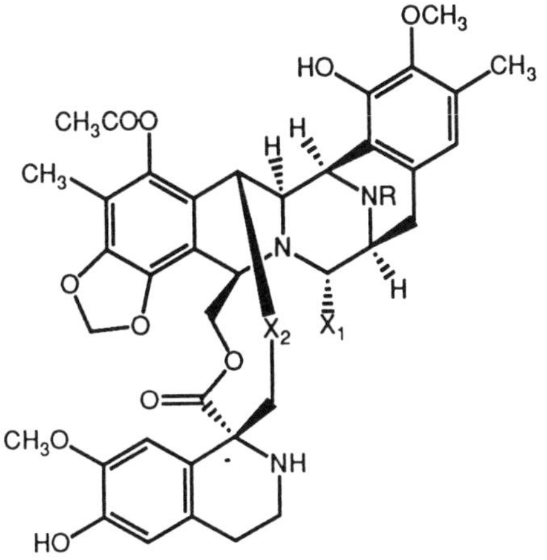

117 : R = CH_3, X_1 = OH, X_2 = S
118 : R = H, X_1 = OH, X_2 = S
119 : R = CH_3, X_1 = H, X_2 = S
121 : R = CH_3, X_1 = OH, X_2 = SO
122 : R = CH_3, X_1 = CN, X_2 = S
123 : R = H, X_1 = CN, X_2 = S

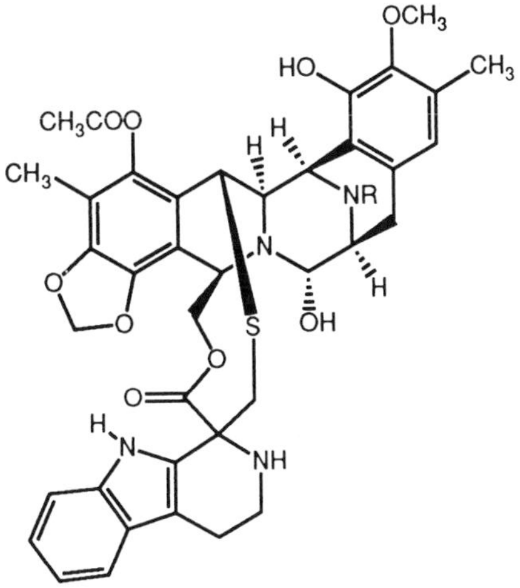

124 : R = H
125 : R = CH_3

During further isolation of **118** for more biological testing, Sakai (26) discovered two new ecteinascidins, 722 (**124**) and 736 (**125**). These compounds possess a tetrahydro-β-carboline unit instead of a third tetrahydroisoquinoline. Sakai found HSCCC to be a very important purification technique for these compounds, since it gives good separation of the ecteinascidins from inactive material without causing decomposition. A partially purified fraction from an extract of *E. turbinata* from Sunshine Key, Florida, was subjected to HSCCC with EtOAc/benzene/cyclohexane/MeOH/H_2O (4:4:3:4:3; MP = LP), separating it into five new fractions. The first fraction was further purified to give **117–119**, while the later fractions were shown to contain **124** and **125** by fast atom bombardment mass spectrometry (FABMS). An extract from a different sample of tunicate was separated by MLCPC using the slightly modified system EtOAc/benzene/cyclohexane/MeOH/H_2O (3:4:4:4:3; MP = UP). The seventh of 10 fractions thus obtained was purified to give **125**, while the ninth gave **124**. The material remaining in the stationary phase was further separated to give **117**, **119**, and **121**, while earlier fractions contained **122**. An extract of a sample of tunicate collected in Puerto Rico was separated by yet another HSCCC solvent system, CH_2Cl_2/toluene/MeOH/H_2O(15:15:23:7; MP = LP). The second of 10 fractions contained **125**, the third **124**, the fifth **117** and **119**, and the sixth **118**.

9.5. CONCLUSION

Overall, HSCCC is a very useful method for the separation of marine natural products. The number of published reports using this purification tool is increasing as more groups acquire their own HSCCC systems. In our lab, at least ten isolation projects have employed this technique. The advantages gained from the lack of a solid support and the different mode of separation as compared to LSC make for a technique that is nearly ideal for many natural product isolations. Its ability to separate closely related compounds and to minimize decomposition of unstable molecules will ensure its continued use. At this point, the main barriers to its use are the initial cost of buying the system and the development of a solvent system for a given separation.

From reviewing the applications described above it seems clear that several solvent systems are becoming relatively standard. The most popular is hexane/EtOAc/MeOH/H_2O (4:7:4:3) with either phase as the mobile phase. This solvent system has been used for a wide variety of compounds (26, 32, 47, 49, 61, 62, 64, 72–75). Several variations in the ratios of these solvents, especially 3:7:5:5, have also been used. The solvent system hexane (or heptane)/CH_2Cl_2/MeCN (10:3:7) has been used for the separation of a number of lipophilic compounds (26, 33, 41–43). These two solvent systems appear to represent good starting points for the development of systems for HSCCC separations of marine natural products.

REFERENCES

1. D. Faulkner, *J. Nat. Prod. Rep.*, **8**, 97 (1991).

2. M. H. G. Munro, R. T. Luibrand, and J. W. Blunt. In *Bioorganic Marine Chemistry*, Scheuer, P. J., Ed. Springer-Verlag: Berlin, 1987; Vol. 1, pp. 93.

3. W. Bergmann and D. C. Burke, *J. Org. Chem.*, **20**, 1501 (1955).

4. W. Bergmann and D. C. Burke, *J. Org. Chem.*, **21**, 226 (1956).

5. W. Bergmann and M. F. Stempien, Jr. *J. Org. Chem.*, **22**, 1575 (1957).

6. K. L. Rinehart, Jr., J. B. Gloer, R. G. Hughes, Jr., H. E. Renis, J. P. McGovren, E. B. Swynenberg, D. A. Stringfellow, S. L. Kuentzel, and L. H. Li, *Science* **212**, 933 (1981).

7. K. L. Rinehart, V. Kishore, K. C. Bible, R. Sakai, D. W. Sullins, and K.-M. Li, *J. Nat. Prod.*, **51** 1 (1988).

8. K. L. Rinehart, V. Kishore, S. Nagarajan, R. J. Lake, J. B. Gloer, F. A. Bozich, K.-M. Li, R. E. Maleczka, Jr., W. L. Todsen, M. H. G. Munro, D. W. Sullins, and R. Sakai, *J. Am. Chem. Soc.*, **109**, 6846 (1987).

9. G. R. Pettit, C. L. Herald, D. L. Doubek, D. L. Herald, E. Arnold, and J. Clardy, *J. Am. Chem. Soc.*, **104**, 6846 (1982).

10. A. S. Kraft, J. B. Smith, and R. L. Berkow, *Proc. Natl. Acad. Sci. USA*, **83**, 1334 (1986).

11. S. M. Kupchan, R. W. Britton, M. F. Ziegler, and C. W. Sigel, *J. Org. Chem.*, **38**, 178 (1973).

12. W. C. Still, M. Kahn, and A. Mitra, *J. Org. Chem.*, **43**, 2923 (1978).

13. J. W. Blunt, V. L. Calder, G. D. Fenwick, R. J. Lake, J. D. McCombs, M. H. G. Munro, and N. B. Perry, *J. Nat. Prod.* **50**, 290 (1987).

14. A. P. Foucault, *Anal. Chem.* **63**, 569A (1991).

15. J. B. McAlpine, and J. E. Hochlowski, In *Natural Products Isolation*, Wagman, G. H., Cooper, R., Eds. Journal of Chromatography Library 43; Elsevier: New York, 1989, Chapter 1.

16. N. B. Mandava and Y. Ito, Eds., Chromatography Science 44; Dekker; New York, 1988.

17. W. D. Conway, *Countercurrent Chromatography: Apparatus, Theory, and Applications*; VCH: New York, 1990.

18. A. Marston, I. Slacanin, and K. Hostettmann, *Phytochem. Anal.*, **1**, 3 (1990).

19. D. E. Schaufelberger, *J. Chromatogr.*, **538**, 45 (1991).

20. Y. Ito, *J. Chromatogr*, **214**, 122 (1981).

21. G. R. Pettit, F. Gao, D. Sengupta, J. C. Coll, C. L. Herald, D. L. Doubek, J. M. Schmidt, J. R. Van Camp, J. J. Rudloe, and R. A. Nieman, *Tetrahedron*, **47**, 3601 (1991).

22. D. E. Schaufelberger, G. N. Chmurny, J. A. Beutler, M. P. Koleck, A. B. Alvarado, B. W. Schaufelberger, and G. M. Muschik, *J. Org. Chem.*, **56**, 2895 (1991).

23. D. E. Schaufelberger, AG Sandoz Pharma, personal communication, 1991.

24. G. R. Pettit, F. Gao, D. L. Herald, P. M. Blumberg, N. E. Lewin, and R. A. Neiman, *J. Am. Chem. Soc.*, **113**, 6693 (1991).

25. M. R. Kernan, T. F. Molinski, and D. J. Faulkner, *J. Org. Chem.*, **53**, 5014 (1988).

26. R. Sakai and K. L. Rinehart, *J. Nat. Prod.*, **58**, 773 (1995).

27. M. Murata, A. M. Legrand, Y. Ishibashi, M. Fukui, and T. Yasumoto, *J. Am. Chem. Soc.*, **112**, 4380 (1990).

28. P. J. Scheuer, University of Hawaii at Manoa, personal communication, 1991.

29. S. Sakemi, T. Ichiba, S. Kohmoto, G. Saucy, and T. Higa, *J. Am. Chem. Soc.*, **110**, 4851 (1988).

30. S. Sakemi, T. Higa, U. Anthoni, and C. Christophersen, *Tetrahedron*, **43**, 263 (1987).

31. H. H. Sun, S. S. Cross, M. Gunasekera, and F. E. Koehn, *Tetrahedron*, **47**, 1185 (1991).

32. S. Kohmoto, O. J. McConnell, A. Wright, and S. Cross, *Chem. Lett.*, 1687 (1987).

33. S. Kohmoto, O. J. McConnell, A. Wright, F. Koehn, W. Thompson, M. Lui, and K. M. Snader, *J. Nat. Prod.*, **50**, 336 (1987).

34. B. N. Ravi, H. P. Perzanowski, R. A. Ross, T. R. Erdman, P. J. Scheuer, J. Finer, and J. Clardy, *Pure Appl. Chem.*, **51**, 1893 (1979).

35. K. L. Rinehart, Jr., R. Sakai, and J. G. Stroh, U.S. Patent 4 948 791; *Chem. Abs.*, **114**, 214413h (1991).

36. A. C. Stierle, J. H. Cardellina, and F. L. Singleton, *Experentia*, **44**, 1021 (1988).

37. F. J. Schmitz, D. J. Vanderah, K. H. Hollenbeak, C. E. L. Enwall, Y. Gopichand, P. K. SenGupta, M. B. Hossain, and D. van der Helm, *J. Org. Chem.*, **48**, 3941 (1983).

38. R. L. Dillman and J. H. Cardellina, II, *J. Nat. Prod*, **54**, 1159 (1991).

39. A. C. Stierle, J. H. Cardellina, II, and G. A. Strobel, *Proc. Natl. Acad. Sci. USA*, **85**, 8008 (1988).

40. A. A. Stierle, J. H. Cardellina, II, and F. L. Singelton, *Tetrahedron Lett.*, **32**, 4847 (1991).

41. M. F. Raub, J. H. Cardellina, II, M. I. Choudhary, C.-Z. Ni, J. Clardy, and M. C. Alley , *J. Am. Chem. Soc.*, **113**, 3178 (1991).

42. S. Sakemi, L. E. Totton, and H. H. Sun, *J. Nat. Prod.*, **53**, 995 (1990).

43. D. B. Stierle and D. J. Faulkner, *J. Nat. Prod.*, **54**, 1134 (1991).

44. S. Kohmoto, O. J. McConnell, and A. Wright, *Experentia*, **44**, 85 (1988).

45. H. H. Sun and S. Sakemi, *J. Org. Chem.*, **56**, 4307 (1991).

46. R. L. Dillman and J. H. Cardellina, II, *J. Nat. Prod.*, **54**, 1056 (1991).

47. R. Sakai, S. Kohmoto, T. Higa, C. W. Jefford, and G. Bernardinelli, *Tetrahedron Lett.*, **28**, 5493 (1987).

48. S. C. Bobzin and D. J. Faulkner, *J. Org. Chem.*, **56**, 4403 (1991).

49. S. Sakemi and H. H. Sun, *J. Org. Chem.*, **56**, 4304 (1991).

50. S. Tsujii, K. L. Rinehart, S. P. Gunasekera, Y. Kashman, S. S. Cross, M. S. Lui, S. A. Pomponi, and M. C. Diaz, *J. Org. Chem.*, **53**, 5446 (1988).

51. S. Sakemi, Pfizer Central Research, Nagoya, personal communication, 1991.

52. J. N. McGuire and K. L. Rinehart, University of Illinois at Urbana-Champaign, unpublished results.

53. P. A. Keifer, R. E. Schwartz, M. E. S. Koker, R. G. Hughes, Jr., D. Rittschoff, and K. L. Rinehart, *J. Org. Chem.*, **56**, 2965 (1991).

54. D. E. Schaufelberger and G. R. Pettit, *J. Liq. Chromatogr.*, **12**, 1909 (1989).

55. G. R. Pettit, C. L. Herald, J. E. Leet, R. Gupta, D. E. Schaufelberger, R. B. Bates, P. J. Clewlow, D. L. Doubek, K. P. Manfredi, K. Rützler, J. M. Schmidt, L. P. Tackett, F. B. Ward, M. Bruck, and F. Camou, *Can. J. Chem.*, **68**, 1621 (1990).

56. F. J. Schmitz, F. S. DeGuzman, Y.-H. Choi, M. B. Hossain, S. K. Rizvi, and D. van der Helm, *Pure Appl. Chem.*, **62**, 1393 (1990).

57. F. J. Schmitz, F. S. DeGuzman, M. B. Hossain, and D. van der Helm, *J. Org. Chem.*, **56**, 804 (1991).

58. G. P. Gunawardana, S. Kohmoto, S. P. Gunasekera, O. J. McConnell, and F. E. Koehn, *J. Am. Chem. Soc.*, **110**, 4856 (1988).

59. G. P. Gunawardana, F. E. Koehn, A. Y. Lee, J. Clardy, H.-y. He, and D. J. Faulkner, *J. Org. Chem.*, **57**, 1523 (1992).

60. G. P. Gunawardana, S. Kohmoto, and N. S. Burres, *Tetrahedron Lett.*, **30**, 4359 (1989).

61. S. Sakemi, H. H. Sun, C. W. Jefford, and G. Bernardinelli, *Tetrahedron Lett.*, **30**, 2517 (1989).

62. H. H. Sun, S. Sakemi, N. Burres, and P. McCarthy, *J. Org. Chem.*, **55**, 4964 (1990).

63. N. B. Perry, J. W. Blunt, M. H. G. Munro, T. Higa, and R. Sakai, *J. Org. Chem.*, **53**, 4127 (1988).

64. E. A. Jares-Erijman, R. Sakai, and K. L. Rinehart, *J. Org. Chem.*, **56**, 5712 (1991).

65. R. G. S. Berlinck, J. C. Braekaman, D. Daloze, K. Hallenga, R. Ottinger, I. Bruno, and R. Riccio, *Tetrahedron Lett.*, **31**, 6531 (1990).

66. Y. Kashman, S. Hirsh, O. J. McConnell, I. Ohtani, T. Kusumi, and H. Kakisawa, *J. Am. Chem. Soc.*, **111**, 8925 (1989).

67. K. L. Rinehart, T. G. Holt, N. L. Fregeau, J. G. Stroh, P. A. Keifer, F. Sun, L. H. Li, and D. G. Martin, *J. Org. Chem.*, **55**, 4512 (1990).

68. T. G. Holt, The isolation and structural characterization of the ecteinascidins. Ph.D. Thesis, University of Illinois, Urbana, 1986; *Chem. Abstr.*, **106**, 193149u (1987); *Diss. Abstr. Int. B.* **47**, 3771 (1987).

69. N. L. Fregeau, Biologically active compounds from a clam and a tunicate. Ph.D. Thesis, University of Illinois, Urbana, 1992.

70. K. L. Rinehart and T. G. Holt, U.S. Patent Application Serial No. 872 189, June 9, 1986; PCT Int. Appl. WO8707610, Dec. 17, 1987; *Chem. Abstr.*, **109**, 811j (1988).

71. A. E. Wright, D. A. Forleo, G. P. Gunawardana, S. P. Gunasekera, F. E. Koehn, and O. J. McConnell, *J. Org. Chem.*, **55**, 4508 (1990).

72. R. Sakai, K. L. Rinehart, Y. Gaun, and A. H.-J. Wang, *Proc. Natl. Acad. Sci. USA*, **89**, 11456 (1992).

73. E. A. Jares-Erijman, C. P. Bapat, A. Lithgow-Bertelloni, K. L. Rinehart, and R. Sakai, *J. Org. Chem.*, **58**, 5732 (1993)

74. E. A. Jares-Erijman, A. L. Ingrum, J. R. Carney, K. L. Rinehart, and R. Sakai, *J. Org. Chem.*, **58**, 4805 (1993).

75. E. A. Jares-Erijman, A. A. Ingrum, F. Sun, and K. L. Rinehart, *J. Nat. Prod.*, **56**, 2186 (1993).

CHAPTER

10

SEPARATION OF COMPLICATED ANTIBIOTICS BY HIGH-SPEED COUNTERCURRENT CHROMATOGRAPHY

KEN-ICHI HARADA

Faculty of Pharmacy, Meijo University, Tempaku, Nagoya 468, Japan

10.1. INTRODUCTION

It is both essential and important to isolate and purify a desired compound from a complicated matrix such as a natural source and a biological sample, but usually various complications are present. Recent advances in equipment and adsorbents for high-performance liquid chromatography (HPLC) seems to highly facilitate the isolation work. However, there have been some drawbacks in HPLC, such as sample loss, tailing of solute peaks, and contamination, all arising from the use of solid supports. Although counter-current chromatography (CCC) has many desirable features and various types of CCC have been developed to complement chromatography with solid supports, they have been mainly used for long-term preparative separation of natural products (1). High-speed countercurrent chromatography (HSCCC), which was developed by Ito, has been a very desirable tool for natural product chemists (2–4) and in fact, it has begun to be applied as a rapid separation technique, which can be completed within several hours. However, there has been little use of HSCCC (5, 6), probably because few appropriate methodologies have been established. To widen the use of HSCCC, first, many experimental results must be accumulated and second, some advantages for HSCCC must be emphasized so that its potential can be realized. Third, a systematic methodology should be established.

This chapter focuses on the establishment of a procedure for effective separation of natural products by HSCCC, as illustrated by the study of two complicated antibiotics, sporaviridins (SVD) and bacitracins (BC). The procedure includes development of HPLC, selection of a two-phase solvent

High-Speed Countercurrent Chromatography, Edited by Yoichiro Ito and Walter D. Conway.
Chemical Analysis Series, Vol. 132.
ISBN 0-471-63749-1   © 1996 John Wiley & Sons, Inc.

system, separation of these antibiotics by HSCCC, and comparison of separation behavior by HPLC and HSCCC.

10.2. SELECTED ANTIBIOTICS

An important advantage of HSCCC is the absence of solid supports, so that decomposition of samples is greatly reduced during the course of chromatography. There are many labile natural products showing biological activities. For the isolation of these compounds HSCCC is more suitable than HPLC. Additionally, since HSCCC has high-resolution power (7), we attempted to achieve precise separation in these experiments. For these reasons the following polar compounds were selected to illustrate the potential of HSCCC: SVD and BC that have closely related components and are susceptible to degradation during the course of adsorption chromatography.

SVD (Fig. 10.1) are basic and water-soluble antibiotics produced by *Streptosporangium viridogriseum*. These antibiotics were isolated in 1963 and first reported in 1966 (8). They are active against gram-positive bacteria, acid-fast bacteria, and trichophyton. Since the intact antibiotics are very unstable, particularly under basic conditions, the isolation and structural determination were carried out using more stable derivatives, N-acetylsporaviridins (N-Ac-SVD) (9). They consist essentially of six components, as each has a 34-membered lactone and 7 monosaccharide units, a pentasaccharide (viridopentaose) (10), and 2 monosaccharides. Because the structural differences among them are derived from those of the methyl and ethyl groups at C2 and three viridopentaoses A, B, and C at C13, six components (A_1–C_2) of N-Ac-SVD exist. Because these derivatives show no biological activity, it was required to separate and purify the intact components. To obtain each component a preparative HPLC method was optimized in consideration of their properties. As will be shown later, a mobile phase containing aqueous ammonium chloride depressed the unfavorable chemical behavior and gave a good separation. Although the six components were successfully isolated and each component isolated had the expected antimicrobial activity, it was laborious to do preparative HPLC. It took many runs to obtain about 10 mg of each component from the complex mixture. In addition, it was difficult to remove ammonium chloride from the desired fraction.

BC, peptide antibiotics produced by *Bacillus subtilis* and *B. licheniformis*, exhibit an inhibitory activity against gram-positive bacteria (11) and are among the most commonly used antibiotics as animal feed additives (12, 13). These compounds were discovered in 1943 and were originally considered to be one substance (14). More recently, it has been demonstrated that they consist of over 20 components of differing antimicrobial activities (15, 16). The major antimicrobial components are BC-A and BC-B, and BC-F is a degrada-

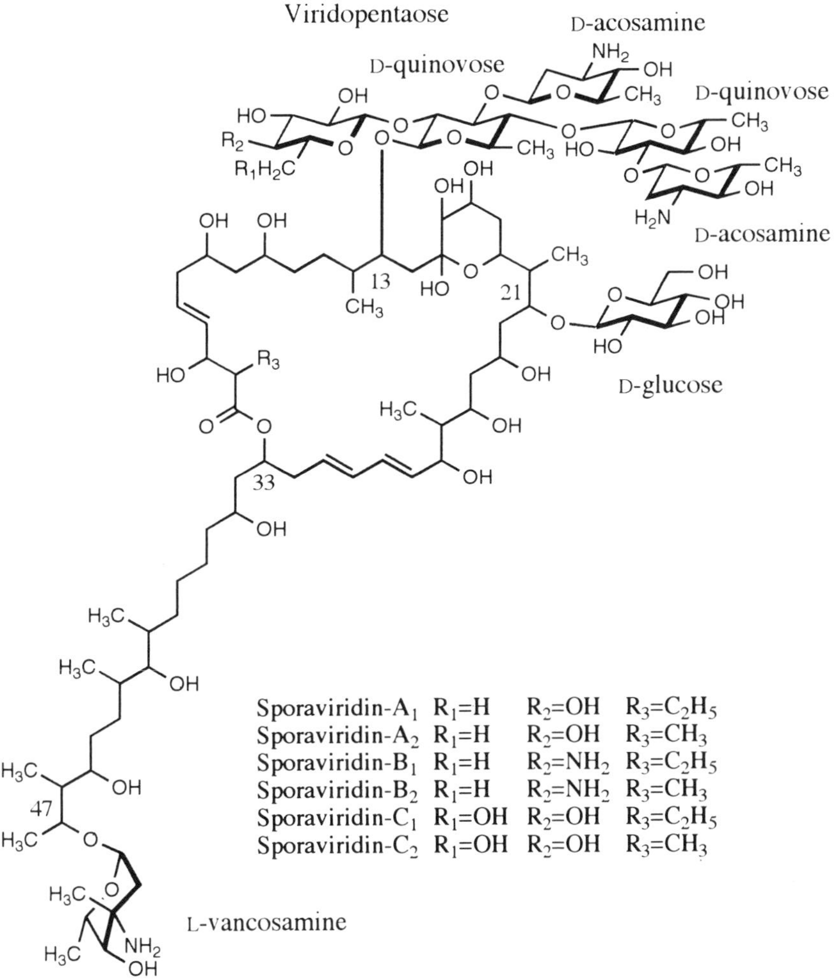

Figure 10.1. Structures of SVD.

tion product that shows nephrotoxicity (14, 15, 17). Only the structures of BC-A and BC-F (Fig. 10.2) have been determined to date (14, 18–20). In order to establish appropriate chromatographic methods for the separation of the components, it was necessary to consider the following various properties. Bacitracins are highly polar compounds composed of a cyclic heptapeptide and a branch containing a thiazole ring. The BC complex has low solubility in benzene, chloroform, ethyl acetate, and acetone and is soluble to the approximate limit of 20 mg/mL in water or methanol. When silica gel and Sephadex

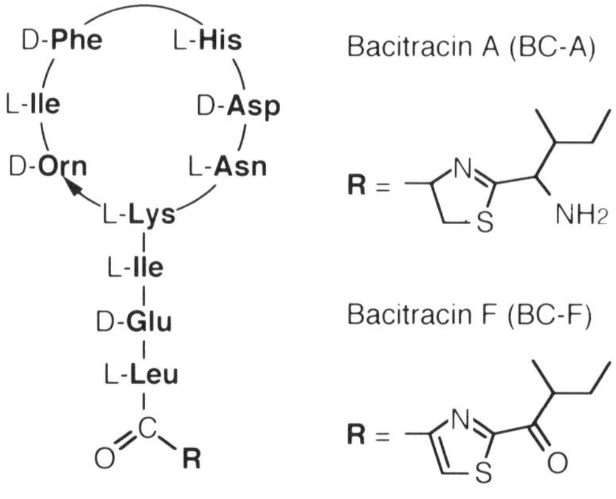

Figure 10.2. Structures of BC-A and BC-F.

are used for preparative separation, the recovery is very poor, probably because polar functional groups interact with silanol and other polar groups of the adsorbents. These components have previously been separated by countercurrent distribution methods without complete resolution (15, 21–24). Additionally, the antibiotics are thermally unstable and labile under basic conditions. It has been demonstrated that BC-A decomposes thermally to give BC-F.

10.3. OPERATION OF HSCCC

The apparatus used was a Shimadzu (Kyoto, Japan) prototype of the coil planet centrifuge (CPC) (HSCCC-1A) for HSCCC with a 160 m length of 1.6-mm i.d. polytetrafluoroethylene (PTFE) tubing around a 10-cm diameter column holder making multiple coil layers. The total capacity was 325 mL and the column was revolved at 800 rpm. Solvents were delivered with a Shimadzu LC-6A pump. After filling with the stationary phase, followed by sample injection, the mobile phase was applied at 3 mL/min (25, 26). In the case of BC ultraviolet (UV) detection was performed at 254 nm using a Shimadzu SPD-6A (26). Eluates were collected in a Pharmacia (Uppsala, Sweden) FRAC-100 fraction collector.

10.4. ESTABLISHMENT OF HPLC FOR SVD AND BC

During isolation and structural determination of N-Ac-SVD, it was found that they are unstable under basic conditions and the characteristics behavior was used for the degradative studies. Treatment of N-Ac-SVD with a base such as aqueous ammonia and 1,8-diazabicyclo[5.4.0]undec-7-ene (DBU) resulted in the cleavage of a glycosidic bond to liberate the pseudoaglycones and the constituent pentasaccharides, viridopentaoses (27). These degradation products played important roles in determining the total structures of N-Ac-SVD. This chemical reaction is more facile in the case of the intact SVD. For example, an aqueous solution of SVD has pH 9, and the half-lifetimes in aqueous and methanol solutions were about 2 days and 1 week, respectively. Thus they decompose quickly to their degradation products due to their basicity in aqueous and methanol solutions.

For these reasons, a mobile phase for the HPLC separation of SVD components had to be carefully chosen. It has already been shown that reversed-phase HPLC (RP-HPLC) using an ammonium chloride containing mobile phase is suitable for analysis of aculeximycin which is also a basic glycoside antibiotic and is similar to SVD (28). The stability of SVD was examined in the mobile phase, methanol/1 M ammonium chloride and has proven to be considerably more stable in this mobile phase than those in aqueous and methanol solutions, indicating that this solvent system depresses the chemical decomposition of SVD and is also suitable for the analysis of the intact SVD. After optimization of the operating conditions for the HPLC, they were fixed as follows: column contains ODS-silica gel; mobile phase is methanol/1 M ammonium chloride (70–76: 30–24); flow-rate is 1 mL (analysis) or 2 mL (preparative separation)/min; detection is UV (232 nm) (29).

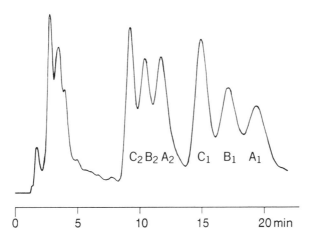

Figure 10.3. High-performance liquid chromatogram of SVD. Column consists of Cosmosil 5C18 (4.6 × 150 mm); mobile phase is methanol/1 N ammonium chloride (74:26); flow-rate = 1 mL/min; detection is UV (232 nm).

Figure 10.3 shows the HPLC of SVD components under the analysis conditions. Although this is not a baseline separation, the six components can be resolved to a considerable extent. Several peaks around the retention time of 3 min were found to be the pseudoaglycones.

As mentioned earlier, BC consist of over 20 components. All of the previously reported RP-HPLC methods used complicated gradient elution systems having long analysis times and these methods did not provide reproducible results (16, 30, 31). Also ion exchange HPLC gave low resolution of the BC components (33). According to our preliminary studies, BC appeared as tailing peaks on a conventional ODS silica gel column, probably due to interaction between the amino group of BC and residual silanol groups on the column. Use of a silanol-free stationary phase would prevent tailing of solute peaks. Capsule (32) and polymer (34) type RP-HPLC columns are made from silica gel coated with a silicone polymer and a porous polymer gel modified with hydrophobic materials such as C18, so that they are free

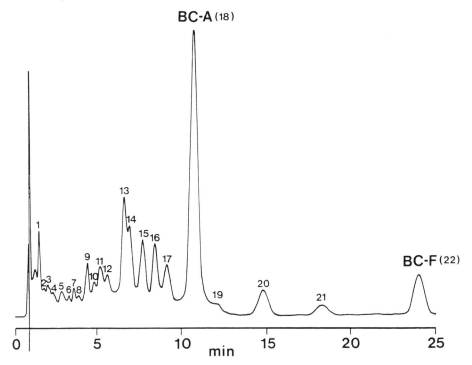

Figure 10.4. High-performance liquid chromatogram of BC. Column consists of Capcell pak 5C18 (4.6 × 150 mm); mobile phase is methanol/0.04 M disodium phosphate (6:4); flow-rate = 1.3 mL/min; detection is UV (234 nm).

from silanol groups. After extensive optimization studies, the following conditions were proposed: column used is Capcell Pak C18; mobile phase is methanol/0.04 M aqueous disodium hydrogenphosphate solution (6:4); flow-rate 1.3 mL/min; detection is UV (234 nm) for analysis and column, Capcell Pak C18; mobile phase is methanol/0.05 M sodium sulphate solution (6:4); flow-rate is 1.3 mL/min; detection is UV (234 nm) for preparative separation (35). Figure 10.4 shows the HPLC of BC, which are clearly separated into 22 peaks.

10.5.   SELECTION OF PROPER TWO-PHASE SOLVENT SYSTEM

Precise separation by HSCCC is mainly dependent on the selection of a proper two-phase solvent system, because it is based on real liquid–liquid partition. Although a systematic search for suitable two-phase solvent systems for HSCCC has been reported (4, 5, 36), it has been frequently difficult to obtain an optimum solvent system. To select a suitable solvent system, settling time of the two-phase solvent system is used, and partition coefficients of the individual components and solubility of the SVD complex were carefully considered. A two-phase solvent system with less than 30 s of settling time is desirable for high retention of stationary phase (2). The SVD complex is only soluble in polar solvents such as water, methanol, and n-butanol. In fact, it was extracted with n-butanol (8). Additionally, a mobile phase, chloroform/methanol/7% ammonium hydroxide, was used for separation on normal phase thin-layer chromatography (TLC). So two systems, chloroform/water and n-butanol/water were examined mainly for SVD.

The partition coefficient (K value) was estimated by the following simple test tube experiment. The SVD complex was partitioned with a two-phase solvent system, which had already been equilibrated, and the resulting upper and lower phases were analyzed by the HPLC, which had been optimized. The total of the upper and lower phase peak areas was calculated and each partition coefficient was determined by dividing the corresponding peak area of the upper phase by that of the lower phase. Table 10.1 shows the totals (averages) and the individual partition coefficients of the SVD components with chloroform/ethanol/water and chloroform/ethanol/methanol/water systems together with their settling times. The partition coefficient can predict approximate elution volume and the separation among the six components depends on the degree of dispersion of the six values. Ideally, these values should be equally dispersed around $K = 1$. To quickly express and understand these results Fig. 10.5 was devised. It is easy to predict that no chloroform-containing solvent system will completely separate all components, but it probably could clearly separate between SVD-C_1 and SVD-C_2. In fact, it was

Table 10.1. Partition Coefficients of SVD Components with Chloroform Systems

Solvent System[a]	s.t.[b]	Total	C_2	B_2	A_2	C_1	B_1	A_1
C:E:M:W								
5:2:2:4	17	1.49	2.89	1.54	1.06	2.03	1.12	1.10
5:2:3:4	21	1.32	2.38	1.40	1.08	1.53	0.97	1.01
5:3:3:4	34	0.73	1.41	0.77	0.64	1.05	0.67	0.71
5:4:3:4	60	0.84	0.69	1.14	0.89	0.94	0.69	0.67
5:3:3:3	51	1.16	1.85	1.19	1.03	1.29	1.09	0.84
C:E:W								
7:13:8	110	0.33	0.38	0.34	0.29	0.34	0.33	0.32
5:2:3	18	4.22	14.00	4.25	3.41	5.66	2.99	2.63
5:3:3	20	0.57	1.80	0.57	0.46	0.77	0.35	0.28
5:4:3	26	0.80	1.43	0.68	0.65	1.06	0.55	0.61
5:3:4	16	1.25	2.93	1.33	1.04	1.91	0.73	0.66
5:4:2	24	0.69	1.53	0.72	0.56	0.86	0.43	0.33
5:4:1	127	0.46	0.91	0.40	0.37	0.58	0.33	0.32
4:3:3	23	0.71	1.92	0.60	0.55	0.98	0.42	0.37
3:3:3	32	0.39	0.86	0.36	0.31	0.47	0.26	0.29

[a]Chloroform = C, ethanol = E, methanol = M, water = W.
[b]Settling time = s.t. (s).

impossible to attain the separation of the six components using any chloroform systems.

As mentioned above, SVD complex was extracted with n-butanol from the fermentation broth (8). Indeed, when SVD complex was partitioned with n-butanol and water, it passed almost completely into the upper phase (Table 10.2). This result indicates that the solubility of the n-butanol phase must be decreased and a nonpolar solvent, such as n-hexane and diethyl ether, was added to the n-butanol and water solvent system as the modifier. The optimization for obtaining a suitable solvent system was done in the following way. Initially, volumes of n-butanol and water were fixed at 10 mL and that of diethyl ether was varied, so that a solvent system, n-butanol/diethyl ether/water (10:4:10), was selected. Next, volumes of n-butanol and diethyl ether were fixed and that of water was changed from 11 to 15 mL. In the case of the 10:4:12 ratio, an almost equal dispersion of partition coefficients for the six components was obtained, as shown in Fig. 10.6. This solvent system is most suitable for the separation of the six components at present (25).

In the same manner as for SVD a suitable two-phase solvent system has been optimized for separation of the BC components by HSCCC. The following organic solvents, n-butanol, ethyl acetate, and chloroform, were selected and then two-phase solvent systems composed of one of these three

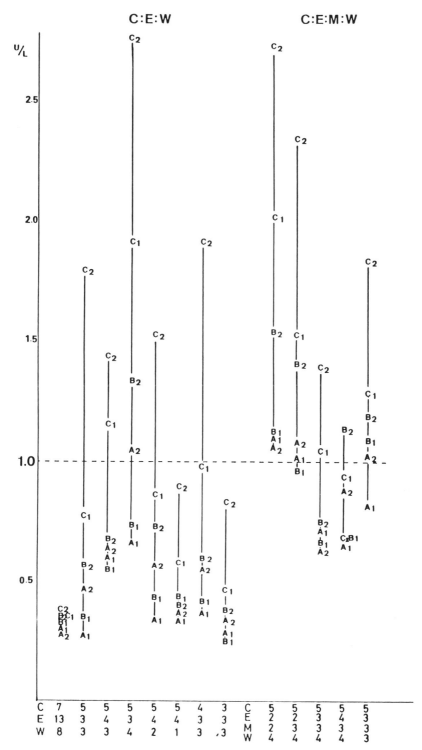

Figure 10.5. Partition coefficients of the six components of SVD with chloroform/ethanol/water (C:E:W) and chloroform/ethanol/methanol/water (C:E:M:W).

Table 10.2. Partition Coefficients of SVD Components with *n*-Butanol Systems

Solvent System[a]	s.t.[b]	Total	C_2	B_2	A_2	C_1	B_1	A_1
n-Bu:W								
10:10	11	5.66	2.96	6.41	6.65	4.87	8.81	9.09
10:15	12	4.94	2.35	5.33	6.30	4.24	7.45	9.61
10:20	14	6.13	2.77	7.03	7.19	5.17	10.62	12.57
10:25	15	4.57	2.01	5.15	6.43	3.80	7.17	8.59
n-Bu:BE:W								
10:1:10	13	3.61	1.69	3.60	4.55	3.11	5.03	7.13
10:2:10	12	3.34	1.58	3.45	4.08	2.86	4.92	5.71
10:3:10	10	2.21	0.96	2.17	2.78	1.84	3.34	4.19
10:4:10	10	1.29	0.50	1.12	1.59	1.04	1.85	2.91
10:5:10	10	0.94	0.38	0.78	1.11	0.74	1.25	2.12
10:6:10	10	0.98	0.39	0.90	1.19	0.81	1.39	1.70
10:7:10	10	0.80	0.24	0.63	1.10	0.59	1.08	1.82
10:8:10	11	0.68	0.44	0.51	0.78	0.59	0.95	1.37
10:9:10	11	0.60	0.21	0.49	0.78	0.46	0.78	1.31
10:10:10	10	0.41	0.12	0.35	0.45	0.29	0.63	0.96
10:4:11	13	0.95	0.31	0.89	1.24	0.70	1.50	2.00
10:4:12	15	1.13	0.38	1.09	1.41	0.80	1.85	2.32
10:4:13	15	1.09	0.37	1.05	1.51	0.73	1.58	2.10
10:4:14	15	0.88	0.29	1.09	1.17	0.57	1.22	1.74
10:4:15	14	1.32	0.49	1.22	1.84	1.01	1.77	2.55

[a] *n*-Butanol = *n*-Bu, diethyl ether = DE, water = W.
[b] Settling time = s.t. (s).

solvents and water and/or methanol, which dissolve the BC complex, were prepared. Settling times and individual partition coefficients of peaks 13–22 (see Fig. 10.4) using *n*-butanol and ethyl acetate systems are summarized in Table 10.3. For *n*-butanol solvent systems, peaks 13–18 have suitable partition coefficients, but those of peaks 20–22 are too high. These solvent systems also have the disadvantage of being laborious to evaporate. Another disadvantage is that the ethyl acetate, alcohol, and water system showed too long settling times. Addition of inorganic salts, such as sodium chloride, potassium chloride, ammonium chloride, and ammonium acetate, to these systems did not improve the K values. Chloroform, methanol, and water systems have been extensively used in droplet CCC (37). Direct application of these solvent systems to HSCCC was unsuccessful in BC separation, but the complete or partial replacement of methanol with ethanol gave good results. Other systems are summarized in Table 10.4. The systems using chloroform/ethanol/

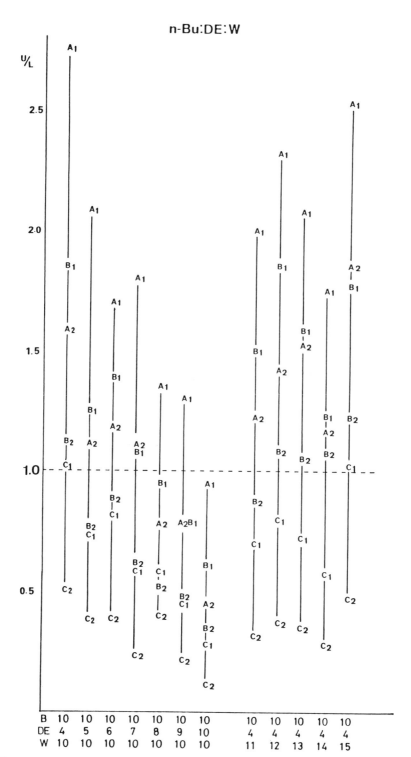

Figure 10.6. Partition coefficients of the six components of SVD with *n*-butanol/diethyl ether/water (*n*:Bu:DE:W).

Table 10.3. Partition Coefficients of Peaks 13–22 of BC with *n*-Butanol and Ethyl Acetate Systems

Solvent System[a]	Ratio	s.t. (s)[b]	13, 14[c]	15[c]	16[c]	17[c]	18[c]	20[c]	21[c]	22[c]
n-Bu:DE:W	10:0:10	15	0.63	1.52	0.63	1.48	0.75	13.86	Up	23.14
n-Bu:DE:W	10:1:10	15	0.49	1.25	0.51	1.24	1.03	11.13	Up	15.47
n-Bu:DE:W	10:3:10	14	0.22	0.59	0.21	0.51	0.53	6.37	Up	9.95
EA:E:W	5:2:3	50	0.10	0.12	0.08	0.10	0.16	0.56	0.49	0.60
EA:*i*-P:W	4:2:3	60	0.07	0.14	0.06	0.13	0.15	0.74	0.68	0.85
EA:M:A:W	5:1:1:2	60	Lo	Lo	Lo	Lo	0.01	0.08	Lo	0.12
EA:E:*i*-P:W	4:1:1:4	60	0.06	0.13	0.05	0.10	0.12	0.57	Lo	0.60
EA:*n*-Bu:M:W	4:1:1:4	40	Lo	Lo	Lo	Lo	0.03	0.50	Lo	0.76
EA:*n*-Bu:M:W	5:1:1:4	30	Lo	Lo	Lo	Lo	0.02	0.29	Lo	0.50
EA:E:*i*-P:W	4:1:1:4	60	0.06	0.13	0.05	0.10	0.12	0.57	Lo	0.61
EA:E:*i*-P:10% NaCl(aq.)	4:1:1:4	30	0.07	(0.08)		0.08	0.11	0.60	0.50	0.67
EA:E:*i*-P:10% KCl(aq.)	4:1:1:4	30	0.21	(0.22)		0.15	0.33	0.88	0.85	1.25
EA:E:*i*-P:10% NH$_4$Cl(aq.)	4:1:1:4	32	Lo	Lo	Lo	Lo	0.07	0.45	0.64	0.64
EA:E:*i*-P:10% NH$_4$OAc(aq.)	4:1:1:4	35	—	—	—	—	—	—	—	—
EA:M:A:W	5:1:1:2	60	Lo	Lo	Lo	Lo	0.01	0.08	Lo	0.12
EA:M:A:10% NaCl(aq.)	5:1:1:2	30	Lo	Lo	Lo	Lo	Lo	Lo	Lo	Lo
EA:M:A:10% KCl(aq.)	5:1:1:2	30	Lo	Lo	Lo	Lo	0.01	Lo	Lo	0.15
EA:M:A:10% NH$_4$Cl(aq.)	5:1:1:2	30	Lo	Lo	Lo	Lo	Lo	Lo	Lo	Lo
EA:M:A:10% NH$_4$OAc(aq.)	5:1:1	40	Lo	Lo	Lo	Lo	Lo	Lo	Lo	Lo

[a] *n*-Butanol = *n*-Bu, diethyl ether = DE, water = W, ethyl acetate = EA, ethanol = E, *i* = propanol = *i*-P, methanol = M, acetone = A.

[b] Settling time = s.t. (s).

[c] Up: Bacitracins were exclusively partitioned into the upper phase.
Lo: Bacitracins were exclusively partitioned into the lower phase.
( ): Both components were not separated.
—: Decomposition.

Table 10.4. Partition Coefficients of Peaks 13–22 of BC with Chloroform Systems

Solvent System[a]	Ratio	s.t. (s)[b]	13, 14[c]	15[c]	16[c]	17[c]	18[c]	20[c]	21[c]	22[c]
C:E:M:W	5:2:3:4	18	7.20	(4.92)		2.46	4.17	0.64	0.65	0.48
C:E:M:W	5:2:1:4	19	Up	Up	Up	Up	33.27	1.62	1.38	0.75
C:E:M:W	5:3:3:4	27	3.35	1.59	4.43	1.40	2.37	0.57	0.57	0.45
C:E:W	5:3:3	15	11.14	(6.37)		3.20	5.34	0.32	0.35	0.27
C:E:W	5:4:2	20	3.19	(2.19)		1.05	2.00	0.25	0.26	0.21
C:E:W	5:4:3	26	5.49	1.78	6.74	1.46	2.20	0.16	Lo	0.16
C:E:W	5:4:4	37	6.10	2.25	6.45	2.04	2.68	0.14	Lo	0.10

[a]Chloroform = C, ethanol = E, methanol = M, water = W.
[b]Settling time = s.t. (s).
[c]Up: Bacitracin components were exclusively partitioned into the upper phase.
Lo: Bacitracin components were exclusively partitioned into the lower phase.
( ): Both components were not separated.

313

methanol/water (5:3:3:4) and chloroform/ethanol/water (5:4:3) gave the best combination of partition coefficients (26).

10.6. SEPARATION OF SVD AND BC COMPONENTS BY HSCCC

The preparative separation of the six components of SVD by HSCCC was performed under the optimized conditions. In this experiment the retention of the stationary phase was 75.4%, and the total elution time and elution volume were 3.5 h and 500 mL, respectively. The six components were eluted in the order of their partition coefficients, SVD-C_2, C_1, B_2, A_2, B_1, and A_1. The almost purified components, A_1 (1.3 mg), A_2 (0.6 mg), B_1 (0.7 mg), B_2 (0.5 mg), C_1 (1.1 mg), and C_2 (1.4 mg), were obtained from 15 mg of the complex mixture. Their HPLCs are shown in Fig. 10.7. Two or three peaks around the retention time of 3 min correspond to the pseudoaglycones produced by the cleavage of the glycosidic linkage at C13 of the aglycone. The ratio of the purified components to the starting SVD complex was about 40%. In the case of 100 mg of the complex mixture similar results were obtained (25).

Figure 10.8 shows the CCC of the BC components using chloroform/ethanol/water (5:4:3) as the two-phase solvent system. The settling time was 26 s and the lower phase was delivered at 3 mL/min. A 50 mg amount of the complex was introduced into the apparatus and rotated at 800 rpm. The retention of the stationary phase was 74.1% and the experiment run was about 3 h. These components were eluted in order of their partition coefficients and peaks 18 and 22 could be almost completely separated. After the mass spectrometric analysis, they were identified as BC-A and BC-F, respectively. Table 10.5 shows the comparison of HPLC with HSCCC on preparative separation of BC-A (26). The semipure BC-A separated initially from the complex by HSCCC, was subjected to preparative HPLC and re-HSCCC. Since the injected amount was limited due to the solubility in the case of HPLC, the number of the injection was 82 and it took about 10 days. The separation by HSCCC was finished in 3 h and the HPLC of the BC-A obtained is shown in Fig. 10.9. The BC-A from HPLC was not always pure, whereas HSCCC gave it in a pure state and the recovery was much better than that by HPLC. A large amount (5 g) of the BC complex mixture was also separated by a cross-axis synchronous flow-through CPC using the same two-phase solvent system (38, 39).

10.7. COMPARISON OF ELUTION BEHAVIOR OF SVD AND BC BY BETWEEN HPLC AND HSCCC

As mentioned above, separation by HSCCC is based on real liquid–liquid partition. Thus it would be interesting to compare the separation behavior by

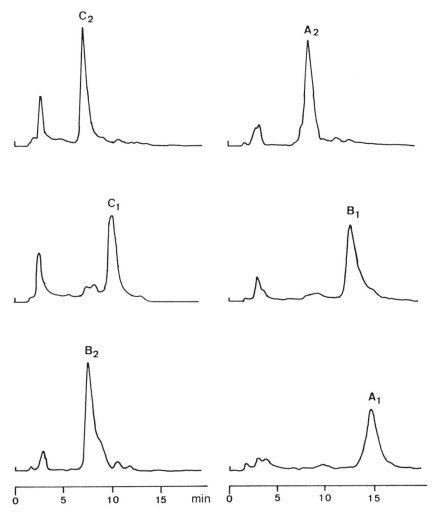

Figure 10.7. High-performance liquid chromatograms of each component of SVD. Column consists of Cosmosil 5C18 (4.6 × 150 mm); mobile phase is methanol/1 N ammonium chloride (76:24), flow-rate = 1 mL/min; detection is UV (232 nm).

HSCCC with that of HPLC on ODS silica gel. An interesting difference was found between HSCCC and HPLC in the elution behavior of SVD complex. In the case of HSCCC their order of elution was C_2, C_1, B_2, A_2, B_1, and A_1 (25), whereas C_2, B_2, A_2, C_1, B_1, and A_1 were eluted in order in the HPLC case (29). Although both methods can be characterized as reversed-phase separation modes, the substituent R_1 operated mainly on the elution order in the

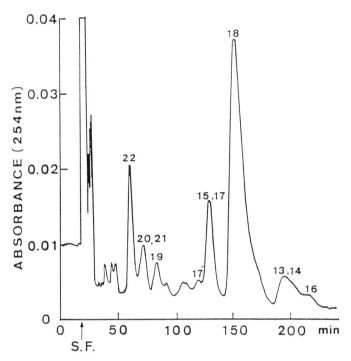

Figure 10.8. High-speed countercurrent chromatogram of BC components. Chromatographic conditions as in "Operation of HSCCC."

Table 10.5. Comparison of HPLC with HSCCC on Preparative Separation of BC-A

	Solvent System	Sample Size (mg)	Time (h)	Yield (%)
HPLC	MeOH/0.04 M Na$_2$SO$_4$ = 68–32	26.8	192	30.6
HSCCC	CHCl$_3$/EtOH/H$_2$O = 5–4–3	69.4	3	56.8

HSCCC, but the substituent R_3 considerably affected the elution in the HPLC. In other words, while the oligosaccharide moiety plays an important role in the separation by HSCCC, the aglycone moiety contributes mainly to the elution in the HPLC.

Bacitracin complex was separated under normal phase conditions. The components were eluted in order of their partition coefficients and peak 22 (BC-F) is the most nonpolar component in the mixture (26). Since the

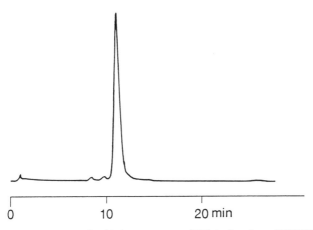

Figure 10.9. High-performance liquid chromatogram of BC-A after the re-HSCCC. Chromatographic conditions as in Fig. 10.2.

separation by HSCCC under reversed-phase conditions has not been done yet, it is difficult to compare both elution behaviors precisely. However, the elution order under normal phase conditions was not consistent with that by RP-HPLC (35). Very recently, the total structures of 13 minor components of BC have been proposed (40).

10.8. CONCLUSION

As stated earlier, SVD and BC are complex mixtures and are highly polar compounds. To examine the potential of HSCCC, it was applied to rapid preparative separation of such compounds. To effectively accomplish this, first HPLC was established for observation of separation by HSCCC and selection of a two-phase solvent. Second, an appropriate two-phase solvent system was selected by considering the partition coefficients estimated by HPLC, because optimizing the two-phase solvent system for each application is critical for successful resolution by HSCCC. Third, preparative separation was performed under the established conditions. The operations were finished in several hours and the HSCCC provided high resolution of the components of these labile antibiotics. While involatile inorganic salts are usually added to a mobile phase to increase resolution in RP-HPLC and it frequently takes much time to eliminate them, HSCCC gives considerable resolution without addition of inorganic salts. These results strongly indicate that a procedure including these three steps is suitable for separation of natural products by

HSCCC and a combination of HPLC for analysis and HSCCC for preparative separation greatly facilitates development of the needed methodology.

The experimental results suggest, as an additional aspect of HSCCC, that its separation mechanism is slightly different from that of HPLC. For the natural product chemist, it is advantageous to have as many chromatographies with different separation mechanisms as possible. An effective combination of them would be promising for rapid separation and purification of a desired compound from a complicated matrix.

ACKNOWLEDGMENT

The author would like to thank the following co-workers: M. Suzuk: I, Kimura, A Yoshikawa, Y. Yamazaki (Meijo University), H. Oka, Y. Ikai (Aichi Prefectural Institute of Public Health), S. Hattori, K. Komori (Shimadzu Co.), H. Nakazawa (The Institute of Public Health) and Y. Ito (NIH).

REFERENCES

1. Y. Ito and W. D. Conway, *Anal. Chem.*, **56**, 534A (1984).

2. Y. Ito *CRC Crit. Rev. Anal. Chem.*, **17**, 65 (1986).

3. Y. Ito, "High-Performance CCC," in N. Mandava and Y. Ito, Eds., *Countercurrent Chromatography: Theory and Practice*, Marcel-Dekker, New York, 1988, pp. 330–414.

4. A. P. Foucault, *Anal. Chem.*, **63**, 569A (1991).

5. D. G. Martin, "Countercurrent Chromatography for Drug Discovery and Development," in N. B. Mandava and Y. Ito, Eds., *Countercurrent Chromato-graphy: Theory and Practice*, Marcel-Dekker, New York, 1988, pp, 565–581.

6. D. G. Martin, R. E. Peltonen, and J. W. Nielson, *J. Antibiotics*, **34**, 721 (1986).

7. H. Oka, Y. Ikai, N. Kawamura, M. Yamada, J. Hayakawa, K.-I. Harada, K. Nagase, H. Murata, M. Suzuki, and Y. Ito, *J. High Resol. Chromatogr.*, **14**, 306 (1991).

8. T. Okuda, Y. Ito, Yamaguchi, T. Furumai, M. Suzuki, and M. Tsuruoka, *J. Antibiotics, Ser. A*, **19**, 85 (1966).

9. I. Kimura, Y. Ota, R. Kimura, T. Ito, Y. Yamada, Y. Kimura, Y. Sato, H. Watanabe, Y. Mori, K.-I. Harada, M. Suzuki, and T. Iwashita, *Tetrahedron Lett.*, **23**, 1921 (1987).

10. K.-I. Harada, S. Ito, M. Suzuki, and T. Iwashita, *Chem. Pharm. Bull.*, **31**, 3829 (1983).

11. G. T. Barry, J. D. Gregory, and L. C. Craig, *J. Biol. Chem.*, **175**, 485 (1948).

12. D. J. Hanson, *Chem. Eng. News*, **63**, 7 (1985).

13. T. Yagasaki, *J. Food Hyg. Soc. Jpn.*, **27**, 451 (1986).

14. G. A. Brewer, *Anal. Profiles Drug Subst.*, **9**, 1 (1980).

15. G. G. F. Newton and E. P. Abraham, *Biochemistry*, **53**, 597 (1953).

16. K. Tsuji and L. H. Robertson, *J. Chromatogr.*, **112**, 663 (1975).

17. L. C. Craig, W. Hauseman, and J. R. Weisiger, *J. Biol. Chem.*, **199**, 865 (1952).

18. W. Stoffel and L. C. Craig, *J. Am. Chem. Soc.*, **83**, 145 (1961).

19. C. Ressler and D. V. Kashelikar, *J. Am. Chem. Soc.*, **88**, 2025 (1966).

20. Y. Hirotsu, Y. Nishiuchi, and T. Shiba, *Peptide Chemistry 1978*, in N. Izumiya, Ed., Protein Research Foundation, Osaka, Japan 1978, p. 171.

21. W. Hausmann, J. R. Weisiger, and L. C. Craig, *J. Am. Chem. Soc.*, **77**, 723 (1955).

22. L. C. Craig and W. Konigsberg, *J. Org. Chem.*, **22**, 1345 (1957).

23. W. Konigsberg and L. C. Craig, *J. Org. Chem.*, **27**, 934 (1962).

24. L. C. Craig, W. F. Philips, and M. Burachik, *Biochemistry*, **8**, 2348 (1969).

25. K.-I. Harada, I. Kimura, A. Yoshikawa, M. Suzuki, H. Nakazawa, S. Hattori, K. Komori, and Y. Ito, *J. Liq. Chromatogr.*, **13**, 2373 (1990).

26. K.-I. Harada, Y. Ikai, Y. Yamazaki, H. Oka, H. Suzuki, H. Nakazawa, and Y. Ito, *J. Chromatogr.*, **538**, 203 (1991).

27. I. Kimura, K. Yamamoto, K.-I. Harada, and M. Suzuki, *Tetrahedron Lett.*, **28**, 1917 (1987).

28. H. Murata, N. Kojima, K.-I. Harada, M. Suzuki, T. Ikemoto, T. Shibhuya, T. Haneishi, and A. Torikata, *J. Antibiotics*, **42**, 691 (1989).

29. K.-I. Harada, I. Kimura, T. Sakazaki, H. Murata, and M. Suzuki, *J. Antibiotics*, **42**, 1056 (1989).

30. K. Tsuji, J. H. Robertson, and J. A. Bach, *J. Chromatogr.*, **99**, 597 (1975).

31. J. B. Gallagher, P. W. Love, and L. Knotts, *J. Assoc. Off. Anal. Chem.*, **65**, 1178 (1982).

32. S. Gupta, E. Pfannkoch, and F. E. Regnier, *Anal. Biochem.*, **128**, 196 (1983).

33. Y. Ohtsu, H. Fukui, T. Kanda, K. Nakayama, M. Nakano, O. Nakano, and Y. Fujiyama, *Chromatographia*, **24**, 380 (1988).

34. K. Yasuka, Y. Tamura, T. Uchida, Y. Yanagihara, and K. Noguchi, *J. Chromatogr.*, **410**, 129 (1987).

35. H. Oka, Y. Ikai, N. Kawamura, M. Yamada, K.-I. Harada, Y. Yamazaki, and M. Suzuki, *J. Chromatogr.*, **462**, 315 (1989).

36. F. Oka, H. Oka, and Y. Ito, *J. Chromatogr.*, **538**, 99 (1991).

37. H. Hostettmann and A. Marston, "Natural Products Isolation of Droplet Countercurrent Chromatography," in N. Mandava and Y. Ito, Eds., *Countercurrent Chromatography: Theory and Practice*, Marcel-Dekker, New York, 1988, pp. 465–492.

38. M. Bhatnagar, H. Oka, and Y. Ito, *J. Chromatogr.*, **463**, 317 (1989).

39. Y. Ikai, H. Oka, J. Hayakawa, K.-I. Harada, and M. Suzuki, *J. Antibiotics*, **45**, 1325 (1992).

40. Y. Ikai, H. Oka, J. Hayakawa, M. Matsumoto, M. Saito, K.-I. Harada, Y. Mayumi, and M. Suzuki, *J. Antibiotics*, **48**, 233 (1995).

CHAPTER

11

N-BROMOACETYL-3,3′5-TRIIODO-L-THYRONINE AND *N*-BROMOACETYL-L-THYROXINE: SYNTHESIS AND CHARACTERIZATION

HANS J. CAHNMANN

Genetics and Biochemistry Branch, National Institute of Diabetes, Digestive and Kidney Diseases, National Institutes of Health, Bethesda, Maryland 20892

11.1. INTRODUCTION

The thyroid hormone 3,3′,5-triiodo-L-thyronine (T_3) and its *in vivo* precursor, L-thyroxine (T_4), often also referred to as a thyroid hormone, have many life-sustaining functions. They promote growth, induce tissue differentiation, and regulate many metabolic and developmental processes. These hormones are transported in plasma by carrier proteins to the various peripheral cells where they initiate the processes necessary for carrying out these functions. They attach themselves with high affinity in a reversible fashion to the carrier proteins and, on the cellular level (membrane, cytoplasm, nucleus), to other proteins (receptors). The noncovalent attachment to carrier proteins and receptors takes place at very specific binding sites.

To investigate the mechanisms of thyroid hormone transport and action, it is desirable and often necessary to substitute a covalent (irreversible) bond between the ligand T_3 or T_4 and the carrier or receptor for the noncovalent (reversible) bond (affinity labeling) so that no dissociation can take place under drastic experimental conditions, such as electrophoresis, heat, oxidation, and reduction.

The underivatized hormones can function as affinity labels only after irradiation with light in the near ultraviolet (UV), which converts them to free radicals (photoaffinity labeling) (1). Alternatively, the hormones can serve as affinity labels *without* the use of light after their conversion to derivatives capable of establishing a covalent bond between them and the carrier or receptor protein under mild conditions (physiological pH, low temperature).

High-Speed Countercurrent Chromatography, Edited by Yoichiro Ito and Walter D. Conway.
Chemical Analysis Series, Vol. 132.
ISBN 0-471-63749-1 © 1996 John Wiley & Sons, Inc.

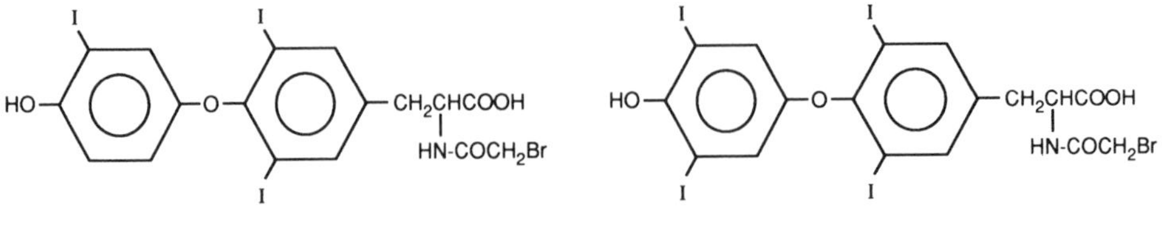

Figure 11.1. Structure of $BrAcT_3$ and $BrAcT_4$.

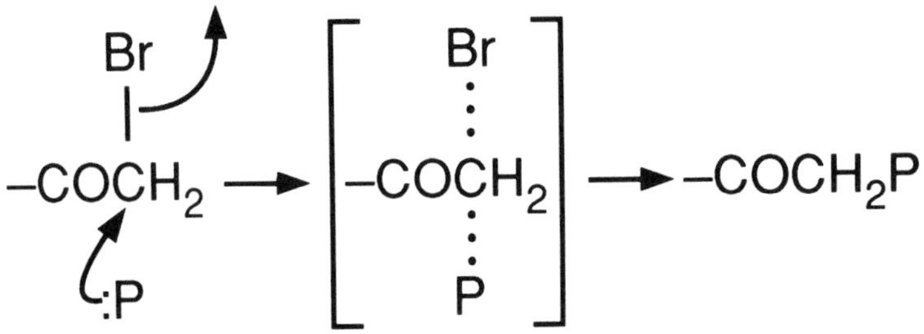

Figure 11.2. Interaction of $BrAcT_3$ and $BrAcT_4$ with a carrier or receptor protein (P).

Figure 11.1 shows the most widely used thyroid hormone derivatives, *N*-bromoacetyl-3,3′,5-triiodo-L-thyronine ($BrAcT_3$) and *N*-bromoacetyl-L-thyroxine ($BrAcT_4$). They are affinity labels by virtue of their $COCH_2Br$ group with its electron-deficient methylene carbon. A nucleophilic group in the carrier or receptor protein, for example an ε-Lys residue, reacts with that carbon to establish a covalent bond (Fig. 11.2). The elimination of Br^- (leaving group) and the attachment of the electron-deficient carbon to the protein (P) probably takes place simultaneously in coordinated manner, possibly with formation of a tetrahydral intermediate.

11.2. SYNTHESIS

One- and two-step syntheses have been used in various laboratories. In the one-step procedure (Fig. 11.3A), T_3 or T_4 is heated with bromoacetyl bromide (BrAcBr) in a suitable solvent to form the bromoacetylated hormones $BrAcT_3$ or $BrAcT_4$ with the elimination of HBr. Minor modifications of this method have also been described.

In the two-step procedure (Fig. 11.3B), bromoacetic acid is coupled with *N*-hydroxysuccinimide by means of the coupling agent dicyclohexylcarbodiimide (CCD) to form *N*-hydroxysuccinimide bromoacetate (BrAc-HSI) and dicyclohexylurea, which may be removed by filtration or centrifugation. Then

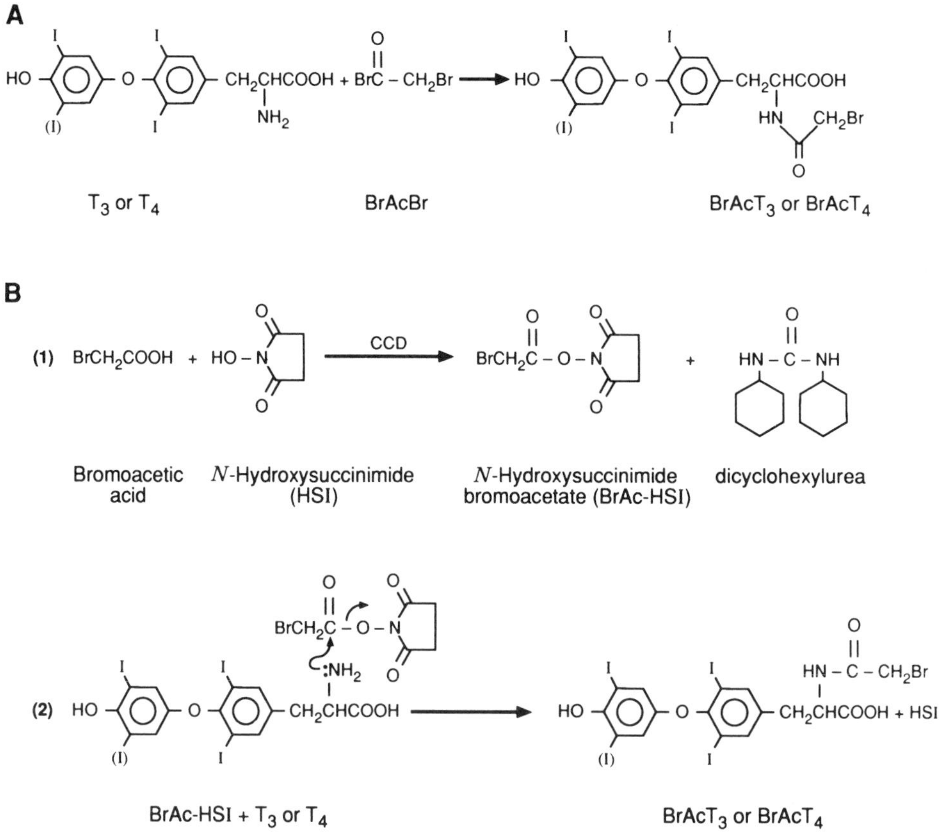

Figure 11.3. One-step (A) and two-step (B) synthesis of BrAcT$_3$ or BrAcT$_4$. CCD, cyclohexylcarbodiimide, (coupling agent).

T$_3$ or T$_4$ is added to the solution of BrAc-HSI, which results in the formation of the bromoacetylated hormones BrAcT$_3$ or BrAcT$_4$.

The one-step method is simpler and also gives superior yields and will, therefore, be described here in detail. The procedures for the synthesis of BrAcT$_3$ and that of BrAcT$_4$ differ somewhat on account of their differences in reactivities (BrAcT$_4$ is more prone to decomposition on heating than BrAcT$_3$) as well as on account of differences in solubilities of T$_3$ and T$_4$ in various solvents. However, the experimental setup is the same (Fig. 11.4).

The reaction is carried out in a round-bottom flask of appropriate size to allow refluxing, except in small-scale experiments ($<10\,\mu$mol T$_3$ or T$_4$, including the carrier-free ^{125}I-labeled hormones), in which case it is more convenient to replace the round-bottom flask with a glass tube fashioned from

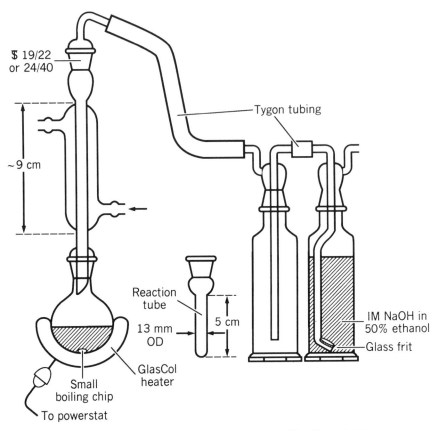

Figure 11.4. Experimental setup for the synthesis of $BrAcT_3$ or $BrAcT_4$.

a shortened Corning 99447 culture tube. The reaction flask or tube is connected to a short water-cooled reflux condenser that in turn is connected via a safety bottle to a wash bottle containing 1 M NaOH in 50% ethanol to absorb any escaping BrAcBr or HBr. The round-bottom flask may be heated with a GlasCol heater or an oil bath.

11.2.1. Synthesis of $BrAcT_3$

The procedure described below is for the bromoacetylation of 100 μmol of T_3 with the addition of a trace amount of iodine-labeled T_3 as an internal marker. This addition is convenient, but not required, since purification of the crude reaction product obtained by bromoacetylation can also be monitored by absorbance. Ethyl acetate is recommended as the solvent because in the purification of crude $BrAcT_3$ by high-speed countercurrent chromatography (HSCCC) (Chapter 6), $BrAcT_3$ is eluted as a very sharp peak (2, 3) so that the entire $BrAcT_3$ can be collected in only a few eluted fractions.

Procedure: A trace amount of iodine-labeled T_3 (e.g., 10–20 µL of $[3'-^{125}I]T_3$, Dupont/NEN 110X, 2200 Ci/mmol) is placed in a round-bottom flask (Fig. 11.4) and the small amount of solvent (50% aqueous propanol) is evaporated by brief lyophilization (10–20 min). Unlabeled T_3 (100 µmol) and ethyl acetate (25 mL) are added to the residue. Further addition (in a fume hood) of 500 µL of BrAcBr (Aldrich B5,641-2 or Eastman Kodak 42) causes dissolution of the suspended T_3. The solution, after addition of a small boiling chip, is refluxed for 10 min using a preheated GlasCol heater or oil bath. The reaction is terminated by immersion of the reaction flask in ice water, followed by addition of 500 µL of methanol (reagent or HPLC grade) to the cooled solution to decompose unreacted BrAcBr.

For the bromoacetylation of other amounts of T_3, ranging from about 10–500 µmol, the amounts of solvent and reagent and the size of the reaction flask are changed proportionately. The solution of the bromoacetylated T_3 is concentrated under reduced pressure (< 25 Torr) to about 0.5 mL in a rotating evaporator such as a Büchi Rotovapor R, using a water bath not exceeding 30°C. The concentrate of crude $BrAcT_3$ may be purified by HSCCC or stored at or below -20°C. The same method is used for the concentration of a solution of bromoacetylated T_4 (below).

For the processing of less than 10 µmol of T_3, which includes carrier-free $[3'-^{125}I]T_3$, the amounts of BrAcBr, ethyl acetate, and methanol are always 50 µL, 2.5 mL, and 50 µL, respectively, independent of the amount of T_3 used. Refluxing is reduced from 10 to 5 min. The round-bottom flask is replaced by a cylindrical tube having the dimensions shown in Fig. 11.4, which fits a metal block heater such as a Multi-Blok Heater 2090 of Labline Instruments. It also fits a Savant SpeedVac Concentrator, which may be used for concentration instead of a rotating evaporator. The tube should be precooled *before* Speed-Vac concentration by brief immersion in dry ice to avoid bumping, but no further cooling is required during concentration. On the contrary, slight warming will prevent ice formation and thus speed up concentration.

11.2.2.   Synthesis of $BrAcT_4$

The procedure is the same as that described above for $BrAcT_3$, except that 1,2-dimethoxyethane, DME, (ethylene glycol dimethyl ether, high-performance liquid chromatography (HPLC) grade, Aldrich 30743-2) is substituted for ethyl acetate because T_4, unlike T_3, does not dissolve in an ethyl acetate–BrAcBr mixture. Furthermore, when trace amounts of T_4 (< 10 µmol) are to be processed, refluxing time is shortened to 2–2.5 min, including about 1 min required for heating the reaction mixture to incipient boiling. (The metal block heater must be preheated to 90–100°C.)

11.3. PURIFICATION

The bromoacetylated thyroid hormones, synthesized as described above (see Section 11.2) require purification. Thin-layer chromatography (TLC) (silica gel), HPLC (reverse-phase), and occasionally gel column chromatography have been used for purification. Analytical HPLC can provide excellent resolution on a microgram or submicrogram scale and therefore is an excellent tool for checking identity and purity of the bomoacetylated hormones. However, on a larger scale (preparative HPLC) resolution is poor. Analytical TLC not only gives less resolution than analytical HPLC, but also has the additional shortcoming of partial decomposition of $BrAcT_3$, and even more so of $BrAcT_4$, on the silica gel surface. Preparative TLC should be avoided because, in addition to very poor resolution when thick layer (0.5–1.0 mm) plates are being used, irreversible adsorption to the silica gel results in low recoveries upon elution.

11.3.1. High-Speed Countercurrent Chromatography

This method has been used recently for the purification of the crude BrAc-hormones (3, 4). It has the virtue of being applicable to a wide range of amounts of bromoacetylated hormones (prepared from a few hundred pico-grams to several grams of T_3 or T_4, which includes the carrier-free iodine-labeled hormones). The equipment used in Ref. 3 and 4 (Ito Multi-Layer Coil Separator-Extractor, P.C. Inc., Potomac, MD), which has a medium capacity, gives good resolution of $BrAcT_3$ or $BrAcT_4$ prepared from up to about 400 mg.

An additional advantage of the method is the absence of a solid support material, such as silica gel, which could be the cause of degradation of the

--▶

Figure 11.5. Elution profiles of crude $BrAcT_3$ obtained in a one-step synthesis (see Figure 3A). A. HPLC elution profile: Waters chromatograph; 15-cm C_{18} Nova column (Waters); flow rate, 1 mL/min; sensitivity, 0.5 absorbance units/chart width; solvents, 15 mM aqueous ammonium acetate pH 4.0 (solvent A) and acetonitrile (solvent B); linear gradient (started 1 min after sample injection), (20–60% B in 40 min; internal standard, 20 μL of 10^{-3} M T_4. B. HSCCC elution profile of the same preparation of crude $BrAcT_3$: Ito Multi-Layer Coil Separator-Extractor, P.C. Inc., Potomac, MD 20854; eluates were monitored by UV absorption at 280 nm; two-phase solvent system, hexane/ethyl acetate/methanol/15 mM aqueous ammonium acetate, pH 4 (1:1:1:1); flow rate, 3 mL/min; fraction size, 3 mL; revolution, 800 rpm; retention of stationary phase, 77% of column capacity (330 mL); maximal column pressure, 65 psi; SF, solvent front.

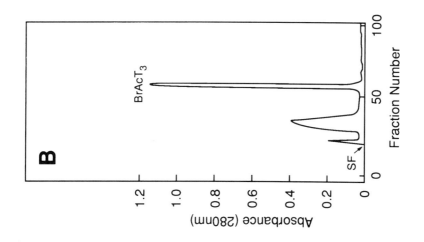

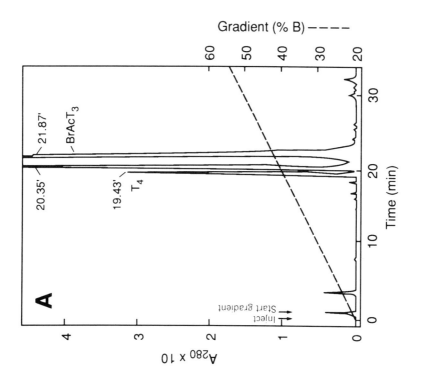

solute or of its irreversible adsorption. The only solid material with which the solution comes in contact is the Teflon tubing that contains the liquid column. HSCCC can be used in conjunction with HPLC since the two methods provide different elution profiles and different resolutions of certain compounds. Hence, two or more solutes not easily resolved by one method may be well separated by the other. An example of this is shown in Fig. 11.5, where the HPLC elution profile of crude $BrAcT_3$ (A) is compared with the HSCCC elution profile of the same $BrAcT_3$ preparation (B). A satellite peak immediately precedes the $BrAcT_3$ peak in A. In contrast, the same material (shown by mass spectrometry to consist largely of acetyl-T_3) emerges from the HSCCC column more than twenty 3-mL fractions earlier.

Various aspects of HSCCC are described elsewhere in this volume. Therefore, only those aspects that pertain to the purification of $BrAcT_3$ and $BrAcT_4$ will be discussed in this chapter.

11.3.2. Solvents

A good two-phase solvent system for the purification of the crude BrAc hormones consists of hexane/ethyl acetate/methanol/15 mM aqueous ammonium acetate, pH 4. The last component is prepared by adjusting the pH of an aqueous solution of 1.2 g of ammonium acetate (reagent grade) to 4 with acetic acid and then adding water to a total volume of 1 L. The use of Milli-Q water (Millipore) or double-distilled water is recommended. The volume ratio of the four components of the solvent system must be adjusted so that good resolution is achieved within a reasonable time (2 h or less). In most instances, a ratio of 4:5:4:5 is appropriate. Only for large amounts of $BrAcT_3$ (prepared from more than $\sim 10\,\mu mol\,T_3$) a ratio of 1:1:1:1 is preferable.

The concentrate of a solution of crude $BrAcT_3$ or $BrAcT_4$ (see Section 11.2) is brought to 0.5–1.5 mL with ethyl acetate and then equal volumes of the other three components of the solvent system are added. The resulting two-phase solution (which, under some circumstances, e.g., overloading, may contain undissolved material) is injected into the chromatographs. The lower, water-rich phase is the mobile phase. The upper, water-poor phase is station-

Figure 11.6. HSCCC elution profiles. Same conditions as in Figure 11.5B, except that the solvent ratio in 6B, 6C, and 6D is 4:5:4:5. The profiles 6A and 6B have been generated from crude BrAc-hormones obtained by bromoacetylation of 100 μmol T_3 or T_4 to which 20 μL of ^{125}I-labeled T_3 or T_4 (Dupont/NEN 110X or 111X, respectively) had been added as internal marker. The profiles 6C and 6D have been generated from crude BrAc-hormones obtained by bromoacetylating 20 μL of Dupont/NEN 110X or 111X only (omitting unlabeled T_3 or T_4). Centrifugation was stopped after $BrAcT_3$ or $BrAcT_4$ had been eluted. In 6C and 6D, elution of column contents was then continued using methanol as eluent (6 mL/min). Arrows, solvent front. Not all fractions are marked by dots.

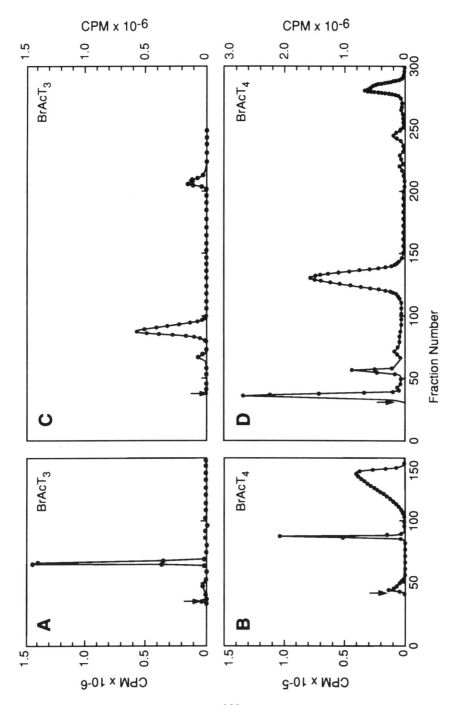

329

ary. After $BrAcT_3$ or $BrAcT_4$ has been eluted, all solutes retained in the stationary phase can be eluted with methanol (see legend to Fig. 11.6).

11.3.3.   Elution

While flow-rate and fraction size may be varied to suit particular needs, it was found that a flow-rate of 3 mL/min and a fraction size of 3 mL are in most instances appropriate with the above-mentioned PC chromatograph. The flow-rate may be increased to 6 mL/min for the elution of solutes retained in the column.

11.3.4.   Eluate Scanning

Eluate fractions are monitored at 280 nm by means of a UV absorbance monitor (e.g., LKB Uvicord S) and radioactivity of the collected eluate fractions is assessed with a gamma counter (e.g., Packard Auto-Gamma 5000 series).

Typical examples of HSCCC elution profiles obtained by bromoacetylation of $100 \mu mol T_3$ or $100 \mu mol T_4$ (containing a small amount of the corresponding iodine-labeled hormone as an internal marker), are shown in Fig. 11.6 (A–D). In these and similar runs, the yields of $BrAcT_3$ or $BrAcT_4$ calculated on the basis of radioactivity and taking into account the radioactivity that is retained in the column varied from 77 to 80% on a 100-μmol scale (Fig. 11.6A and B) and from 49 to 77% when carrier-free iodine-labeled hormones were used (Fig. 11.6C and D). Yields of carrier-free $[3',5'-^{125}I]$ $BrAcT_4$ are considerably lower than those of carrier-free $[3'-^{125}I]BrAcT_3$ presumably due to the fact that $BrAcT_4$ undergoes chemical degradation on heating much more easily than $BrAcT_3$. Figure 11.6A shows the typical needle-shaped peak of $BrAcT_3$, which is caused by the coelution of bromoacetic acid, a side product of the bromoacetylation of T_3 (2). Figure 11.6B shows the equally typical skewed peak of $BrAcT_4$ caused by overloading. All fractions under this peak contain 98 + % pure $BrAcT_4$ as judged by analytical HPLC. When trace amounts of T_4 are bromoacetylated (Fig. 11.6D), the resulting $BrAcT_4$ peak assumes a symmetrical shape. All peaks that are not needle-shaped (Fig. 11.6B–D) can be sharpened to permit collection of the desired BrAc-hormone in only a few eluted fractions (2).

11.3.5.   Storage of HSCCC-Purified $BrAcT_3$ and $BrAcT_4$

While HSCCC fractions containing these hormone derivatives can be stored in stoppered test tubes at $-20°C$ without significant deterioration for at least 2 months ($BrAcT_3$) or 2 weeks ($BrAcT_4$), it is preferable to convert them to dry powders for long-term storage at or below room temperature in well-sealed amber vials or in vials that are otherwise shielded from light. Alternatively, the BrAc-hormones may be stored as solutions in peroxide-free 1,2-dimethoxyethane (ethylene glycol dimethyl ether, HPLC grade) (4). When kept in well-sealed containers at $-20°C$, these solutions show no evidence of significant deterioration (analytical HPLC) after storage for 2–3 months ($BrAcT_3$) or 1 month ($BrAcT_4$), while the dry powders of the BrAc-hormones are stable for at least 1 year at room temperature. These dry powders should not be prepared by evaporation or lyophilization, but by precipitation with water (4).

11.4. VERIFICATION OF IDENTITY AND PURITY

The most useful method for identity and purity checks is analytical HPLC, provided that authentic samples of $BrAcT_3$ and/or $BrAcT_4$ are available. If they are not, pertinent HSCCC fractions must be analyzed by either mass spectrometry or nuclear magnetic resonance or both. (Consult Refs. 3 and 4 for examples). It is sometimes advisable to remove ammonium acetate from the HSCCC fractions prior to these analyses. This can be done by extraction of the BrAc-hormones with peroxide-free ether after partial evaporation of the fractions (4). Thin-layer chromatography, although less precise than HPLC and hence less reliable, may serve as a useful fast method of providing preliminary information, but should not be relied upon as the *only* method of checking identity and/or purity.

11.4.1. HPLC

Figure 11.7 shows the HPLC elution profile of a mixture of HSCCC-purified $BrAcT_3$ and $BrAcT_4$. Since retention times vary not only with the type, age, and dimensions of the column, but also with such variables as flow-rate and operating temperature, one or two internal standards (T_3, T_4) should be used in order to determine whether or not peaks obtained in different HPLC runs represent identical compounds. *Two* standards are preferable because the difference between the retention time of the BrAc-hormone (BrAc-T_3 or $BrAcT_4$) and that of T_4 is strictly proportional to the difference between the retention times of the two standards.

11.4.2. TLC

The relative migration rates of HSCCC-purified $BrAcT_3$ and $BrAcT_4$ on silica gel thin-layer plates are shown in Fig. 11.8. Of the three solvents tested, the

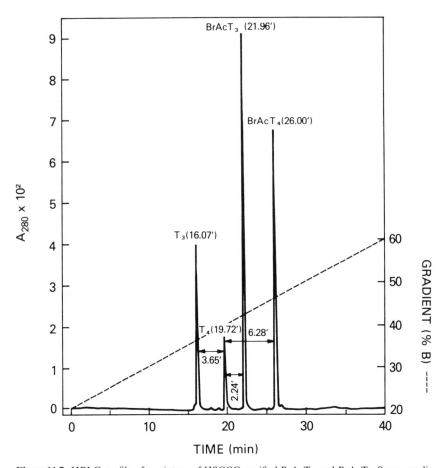

Figure 11.7. HPLC profile of a mixture of HSCCC-purified BrAcT$_3$ and BrAcT$_4$. Same conditions as in Figure 11.5A, except for the following: sensitivity, 0.1 absorbance units/chart width; gradient started 10 min after sample injection; two internal standards, T$_3$ and T$_4$ (2 μL of 1 mg/mL solutions).

widely used solvent A hardly discriminates between the two BrAc-hormones. This solvent also discriminates poorly between either one of the two hormone derivatives and various side products of unknown nature formed in the course of their syntheses. In contrast, solvents B and C exhibit a much greater degree of resolution and are therefore recommended for preliminary identification or purity checks. Their only shortcoming is that they must be prepared fresh every day because they gradually lose resolving power on storage, presumably due to the formation of methyl acetate.

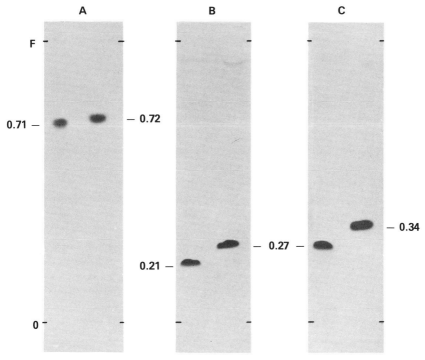

Figure 11.8. TLC of HSCCC-purified $BrAcT_3$ and $BrAcT_4$. Whatman K6F silica gel plates, 5×20 cm, prewashed with solvent system A; solvent systems: A, acetic acid/ethyl acetate, 1:9; B, chloroform/methanol/acetic acid, 45:4:1; C, dichloromethane/methanol/acetic acid, 18:1:1; $BrAcT_3$, left lanes; $BrAcT_4$, right lanes; O, origin; F, solvent front. Numbers denote R_f values.

11.4.3. Absorption Spectra in the Near UV

It is customary to determine the concentration of T_3 and/or T_4 solutions by measuring the absorbance (A) at or above the wavelength of maximal absorption (λ_{max}) which, in alkaline solution, is 320 nm for T_3 and 325 nm for T_4. Corresponding λ_{max} values for $BrAcT_3$ and $BrAcT_4$ are virtually identical, 320 and 326 nm, respectively (4), as expected (Fig. 11.9). However, a determination of the concentration of $BrAcT_3$ or $BrAcT_4$ solution by measuring A at or above these wavelengths is permissible only if it has been established independently by analytical HPLC that the solute causing the absorbance peak in the near UV is pure, uncontaminated with its congener ($BrAcT_4$ or $BrAcT_3$, respectively). These hormone analogs are prone to undergo slow chemical degradation in dilute alkali ($BrAcT_4$ much more so than $BrAcT_3$). Any degradation product that keeps the aromatic portion of the molecule intact,

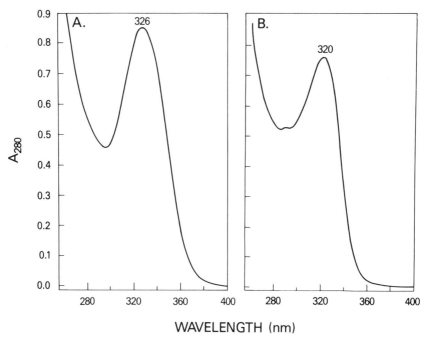

Figure 11.9. Absorption spectra in the near UV of BrAcT$_3$ (A) and BrAcT$_4$ (B), determined in freshly prepared solutions in 0.1 M NaOH.

but has an altered side chain must be expected to produce an absorption peak very close to that of the parent compound (4). No degradation of either BrAcT$_3$ or BrAcT$_4$ is detectible by analytical HPLC within 24 h in solutions in 0.1 M NaOH.

11.5. SUMMARY

A simple one-step method yields the affinity labels BrAcT$_3$ and BrAcT$_4$ in good yield. A solution or suspension of T$_3$ or T$_4$ in an appropriate solvent is refluxed, after addition of BrAcBr, for a specified length of time. Contaminants present in the reaction product are removed by HSCCC.

HSCCC is a very versatile method because it allows efficient purifications of more than 1 g of BrAcT$_3$ or BrAcT$_4$ as well as of trace amounts, such as present in carrier-free iodine-labeled BrAc-hormones.

Analytical HPLC is the method of choice for verifying the degree of purity of HSCCC-generated BrAc-hormones. A useful auxiliary method is analytical

TLC on silica gel plates. Light absorption in the near UV can be used for the determination of the concentration of $BrAcT_3$ or $BrAcT_4$ in freshly prepared alkaline solutions of the pure affinity labels. Mass spectrometry and/or proton nuclear magnetic resonance (^1H NMR) are required for ultimate identification only if authentic samples of $BrAcT_3$ and $BrAcT_4$ are not available.

ACKNOWLEDGEMENT

The work described in this chapter was performed in collaboration with Dr. Yoichiro Ito, National Institutes of Health, Bethesda, Maryland. It could not have been written without the benefit of his expertise in HSCCC. The author thanks Dr. Ito for his help, as well as for many inspiring discussions.

REFERENCES

1. B. van der Walt, V. M. Nikodem, and H. J. Cahnmann, *Proc. Natl. Acad. Sci USA*, **79**, 3508 (1982).

2. Y. Ito, Y. Shibusawa, H. M. Fales, and H. J. Cahnmann, *J. Chromatogr.*, **625**, 177 (1993).

3. H. J. Cahnmann, E. Gonçalves, Y. Ito, H. M. Fales, and E. A. Sokoloski, *J. Chromatogr.*, **538**, 165 (1991).

4. H. J. Cahnmann, E. Gonçalves, Y. Ito, H. M. Fales, and E. A. Sokoloski, *Anal. Biochem.*, **204**, 344 (1992).

CHAPTER

12

SEPARATION AND PURIFICATION OF DYES BY CONVENTIONAL HIGH-SPEED COUNTERCURRENT CHROMATOGRAPHY AND pH-ZONE-REFINING COUNTERCURRENT CHROMATOGRAPHY

ADRIAN WEISZ

Office of Cosmetics and Colors, Food and Drug Administration, Washington, DC 20204

12.1. INTRODUCTION

High-speed countercurrent chromatography (HSCCC) has been applied to the separation of dyes since the mid-1980s when Fales et al. (1) reported their separation of the triphenylmethane color methyl violet 2B. The subsequent applications of HSCCC to this area were demonstrated by Freeman and co-workers (2, 3) through their purification of azo-acid, direct, and disperse dyes. These works were reviewed in-depth by Conway (4) and Ruth and Mandava (5); therefore, further discussion will not be included here. Since the publication of Conway's book, the only reported HSCCC separations of dyes pertain to those of the xanthene class, which are used primarily as color additives in food, drugs, and cosmetics, and as biological stains. The literature documents the presence of contaminants that vary in type and level among batches of these dyes (6, 7). Excess levels of specified contaminants may result in the failure of a batch of color additive to meet certification requirements established by the U.S. Food and Drug Administration (FDA) (8). The presence of contaminants in the dyes can also cause variations in their staining properties. This problem has been documented as the cause of anomalous histochemical staining results (9–12). This chapter reviews the studies of HSCCC separation of dyes published after Conway's book, as well as recent results of the separation and purification of hydroxyxanthene dyes by pH-zone-refining countercurrent chromatography (pH-zone-refining CCC) (13–15), a modified form of conventional HSCCC.

High-Speed Countercurrent Chromatography, Edited by Yoichiro Ito and Walter D. Conway. Chemical Analysis Series, Vol. 132.
ISBN 0-471-63749-1 © 1996 John Wiley & Sons, Inc.

12.2. INSTRUMENTATION

Most of the separations described below were performed using commercially available HSCCC centrifuges that hold an Ito multilayer-coil separation column (16). The column rotates about its own axis, 10 cm from the central axis of the centrifuge. The multilayer coil used for most of the semipreparative- and preparative-scale separations was constructed (by Ito) of polytetrafluoroethylene (PTFE) tubing, either 160 or 165 m × 1.6 mm i.d., with a total capacity of approximately 300 or 325 mL, respectively. The β value (a centrifugal parameter) (17) ranged from 0.5 at the internal terminal to 0.85 at the external terminal. The columns typically consisted of 16 coiled layers, but in one case (18), only six coiled layers were present. (Similar columns are commercially available from Shimadzu, Kyoto, Japan; from P.C. Inc., Potomac, MD; and from Pharma-Tech Research Corp., Baltimore, MD). The separations were initiated by filling the entire column with the stationary (upper nonaqueous) phase, using a high-performance liquid chromatographic (HPLC) or metering pump (e.g., Beckman Accu-Flo pump or a Milton Roy minipump; LDC Analytical, Riviera Beach, FL), and then loading the sample, dissolved in a minimum volume of the solvent system used, into the column by syringe. The mobile (lower aqueous) phase was then pumped into the column at 2 or 3 mL/min while the column was rotated at 800 rpm. The column effluent was monitored with an ultraviolet (UV) detector (e.g., Uvicord S; LKB Instruments, Stockholm, Sweden) at 206, 254, or 280 nm, and a fraction collector (e.g., Ultrorac, LKB Instruments) was used to obtain 2-, 3-, or 6-mL fractions. The fractions separated by HSCCC were analyzed by analytical HPLC (18, 19). The separations described in Sections 12.4.4.3, 12.4.4.6, and 12.4.5.3 were performed with a Model CCC-1000 high-speed countercurrent chromatograph (Pharma-Tech Research Corp.) that holds three Ito multilayer coils connected in series (20). The basic design of this apparatus is described in Chapter 1 in this volume. The total capacity of the columns is approximately 325 mL. Instrumental details related to the separations performed with this apparatus are described in Section 12.4.4.3.

12.3. CONVENTIONAL HSCCC SEPARATIONS

12.3.1. Sulforhodamine B

Oka et al. (18) used HSCCC to separate the aminoxanthene dye sulforhodamine B, **1** (C.I. Acid Red 52, Colour Index No. 45100), from minor components of the Japanese color additive Food Color Red No. 106 (R-106).

   The sample to be separated was a relatively complex mixture [thin-layer chromatography (TLC) separation resulted in three spots, and HPLC separation yielded nine peaks, Fig. 12.1a]. The major problem was selection of

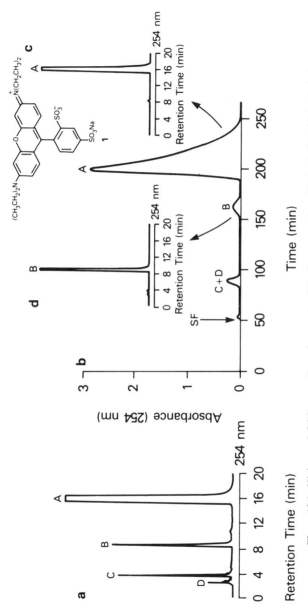

Figure 12.1. High-speed CCC separation of a sample of the Japanese color additive R-106. (a) Analytical HPLC of the sample. (b) High-speed CCC separation of 25 mg of sample. (c) Analytical HPLC of combined HSCCC fractions corresponding to peak A. (d) Analytical HPLC of combined HSCCC fractions corresponding to peak B. [Adapted from Ref. 18.]

339

a two-phase solvent system that would adequately partition the components of interest (peaks A and B, corresponding to **1** and a subsidiary dye, respectively, in Fig. 12.1a).

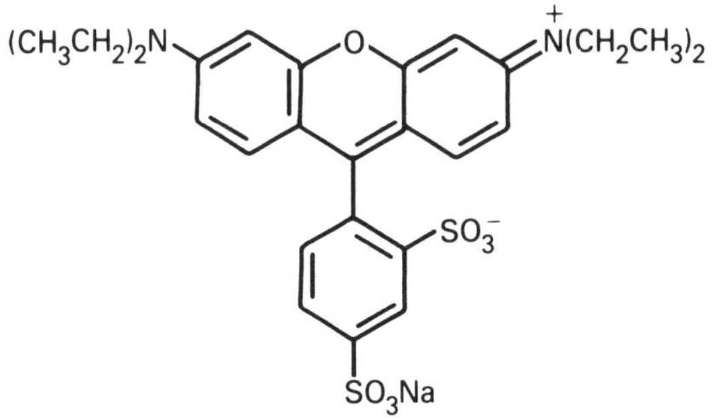

1

Specifically, it was necessary for the respective partition coefficients (K) of the two compounds to be different from one another (and from those of the other components in the mixture), and to be close to unity. This requirement was complicated by the fact that the main component of the mixture, **1**, was highly soluble in water because of its ionic dipolar structure, and relatively insoluble in common organic solvents. The solvent system chosen, n-butanol/0.01 M aqueous trifluoroacetic acid (TFA) (1:1), separated 25 mg of R-106 (Fig. 12.1b), resulting in 21 mg of the main component, **1**, at a purity level of 99.9% (Fig. 12.1c). The major impurity in that specific sample (peak B, Fig. 12.1a) was found to be a desethylated subsidiary color that was isolated at 98% purity (0.9 mg, Fig. 12.1d). The separated compounds were identified by use of fast atom bombardment–mass spectrometry (FAB–MS).

12.3.2. Tetrabromotetrachlorofluorescein and Phloxine B

HSCCC was used to purify 2′,4′,5′,7′-tetrabromo-4,5,6,7-tetrachlorofluorescein, **2** (Colour Index No. 45410:1), and its disodium salt, Phloxine B, **3** (Colour Index No. 45410) (21). Compounds **2** and **3** are the major components of the U.S. certified color additives designated as D&C Red No. 27 and D&C Red No. 28, respectively (8). The color additives are permitted for use in drugs and cosmetics.

The experiments separated 50 mg of commercial dye and yielded approximately 23 mg of **2** or of **3** at a purity level of 99.9% (Fig. 12.2).

Attempts to purify larger quantities (210 mg and 2 g) of commercial Phloxine B in that study were unsuccessful. Trials to separate larger amounts

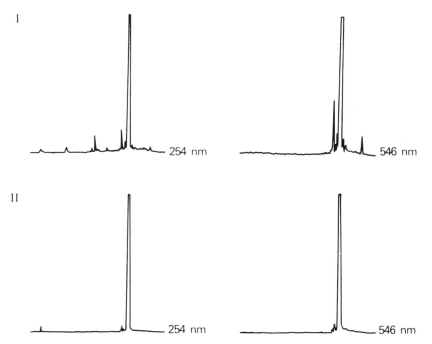

2   **3**

resulted in anomalous double-peak elution curves representing identical, partially purified dye components (Fig. 12.3). This result was attributed to the highly concentration-dependent partition behavior of the dyes in the solvent system used, *n*-butanol/ethyl acetate/0.01 *M* aqueous ammonium acetate

Figure 12.2. Analytical HPLC of a commercial sample of Phloxine B (I) and of the same sample after purification by HSCCC (II). [Adapted from Ref. 21.]

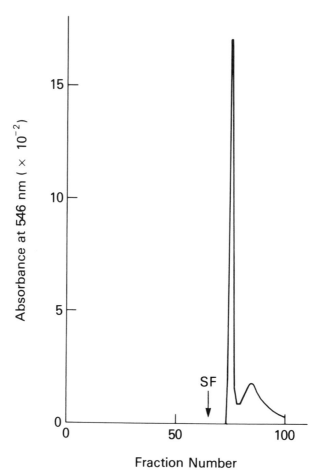

Figure 12.3. Double-peak HSCCC elution curve obtained in the separation of a 2-g sample of commercial Phloxine B. [Reprinted from Ref. 21. SF = solvent front.]

(NH_4OAc) (1:1:2) (Fig. 12.4). The problem was circumvented in a later study, in which **2** was successfully separated from multigram quantities of D&C Red No. 28 by pH-zone-refining CCC (22) (see Section 12.4.4.5).

12.3.3.   Complementary Use of HSCCC and Preparative HPLC in the Separation of a Synthetic Mixture of Dyes

High-speed CCC was effectively implemented as a complement to preparative reversed-phase HPLC (RP-HPLC) for the separation of a complex mixture of

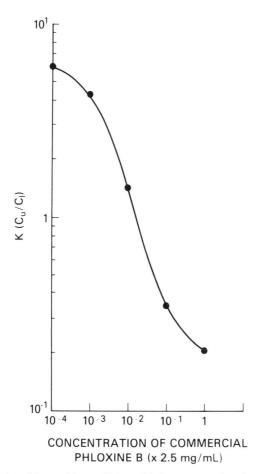

Figure 12.4. Variation of the partition coefficient with the concentration of commercial Phloxine B in *n*-butanol/ethyl acetate/0.01 *M* NH$_4$OAc (1:1:2). [Reprinted from Ref. 21.]

brominated tetrachlorofluorescein dyes (19). An attempt to prepare lower-brominated subsidiary colors of **2** by partial bromination of 4,5,6,7-tetra-chlorofluorescein (TCF), **4** (23) (Scheme 1), yielded a complex mixture shown by RP-HPLC to contain four main components (Fig. 12.5a). This mixture was separated by preparative RP-HPLC into fractions containing pure 4',5'-dibromotetrachlorofluorescein, **5**, and other fractions containing multiple components.

The combined fractions (100 mg), which included one of the major products (peak C, Fig. 12.5b) were further separated by HSCCC. The two-phase solvent

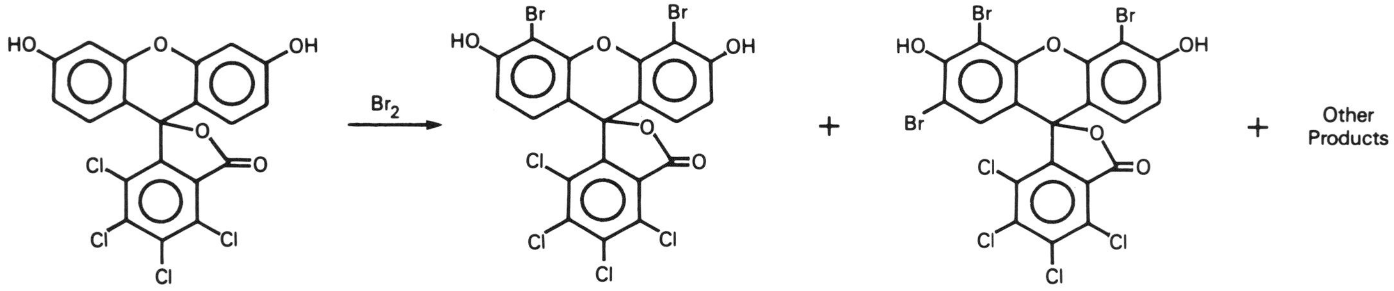

Scheme 1

344

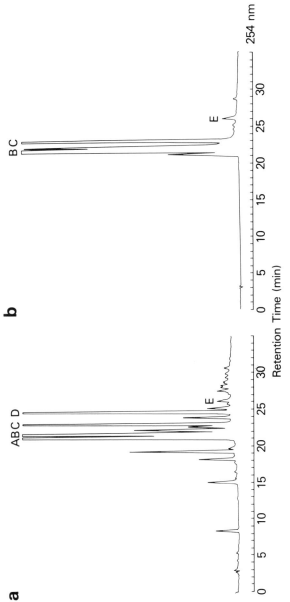

Figure 12.5. Analytical RP-HPLC of (a) the crude brominated tetrachlorofluorescein and (b) the combined preparative RP-HPLC fractions that were further separated by HSCCC. [Adapted from Ref. 19.]

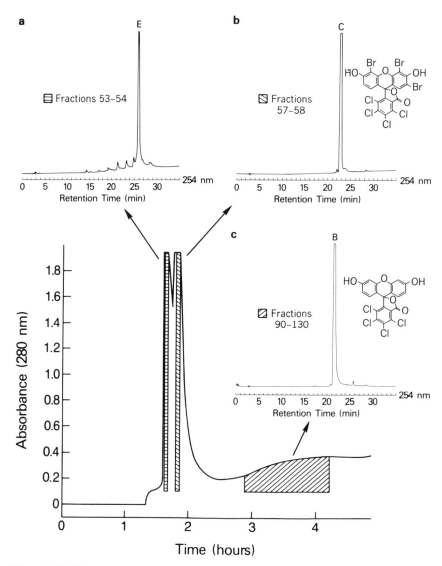

Figure 12.6. High-speed CCC separation of combined preparative RP-HPLC fractions corresponding to peaks B, C, and E (see Fig. 12.5b). (a), (b), and (c) are the chromatograms resulting from analytical RP-HPLC performed on the separated compounds. [Adapted from Ref. 19.]

system. *n*-butanol/ethyl acetate/0.01 *M* NH₄OAc (1:1:2) was chosen for this
separation because it had been previously used successfully to purify the
tetrabrominated analogs **2** and **3** (21). The HSCCC separation of the mixture
resulted in three peaks, and the corresponding eluates were collected in
fractions 53–54, 57–58, and 90–130 (Fig. 12.6), respectively.

The compounds corresponding to fractions 57–58 and 90–130 were isolated as lactones (20 and 12 mg, respectively) and identified by proton nuclear
magnetic resonance (^{1}H NMR) and chemical ionization mass spectrometry
(CIMS) as 2′,4′,5′-tribromotetrachlorofluorescein, **6** (peak C in Fig. 12.5b), and
the unbrominated starting material, TCF, **4** (peak B in Fig. 12.5b). The
HSCCC separation concentrated the component corresponding to the minor
peak E (Fig. 12.5b) into fractions 53–54, as shown in the analytical RP-HPLC
of these fractions (Fig. 12.6a). Peak E remained unidentified.

12.4. pH-ZONE-REFINING COUNTERCURRENT
CHROMATOGRAPHY

12.4.1. Introduction

Recently, Ito et al. (24) reported that addition of a short-chain organic acid to
the sample solution prior to HSCCC separation results in sharpening of the
peaks obtained from elution of compounds containing carboxylic acid groups.
Based on this finding, a new technique for the preparative HSCCC separation
of multigram mixtures of organic acids (that contain one or more —COOH
groups) was developed: pH-zone-refining CCC (13–15). This technique was
applied to the separation of organic bases as well (25). This chapter, however,
discusses applications of this method only to acidic dyes of the hydroxyxanthene type.

12.4.2. Principle

The principles (14) of this technique are described in detail in Chapter 6.
Briefly, pH-zone-refining CCC requires that an acid, such as TFA, be used to
retain the analytes by enhancing their partitioning into the stationary organic
phase.This is achieved by adding acid to the sample solution and/or to the
stationary phase. An aqueous base (e.g., aqueous ammonia) is used as the
mobile phase, and the individual components of the mixture of acids are
isocratically eluted in the order of their pK_a values and hydrophobicities. In
the separation column, each major sample component is concentrated into
a broad band (pH zone) behind the distinct band of the retaining acid. Minor
components present within each band are forced to the boundaries of the pH

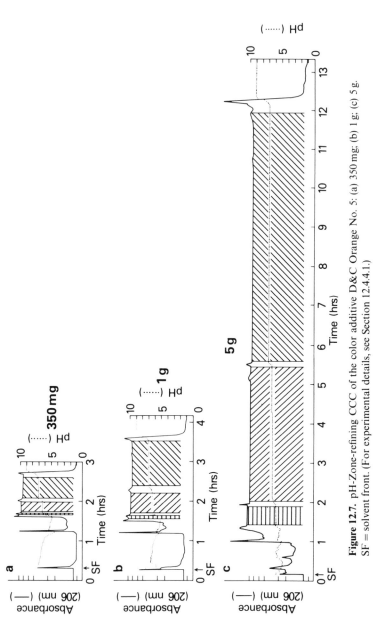

Figure 12.7. pH-Zone-refining CCC of the color additive D&C Orange No. 5: (a) 350 mg; (b) 1 g; (c) 5 g. SF = solvent front. (For experimental details, see Section 12.4.4.1.)

348

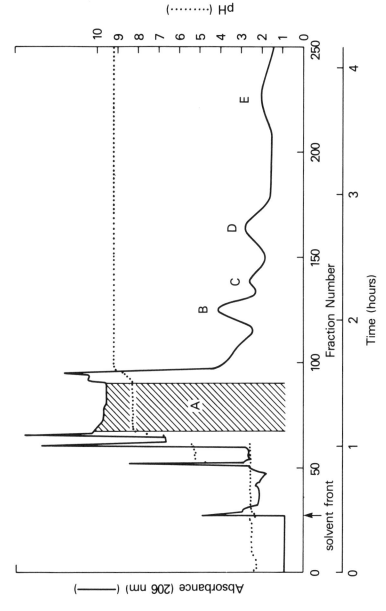

Figure 12.8. pH-Zone-refining CCC of a 350-mg portion of commercial 4,5,6,7-tetrachloro-fluorescein (4).

349

zone, according to their partition coefficients (K). This separation process continues throughout the run, resulting in the consecutive elution of individual major components. Although UV detection may not indicate that separations have occurred (yielding one broad rectangular peak), the pH values of the eluted fractional form a series of plateaus that correspond to the separated components (Fig. 12.7). An increase in sample size results in lengthening each plateau without changing the overall elution profile (Fig. 12.7).

Depending on the composition of the mixture to be separated and on the two-phase solvent system used, the UV elution chromatogram may show one or both of the two distinct patterns: rectangular peaks corresponding to the elution of the separated acidic components and broad Gaussian-shaped peaks that may correspond to nonionizable compounds (e.g., that lack carboxyl groups). In Fig. 12.8, peak A corresponds to elution of TCF (acid form); peaks B, C, D, and E probably correspond to nonionizable compounds.

In the separation of analytes that are only partially soluble in the solvent system, the technique permits loading the sample as a suspension into the separation column. The hydrodynamic mechanism for the separation, along with a simple mathematical treatment of the process, are presented in Chapter 6. Simulated chromatograms for separation of organic acids by pH-zone refining CCC were also obtained by using a computer program (26). The following discussion will concentrate on the preparative-scale separation and purification of hydroxyxanthene dyes by pH-zone-refining CCC.

12.4.3. Preliminary Requirements for pH-Zone-Refining CCC

12.4.3.1. Sample Requirements

pH-Zone-refining CCC may be applied either to purification or to separation of free acids or their derivatives (e.g., lactones) which may be in equilibrium with the corresponding acids under the conditions of the separation experiment. pH-Zone-refining CCC can be effectively used to separate two or more very closely related acids if they differ, even slightly, in their acidity and/or in their hydrophobicity. A study involving the preparative-scale separation (400 mg of crude mixture) of the stereoisomeric acids 9 and 10 (obtained from the coresponding mixture of esters 7 and 8) (27) exemplifies the application of this requirement (Scheme 2).

The separation was possible because of the difference in acidity (~ 0.5 pK units) between an axially oriented (a) and an equatorially oriented (e) carboxyl group in cyclohexanecarboxylic acids (28, 29). The stronger acid, trans isomer **9** (equatorial carboxyl group), eluted before the weaker acid, cis isomer **10**

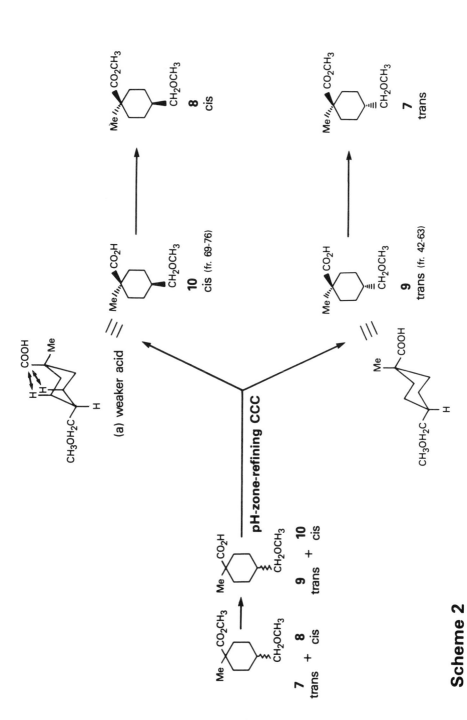

Scheme 2

351

(axial carboxyl group) (27). Experimental work is in progress on the separation of sulfonated dyes which have very low pK_a values.

12.4.3.2.   Selection of the Two-Phase Solvent System

The following steps are recommended for the selection of an appropriate two-phase solvent system for pH-zone-refining CCC.

1. Prepare a two-phase solvent system by adding methyl *tert*-butyl ether (MTBE) [or diethyl ether (Et_2O)] to distilled water, and thoroughly equilibrate it in a separatory funnel at room temperature.

2. Deliver 2-mL aliquots of the upper (U) and the lower (L) phase into a test tube. Add the sample and TFA to the solvents (to produce a concentration of approximately 10 mM). Equilibrate the mixture by vigorous stirring. Dilute an aliquot from each phase with solvent (e.g., methanol) and measure the absorbance at an appropriate wavelength, using a spectrophotometer. Obtain the partition coefficient K_a(U/L) by dividing the absorbance of the upper phase by that of the lower phase.

3. In a similar way, deliver 2-mL aliquots of each phase (from the first step above) into a test tube. Add the sample and a small amount of aqueous ammonia ($\sim 28\%$) to the mixture (to produce a base concentration of approximately 10 mM). Equilibrate the mixture and measure the partition coefficient K_b(U/L) as previously described.

If $K_a \gg 1 \gg K_b$, the above solvent system can be effectively used to separate the sample components. When K_a is 2 or less, the above test should be repeated with a more hydrophobic solvent system, such as n-hexane/ethyl acetate/methanol/water (1:1:1:1). When K_b is greater than 0.5, a more polar solvent system, such as n-butanol/water, may be tested. However, the use of polar solvent systems tends to produce emulsions, resulting in loss of the stationary phase from the column. In this case, the use of a cross-axis coil planet centrifuge (CPC) (30) is recommended to provide enhanced retention of the stationary phase.

12.4.4.   Preparative Separations of Brominated and Chlorinated Fluoresceins

12.4.4.1.   Separation of 4',5'-Dibromo-, 2',4',5'-Tribromo-, and 2',4',5',7'-Tetrabromofluorescein from the Color Additive D&C Orange No. 5

D&C Orange No. 5 (Colour Index No. 45370:1) is a U.S.-certified color additive used in drugs and cosmetics. It is identified as a mixture containing

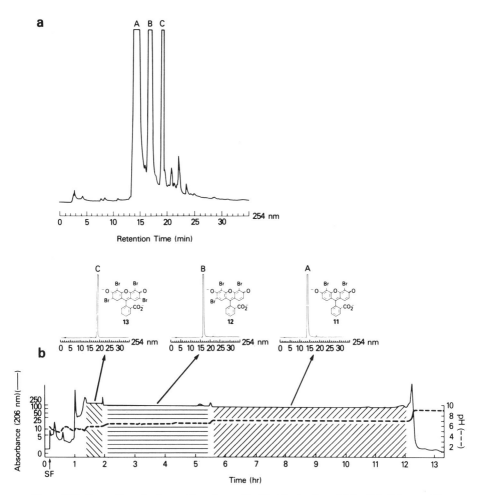

Figure 12.9. Separation of a 5-g portion of a certified batch of color additive D&C Orange No. 5 by pH-zone-refining CCC. (a) RP-HPLC of D&C Orange No. 5. (b) pH-Zone-refining CCC of 5 g of D&C Orange No. 5. *Solvent system* consists of Et_2O/acetonitrile/(0.01 M NH_4OAc adjusted to pH 9 with ammonium hydroxide), 4:1:5 by volume. *Sample* is 5 g of D&C Orange No. 5 in 80 mL of solvent system (40 mL each of upper and lower phases). TFA, 200 μL, was added to the sample solution. The isolated compounds are shown in their acid form. [Adapted from Ref. 14.]

principally three hydroxyxanthene dyes: 50–60% 4′,5′-dibromofluorescein, **11**; 30–40% 2′,4′,5′-tribromofluorescein, **12**; and not more than 10% 2′,4′,5′,7′-tetrabromofluorescein, **13** (8) (Fig. 12.9a, peaks A, B, and C, respectively).

11   **12**   **13**

The pH-zone-refining CCC of a suspension containing 5 g of D&C Orange No. 5 is shown in Fig. 12.9b. The two-phase solvent system consisted of Et_2O/acetonitrile/0.01 M NH_4OAc (4:1:5 by volume) adjusted to pH 9 with ammonium hydroxide. The solvent system was equilibrated, and the two phases were separated shortly before use (14). The sample was prepared by partitioning 5 g of certified D&C Orange No. 5 between the upper and lower phases (40 mL each). Trifluoroacetic acid (200 μL) was added to the sample solution, to yield a partial precipitate. After the dye mixture was sonicated for approximately 1 min, the suspension was injected into the column. The resulting chromatogram has a broad rectangular shape (Fig. 12.9b). The three broad absorbance plateaus correspond to the three pH plateaus. Each plateau represents elution of a pure compound, as illustrated by the associated chromatograms obtained by RP-HPLC of the combined fractions from the three hatched regions. Impurities were concentrated in the few fractions corresponding to the transition zones (not hatched) between plateaus. The sequence of elution of the three main components reflected the order of their decreasing acid dissociation constant, pK_{a2}: **13** < **12** < **11** (14). The pH-zone-refining CCC separations of 0.35, 1, and 5 g of D&C Orange No. 5 resulted in proportionate increases in peak width and pH plateau length. The small degree of overlap between the peaks was maintained (see Fig. 12.7 in Section 12.4.2). The recoveries of the isolated dyes were excellent (Table 12.1).

12.4.4.2.   Separation of 2′,4′,5′,7′-Tetrabromofluorescein and 2′,4′,5′-Tribromofluorescein from the Color Additive D&C Red No. 22 (Eosin Y)

D&C Red No. 22 (Colour Index No. 45380) is a U.S.-certified color additive used in drugs and cosmetics. It is identified as a mixture of mainly tri- and

**Table 12.1. pH-Zone-Refining Countercurrent
Chromatography of D&C Orange No. 5[a]**

Sample Size (mg)	Recovery[b] (%)		
	11	**12**	**13**
350	77.2	87.3	42
5000	82.0	90.3	77

[a]U.S. certified batch. Recoveries were calculated based on the amounts of **11**, **12**, and **13** in the dye as determined in certification by FDA (56.6, 33.7, and 6.8%, respectively).
[b]Isolated as lactones (19) and identified by CIMS and by ^{1}H NMR (14).

tetrabromofluorescein dyes: not more than 25% of the sum of the disodium salts of the tribromofluoresceins and not less than 72% of the disodium salt of 2',4',5',7'-tetrabromofluorescein, **13** (8). An RP-HPLC chromatogram of a certified batch of D&C Red No. 22 is shown in Fig. 12.10a. Eosin Y, which contains the same dye components as those found in D&C Red No. 22, has an additional important use as a biological stain (31). One difference between D&C Red No. 22 and Eosin Y is their total dye content: approximately 90% for D&C Red No. 22 and approximately 80% for Eosin Y (32). The pH-zone-refining CCC of 5 g of D&C Red No. 22 is shown in Fig. 12.10b.

The two-phase solvent system consisted of Et$_2$O/acetonitrile/0.01 M NH$_4$OAc (4:1:5 by volume). The solvent system was equilibrated, and the two phases were separated. The lower phase was adjusted to pH 8.35 by addition of ammonium hydroxide (0.01%). For this separation, TFA (800 μL) was added to the upper phase (500 mL, pH of the upper phase = 1.86) rather than to the sample solution as was the case for the separation of D&C Orange No. 5 (in Section 12.4.4.1). This approach was taken to avoid precipitation of the dye, thus simplifying injection of the sample solution into the column. Addition of the acid to the upper, stationary phase is also recommended when the sample to be separated is unstable in strong acid. The sample was prepared for injection by partitioning 5 g of D&C Red No. 22 between the upper and lower phases (15 mL each). The resulting chromatogram has a broad rectangular shape (Fig. 12.10b). The two broad absorbance plateaus correspond to the two pH plateaus. Before the first absorbance plateau (fractions 10–15), a decrease in intensity can be observed. These fractions contained the main component that eluted as a suspension. Reversed-phase HPLC analyses of these fractions showed that the precipitate was less pure than the compound that eluted later as a concentrated solution (fractions 20–210). Each horizontal plateau represents elution of a single compound as illustrated by the associated RP-

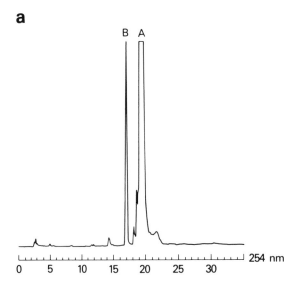

a

B A

254 nm

0   5   10   15   20   25   30

Retention Time (min)

A

13

254 nm

0   5   10   15   20   25   30

Retention time (min)

B

12

254 nm

0   5   10   15   20   25   30

Retention time (min)

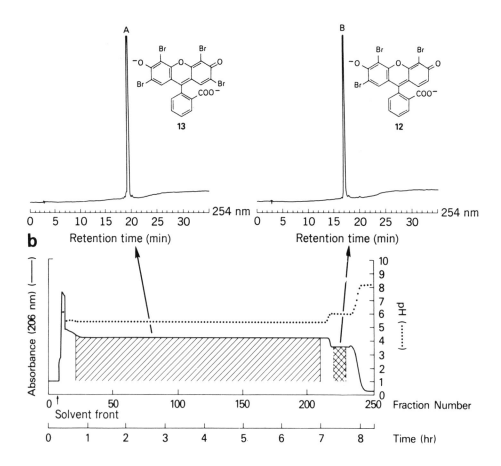

b

Absorbance (206 nm) (——)

pH (·······)

10
9
8
7
6
5
4
3
2
1
0

0   50   100   150   200   250   Fraction Number

Solvent front

0   1   2   3   4   5   6   7   8   Time (hr)

HPLC chromatograms. The dyes were isolated in the lactone form (19) and identified by CIMS and ^{1}H NMR. The separation resulted in 1.46 g of 2′,4′,5′,7′-tetrabromofluorescein, **13**, and 31 mg of 2′,4′,5′-tribromofluorescein, **12**, peaks A and B, respectively, in Fig. 12.10a. The relatively low yield obtained for the pure component **13** is due to the elution of the suspension in fractions 10–15. The elution of the suspension may be attributed to the insufficient amount of retainer acid (TFA) present in the sample solution (i.e., to convert all the sodium salt of the dye into the acid form) (22) (see also Section 12.4.4.5). Optimization of the quantity of the retainer acid (TFA) present in the sample solution may improve the recovery of pure **13** in future separations.

12.4.4.3.   Separation of 4′-Bromo- and 4′,5′-Dibromofluorescein from a Synthetic Mixture

The compound 4′-bromofluorescein (4′BrF) is needed as a reference material for the determination of contaminants in the color additives D&C Orange No. 5 and D&C Red No. 22 described above.

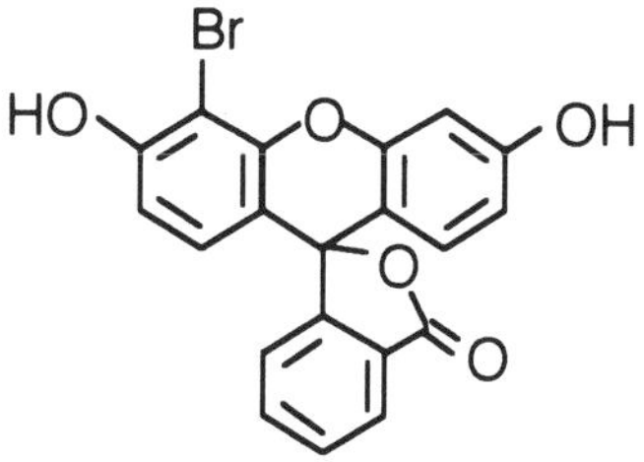

4'BrF

The HPLC analysis of the synthetic mixture [obtained by brominating fluorescein in ethanol (33)], which contained mainly 4′BrF and 4′5′-dibromo-fluorescein (4′5′-diBrF, **11**), is shown in Fig. 12.11a. The pH-zone-refining CCC obtained for the separation of 0.67 g of this mixture is shown in Fig. 12.11b.

Figure 12.10. Separation of a 5-g portion of a U.S.-certified batch of color additive D&C Red No. 22 by pH-zone-refining CCC. (a) RP-HPLC of the dye. (b) pH-Zone-refining CCC of the 5-g sample. *Solvent system* consists of Et$_2$O/acetonitrile/0.01 M NH$_4$OAc (4:1:5 by volume). Lower phase is adjusted to pH 8.35 with ammonium hydroxide (0.01%). The TFA (800 μL) was added to the upper, stationary phase (500 mL). *Sample* consists of 5 g of D&C Red No. 22 in 30 mL of solvent system (15 mL each of upper and lower phases).

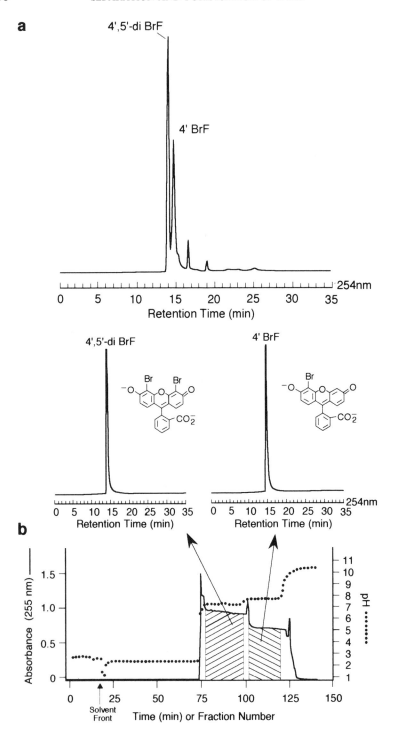

The two-phase solvent system used for this separation consisted of Et_2O/acetonitrile/water (4:1:5 by volume). After equilibration, the separated phases were degassed by sonication. To the upper phase (500 mL) was added $330\,\mu L$ (0.5 g) of TFA, and the pH of the upper phase was thus brought to approximately 2.25. The acidified upper phase was used as the stationary phase. To the lower phase (890 mL) was added 1 mL (0.89 g) of ammonium hydroxide ($NH_3 > 25\%$), and the pH of the lower phase was brought to approximately 10.45. The basified lower phase was used as the mobile phase. The sample solution was prepared by partitioning 0.67 g of the mixture into a solvent system that consisted of 4 mL of mobile phase and 20 mL of upper phase and an additional $100\,\mu L$ of TFA. The HSCCC apparatus used for this separation was a Model CCC-1000 (manufactured by Pharma-Tech Research Corp.) to which several improvements were made to facilitate chromatography (34, 35): the eluant pH was recorded "on-line" using a pH electrode in a flow cell; chromatograms at multiple wavelengths were monitored and stored using a computerized scanning UV–vis detector; and a preparative flow-cell with a pathlength adjustable from 0 to 3 mm was used. Previously, the pH of each fraction was manually measured and the chromatogram was recorded at one UV wavelength (206, 254, or 280 nm) with a fixed pathlength flow-cell. For this separation, the detector cell was set at 0.1 mm and the column was rotated at 1000 rpm in the forward mode. Under a flow-rate of 3 mL/min, the separation was completed in about 2 h. The solvent front (first fraction containing mobile phase) emerged at fraction 19 and the retention of the stationary phase, measured after the separation, was 74.5% of the total column volume. Two components were eluted as broad rectangular peaks that corresponded to two pH plateaus (Fig. 12.11b). The fractions (79–98 and 102–119) that corresponded to these plateaus contained single components (as shown by the associated HPLC in Fig. 12.11b), which were isolated in the lactone form (240 and 212 mg, respectively) and were identified by CIMS and ^1H NMR (33) as 4',5'-diBrF and 4'BrF, respectively.

12.4.4.4 Purification of Commercial 4,5,6,7-Tetrachlorofluorescein

The compound TCF, **4**, is a dye of the hydroxyxanthene class. This compound is used primarily as an intermediate for the preparation of more highly halogenated dyes, such as 2',4',5',7'-tetrabromo-4,5,6,7-tetrachlorofluorescein, **2**, Phloxine B, **3**, and Rose Bengal, **14** (Scheme 3).

Figure 12.11. Separation of 0.67 g of a synthetic mixture of 4'BrF and 4',5'-diBrF by pH-zone-refining CCC. (a) Chromatogram resulting from RP-HPLC analysis. (b) pH-Zone-refining CCC of the separation and the RP-HPLC of the separated components. For experimental details, see the text.

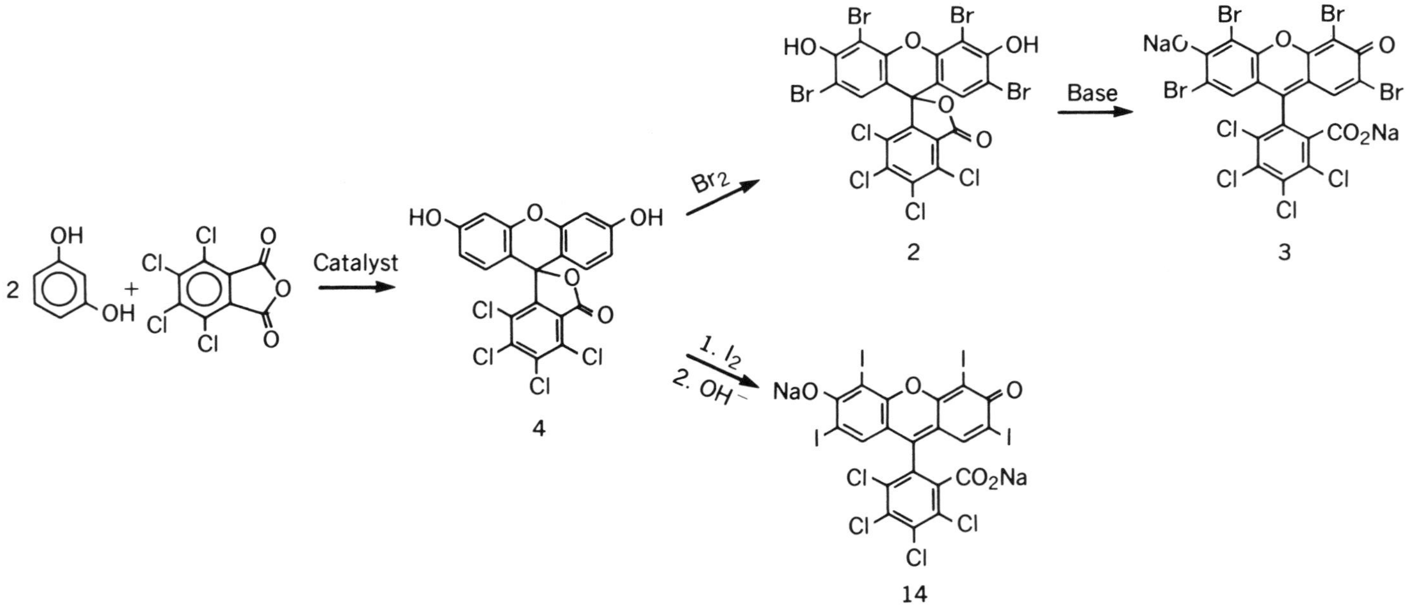

Scheme 3

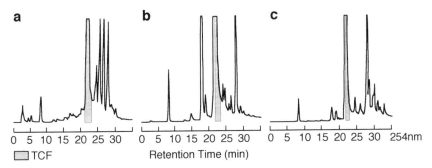

Figure 12.12. Analytical RP-HPLC of batches of TCF obtained from three different distributors.

Compounds **2**, **3**, and **14** are widely used as biological stains. Compounds **2** and **3** have additional importance as main components of U.S.-certified color additives used in drugs and cosmetics (designated as D&C Red No. 27 and D&C Red No. 28, respectively) (8). Commercially available TCF contains a number of contaminants, as shown by RP-HPLC analyses of several batches of TCF obtained from various distributors. The HPLC results further indicated that the specific contaminants varied (Fig. 12.12) (36, 37).

The contaminants of TCF can be carried over to the more highly halogenated dyes during the manufacturing process (Scheme 3). The presence of contaminants in these dyes has significant ramifications in the case of each of the described uses. The pH-zone-refining CCC of a solution containing 350 mg of commercial TCF is shown in Fig. 12.13.

The two-phase solvent system used for this separation was composed of Et_2O/acetonitrile/0.01 M NH_4OAc (4:1:5 by volume). The solvent system was equilibrated and the two phases were separated shortly before use. The lower phase was adjusted to pH 9.06 with ammonium hydroxide (0.1%). The sample mixture was prepared by partitioning 350 mg of commercial TCF between 5.5 mL of each of the upper and lower (not basified) phases. The TFA (200 μL) was added to the sample solution and the mixture was eluted with the basic aqueous mobile phase. The resulting CCC chromatogram (Fig. 12.13b) has five absorbance peaks. The corresponding eluates were collected in fractions 78–80, 92–97, 102–103, 105–109, and 116–168. Each of the absorbance peaks contained a single component that corresponded to one of the RP-HPLC peaks in Fig. 12.13a. These compounds were isolated and identified by two or more of the following methods: CIMS, ^{1}H NMR, high-resolution MS, and RP-HPLC (37). The results are summarized in Fig. 12.14.

Tetrachlorophthalic anhydride or the corresponding free acid, **15**, is used as the starting material for the synthesis of TCF (Scheme 4A) (37). The unreacted

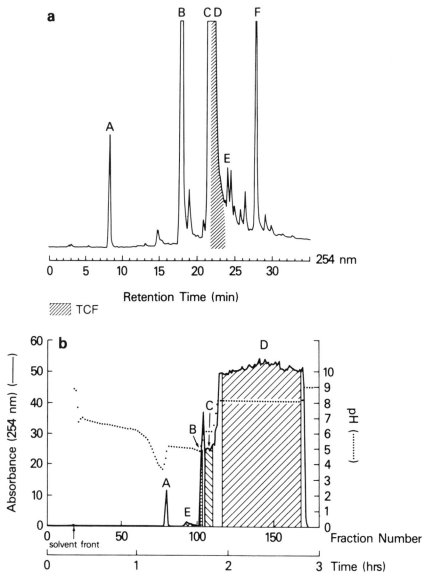

Figure 12.13. Separation of a 350-mg portion of commercial TCF by pH-zone-refining CCC. (a) Reversed-phase HPLC chromatogram of the TCF batch. (b) pH-Zone-refining CCC of the separation. *Solvent system* consists of Et_2O/acetonitrile/0.01 M NH_4OAc (4:1:5 by volume). Lower phase is adjusted to pH 9.06 with ammonium hydroxide (0.1%). *Sample* is a 350-mg TCF suspension in 5.5 mL of each of upper and lower phases. The TFA (200 μL) was added to the sample solution.

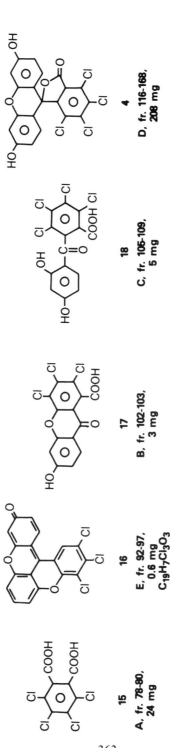

15

A, fr. 78-80,
24 mg

16

E, fr. 92-97,
0.6 mg
$C_{19}H_7Cl_3O_3$

17

B, fr. 102-103,
3 mg

18

C, fr. 105-109,
5 mg

4

D, fr. 116-168,
208 mg

Figure 12.14. Compounds separated by pH-zone-refining CCC from a 350-mg portion of commercial TCF. The letter labels and the indicated fraction numbers correspond to the peaks and the hatched areas, respectively, in Fig. 12.13. The yield for each compound is also given.

363

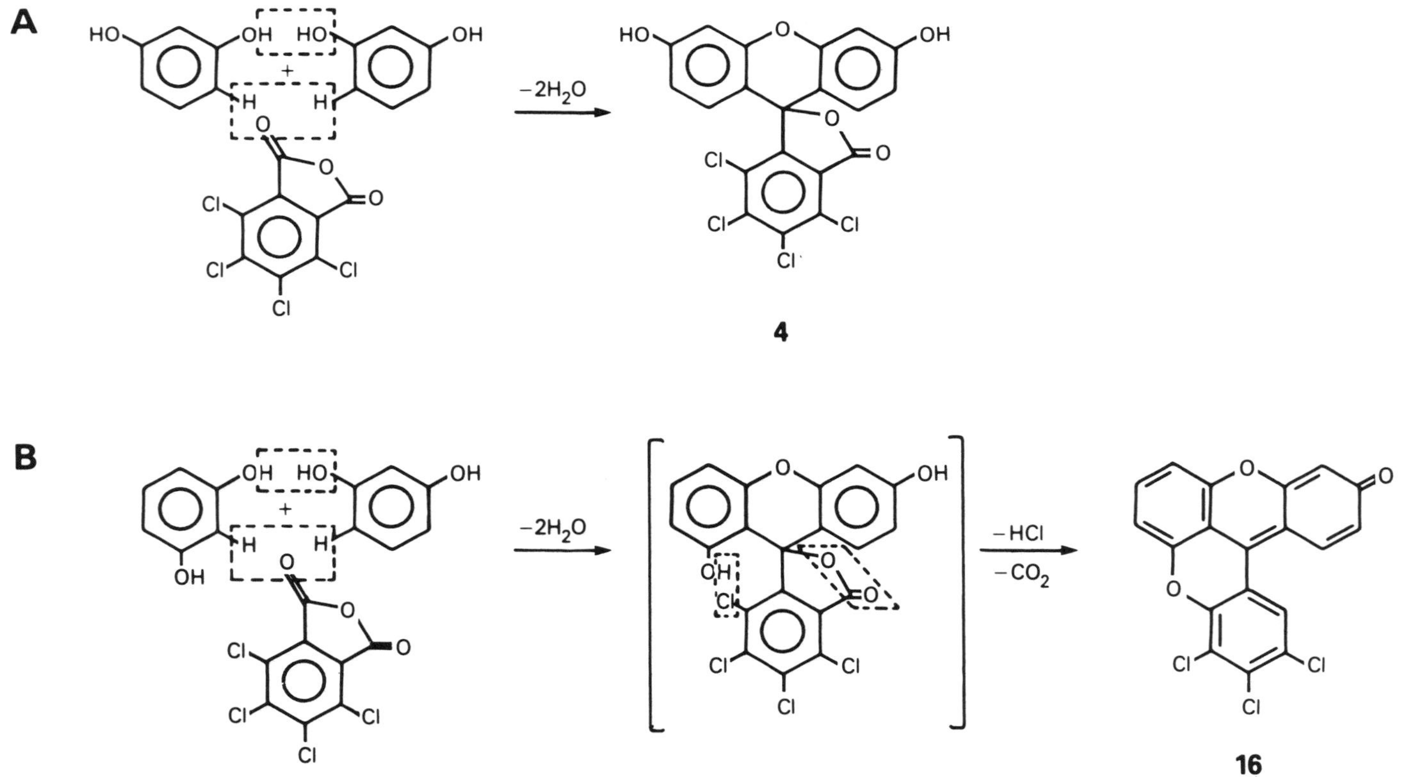

fr. 92-97
peak E
$C_{19}H_7Cl_3O_3$

Scheme 4

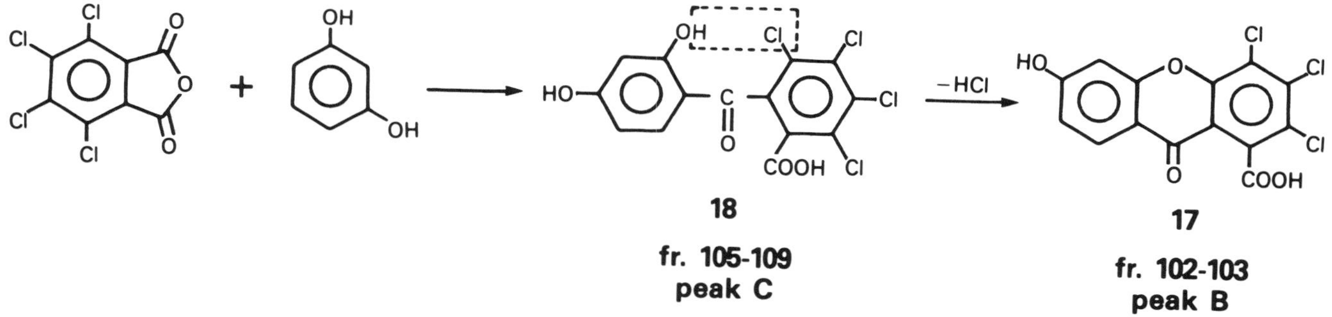

Scheme 5

365

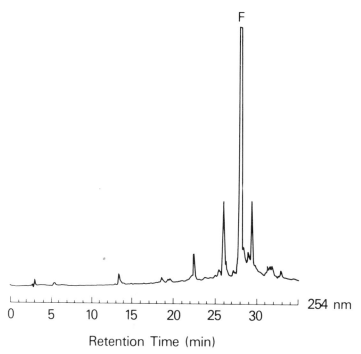

Figure 12.15. Reversed-phase HPLC of the unfractionated column content (stationary phase) after the pH-zone-refining CCC separation shown in Fig. 12.13.

material, if carried over during the manufacturing process, would be present as a contaminant in the final product. Tetrachlorophthalic anhydride is hydrolyzed to tetrachlorophthalic acid, **15**, during the pH-zone-refining CCC separation. Compound **16** (10,11,12-trichloro-3H-[1]benzopyrano[2,3,4-k,l]xanthen-3-one) is a newly found contaminant of TCF. The suggested chemical pathway for its formation is described in Scheme 4B.

The formation of the contaminants 2,3,4-trichloro-6-hydroxyxanthone-1-carboxylic acid, **17**, and 2-(2′,4′-dihydroxybenzoyl)tetrachlorobenzoic acid, **18**, is explained by the reaction of tetrachlorophthalic anhydride with only one molecule of resorcinol (Scheme 5).

The currently described pH-zone-refining CCC separation represents the first reported separation of **17** and **18** from commercial TCF. It is noteworthy that compounds **18** and **4** completely overlap and elute together in the RP-HPLC system used, but are well separated by pH-zone-refining CCC (Fig. 12.13, peaks C and D). Previous references to **17** and **18** in the literature involve only the synthesis of **17** from **18** (38) and the observation of **17** as a decomposition product of TCF in strong alkaline solution (39).

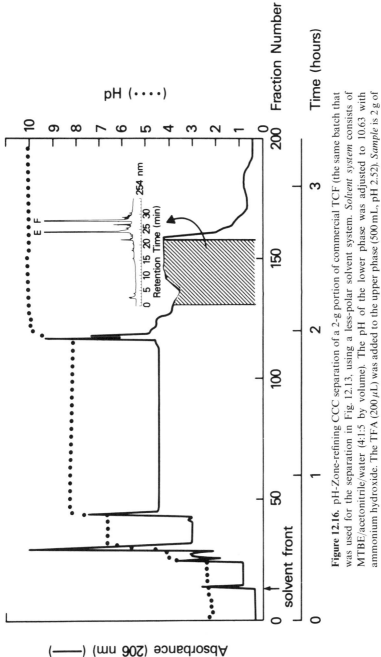

Figure 12.16. pH-Zone-refining CCC separation of a 2-g portion of commercial TCF (the same batch that was used for the separation in Fig. 12.13, using a less-polar solvent system. *Solvent system* consists of MTBE/acetonitrile/water (4:1:5 by volume). The pH of the lower phase was adjusted to 10.63 with ammonium hydroxide. The TFA (200 μL) was added to the upper phase (500 mL, pH 2.52). *Sample* is 2 g of TCF suspended in 100 mL of the acidified upper phase, charged without prior filtration.

367

The compound that corresponds to peak F in Fig. 12.13a is less polar and, under the conditions used for this separation, remained in the stationary phase in the column (Fig. 12.15). The following structure, (11,12,13,14-tetrachloro-3,14b-dihydroxy-[2]benzoxepino[3,4,5-*kl*]xanthene-10(14b*H*)-one, has been proposed for this contaminant on the basis of mass spectral and ^{1}H NMR analyses.

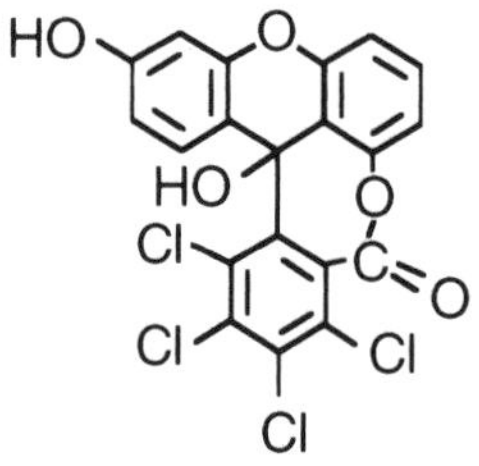

There are several approaches to the collection of nonpolar compounds that remain in the column after the acidic components of the mixture have eluted: (a) stop the rotation of the column and collect the column contents in small fractions by pumping out the stationary phase with the mobile phase; (b) continue the separation for a longer time until the nonpolar compounds elute; and (c) use a less polar solvent system for the separation (see Fig. 12.16).

pH-Zone-refining CCC separations of several 350-mg portions of commercial TCF resulted in similar sharp separations, indicating the high reproducibility of this technique. Separations of gram quantities (1-, 2-, and 5-g portions) of commercial TCF resulted in proportionate increases in peak width and broadening of pH plateaus and in good recoveries of the main compound.

12.4.4.5. Separation of 2′,4′,5′-Tribromo- and 2′,4′,5′,7′-Tetrabromo-4,5,6,7-Tetrachlorofluorescein from the Color Additive D&C Red No. 28 (Phloxine B)

D&C Red No. 28 (Colour Index No. 45410) is a U.S.-certified color additive used in drugs and cosmetics. It is identified as principally **3**. It may contain not more than 4% lower-halogenated subsidiary colors and and not more

→

Figure 12.17. Separation of 3 g of commercial D&C Red No. 28 by pH-zone-refining CCC. (a) Reversed-phase HPLC of the sample. (b) pH-Zone-refining CCC of the separation and the RP-HPLC of the separated components. *Solvent system* consists of Et$_2$O/acetonitrile/0.01 M NH$_4$OAc (4:1:5 by volume). Lower phase adjusted to pH 8.12 with ammonium hydroxide. The TFA (600 µL) was added to the upper stationary, phase (500 mL). *Sample* is 3 g of D&C Red No. 28 in 30 mL of solvent system (20 mL, lower phase, and 10 mL, upper phase).

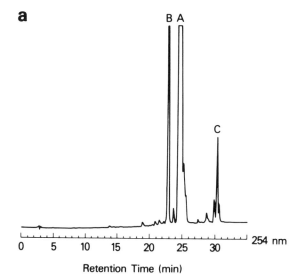

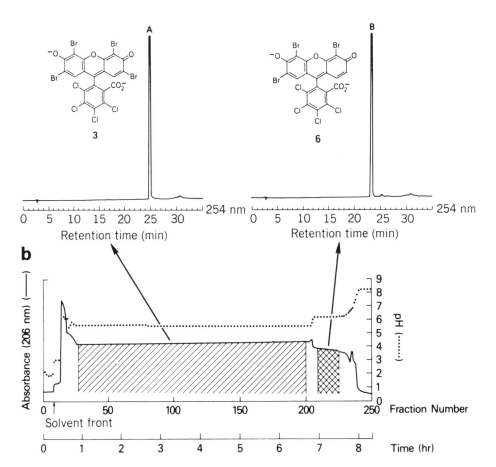

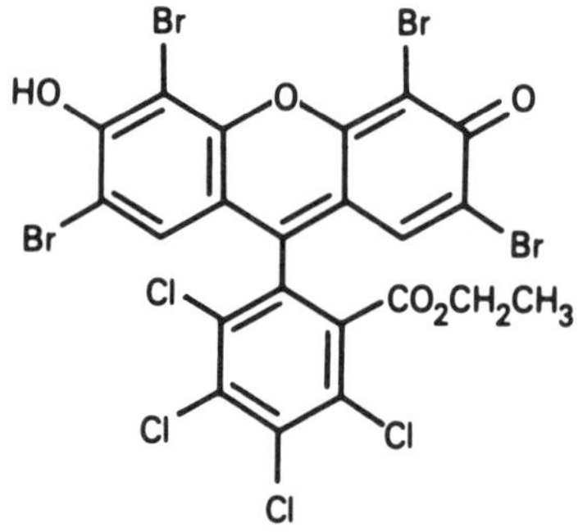

19

than 2% of the ethyl ester of 2',4',5',7'-tetrabromo-4,5,6,7-tetrachloro-fluorescein, **19** (8).

An RP-HPLC chromatogram of a certified lot of D&C Red No. 28 is shown in Fig. 12.17a. Under the name Phloxine B, this dye is used as a biological stain (31, 41). The dye is manufactured by alkaline hydrolysis of **2** (Scheme 3). In a previous study (21) (see Section 12.3.2), it was shown that HSCCC, under the conditions used, could not separate more than 50 mg of commercial D&C Red No. 28. By using a semipreparative CCC column under conditions for pH-zone-refining CCC, multigram quantities of **2** can be separated from the commercial dye (22). The pH-zone-refining CCC of 3 g of commercial D&C Red No. 28 is shown in Fig. 12.17b.

The two-phase solvent system used consisted of Et_2O/acetonitrile/0.01 M NH_4OAc (4:1:5 by volume). The pH of the lower phase was adjusted to 8.12 by addition of ammonium hydroxide. The sample mixture was prepared by distributing 3 g of dye in a solvent consisting of 20 mL of the lower phase and 10 mL of the upper phase. The TFA (600 μL) was added to the upper, stationary phase (500 mL). The resulting chromatogram has a broad rectangular shape (Fig. 12.17b). The two absorbance plateaus correspond to the two pH plateaus. As with the separation of **13** from D&C Red No. 22 (see Section 12.4.4.2), a decrease in absorbance can be observed (fractions 18–25) before the elution of the main component (fractions 27–200). Fractions 18–25

→

Figure 12.18. Separation of a 6-g portion of the color additive D&C Red No. 28 by pH-zone-refining CCC. (a) Reversed-phase HPLC of the sample. (b) pH-Zone-refining CCC of the separation and the RP-HPLC of the separated components. *Solvent system* consists of Et_2O/acetonitrile/0.01 M NH_4OAc (4:1:5 by volume). Lower phase adjusted to pH 9.21 with ammonium hydroxide. *Sample* is 6 g of D&C Red No. 28 dissolved in 50 mL of solvent (20 mL, lower phase, and 30 mL, upper phase). The TFA (1.2 mL) was added to the sample solution.

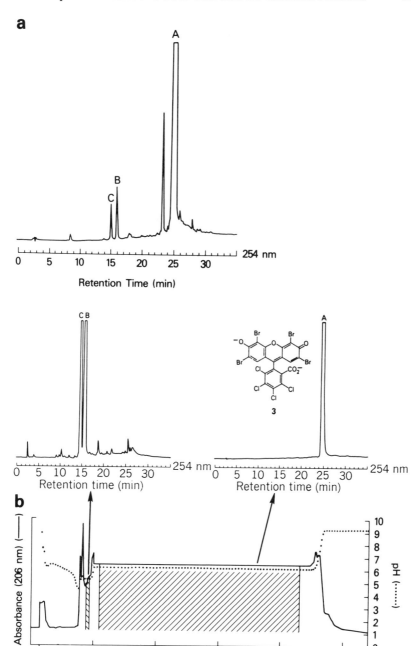

contained the main component **3** as a suspension, slightly contaminated with other impurities. The fractions corresponding to the horizontal plateaus (fractions 27–200 and 209–225) contained pure compounds (Fig. 12.17b, see HPLC chromatograms). The dyes were isolated in the lactone form (**2**, 1.07 g, and **6**, 61 mg) and identified by CIMS and ^{1}H NMR (22). A less polar component of the mixture (**19**), corresponding to peak C in Fig. 12.17a, remained in the stationary phase in the column. The relatively low yield ($\sim 39\%$) obtained for the pure component **2** is apparently due to the elution of the suspension in fractions 18–25. It was suggested (22) that the elution of the suspension was caused by the fact that the amount of retainer acid (TFA) present in the sample solution was insufficient to acidify the sodium salt of the dye present in the sample. In another experiment, a 6-g sample from a different batch of D&C Red. No. 28 was subjected to pH-zone-refining CCC by adding TFA to the sample solution in sufficient quantity to convert all the dye into the acid form (see Fig. 12.18) (22).

In this case, no suspension was eluted and no decrease can be observed in the absorbance intensity before the elution of the main component (fractions 55–215, Fig. 12.18b). The yield ($\sim 80\%$) obtained for the pure main component **2** (4.06 g in the lactone form) was improved in this separation by avoiding elution of the contaminated suspension through the presence of sufficient retainer acid to acidify the sodium salt of the dye. The two minor components corresponding to peaks B and C in Fig. 12.18a eluted in a very concentrated form in fractions 45–47 (Fig. 12.18b).

12.4.4.6. *Separation of 4'-Bromo-4,5,6,7-Tetrachlorofluorescein from a Synthetic Mixture*

For FDA's color additive certification program, pure lower-brominated tetrachlorofluoresceins are needed as reference materials. The compound 4'-bromo-4,5,6,7-tetrachlorofluorescein (4'BrTCF) is needed for the determination of impurities in the color additives D&C Red Nos. 27 and 28.

The chromatogram resulting from HPLC analysis of a synthetic mixture [obtained by partial bromination of TCF in absolute ethanol (34, 35)] that

Figure 12.19. Separation of 4'BrTCF from a synthetic mixture by pH-zone-refining CCC. (a) The HPLC chromatogram of the crude mixture ($\sim 25\%$ 4'BrTCF). (b) pH-Zone-refining CCC of the separation of 5 g of the crude mixture and HPLC chromatogram of the approximately 90% pure 4'BrTCF isolated from the fractions shown by the hatched area. (c) pH-Zone-refining CCC of the separation of 2.4 g of approximately 90% pure 4'BrTCF shown in (b) and the HPLC chromatogram of the over 96% pure 4'BrTCF isolated from the hatched area. For experimental details, see the text.

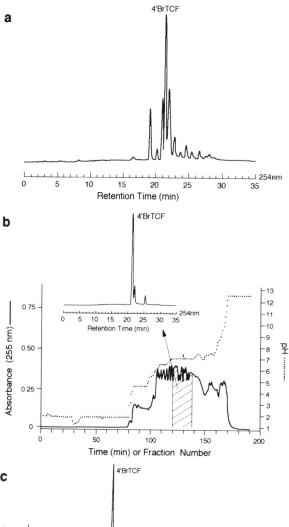

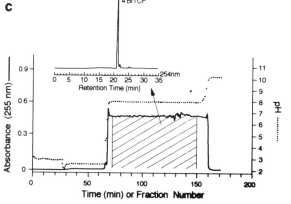

4'BrTCF

contained mainly 4'BrTCF ($\sim 25\%$ by HPLC) is shown in Fig. 12.19a. The pH-zone-refining CCC obtained for the separation of 5 g of this mixture is shown in Fig. 12.19b. The two-phase solvent system consisted of diethyl ether/acetonitrile/water (4:1:5 by volume). After equilibration, the separated phases were degassed by sonication. To the upper phase (500 mL) was added 1 mL (1.41 g, 12.37 mmol) of TFA ($> 98\%$), yielding a solution of 24.7 mM TFA with a pH of approximately 2.3. The acidified upper phase was used as the stationary phase. To the lower phase (850 mL) was added 4.42 g of 50% aqueous NaOH (55.2 mmol), yielding a solution of 65 mM NaOH with a pH of approximately 12.7. The basified lower phase was used as the mobile phase. The sample solution was prepared by dissolving 5 g of crude monobrominated TCF mixture in 60 mL of acidified stationary phase to which was added 10 mL of unbasified lower phase. For this separation, the detector cell was set at 0.05 mm and the column was rotated at 1000 rpm. Under a flow-rate of 3 mL/min, the separation was completed in 3 h. The solvent front (first fraction containing mobile phase) emerged at fraction 27. The retention of the stationary phase measured after the separation was 31%. The 4'BrTCF eluted as a broad rectangular peak that corresponded to a pH plateau (fractions 120–138, hatched area in Fig. 12.19b). The product isolated in lactone form from these fractions (1.28 g) contained approximately 90% pure 4'BrTCF (see associated HPLC chromatogram in Fig. 12.19b). This product and a second batch of approximately 90% pure product (total 2.4 g) were further purified by pH-zone-refining CCC with a different solvent system (methyl tert-butyl ether/water) (Fig. 12.19c), yielding 4'BrTCF (1.63 g, 65% recovery) of approximately 95% purity by HPLC that was characterized by ^{1}H NMR and CIMS (34, 35).

The HSCCC used for these separations was the Model CCC-1000, (Pharma-Tech Research Corp.) with the improvements described in Section 12.4.4.3 (34, 35).

12.4.5.   Preparative Separations of Iodinated Fluoresceins

12.4.5.1.   Separation of 2′,4′,5′-Triiodo-, 2′,4′,7′-Triiodo-, and 2′,4′,5′,7′-Tetraiodofluorescein from the Color Additive FD&C Red No. 3 (Erythrosine)

FD&C Red No. 3 (Colour Index No. 45430) is a U.S.-certified color additive used in food, drugs, and cosmetics. It is identified as a mixture of principally the disodium salt of 2′,4′,5′,7′-tetraiodofluorescein, **20**, and not more than 10% lower-iodinated fluoresceins (8). Under the name Erythrosine, this dye is widely used as a biological stain (31, 32). FD&C Red No. 3 is manufactured by tetraiodination of fluorescein.

20

The pH-zone-refining CCC of the separation of 3 g of commercial FD&C Red No. 3 is shown in Fig. 12.20 (42). The two-phase solvent system used consisted of Et_2O/acetonitrile/0.01 M NH_4OAc (4:1:5 by volume). The pH of the lower phase was adjusted to 7.53 by addition of ammonium hydroxide. The TFA (400 μL) was added to the upper, stationary phase (500 mL). The sample solution was prepared by partitioning 3 g of dye in a solvent consisting of 20 mL of the lower phase and 20 mL of the unacidified upper phase.

The pH-zone-refining CCC has the characteristic broad rectangular shape. The solvent front (first fraction containing mobile phase) emerged at fraction 13. The retention of the stationary phase, measured after the separation, was 66.7% of the total column capacity. The three broad absorbance plateaus that correspond to the three pH plateaus represent elution of pure compounds. The eluates corresponding to these plateaus (fractions 25–79, 86–96, and 106–124) contained single compounds (Fig. 12.20b) that were isolated in the lactone form (19) and characterized by CIMS and ^{1}H NMR. The

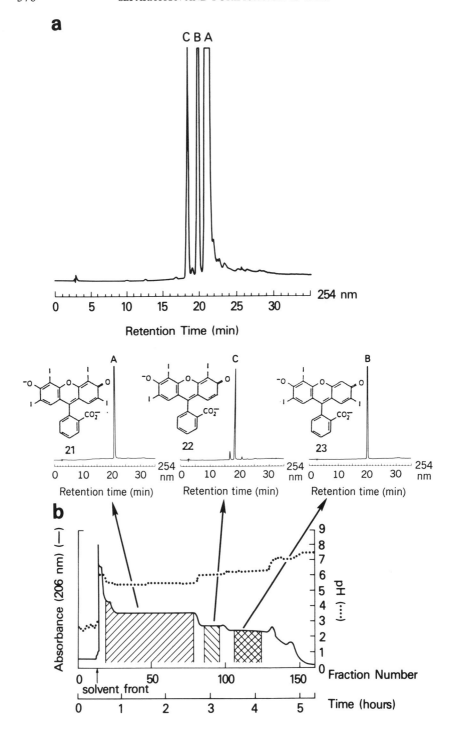

separation resulted in the recovery of 400 mg of 2',4',5',7'-tetraiodofluo-rescein, 21 (peak A in Fig. 12.20a), and of two positional isomers of lower-iodinated subsidiary colors, 2',4',5'-triiodofluorescein, 22 (22 mg), and 2',4',7'-triiodofluorescein, 23 (52 mg) (peaks C and B, respectively, in Fig. 12.20a).

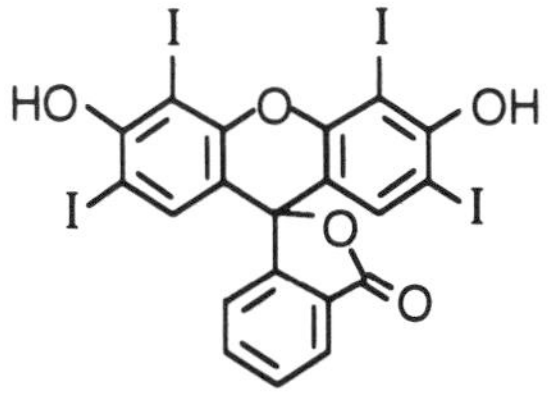

21 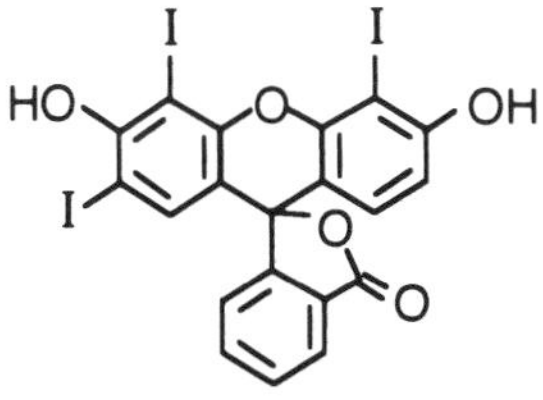   **22** 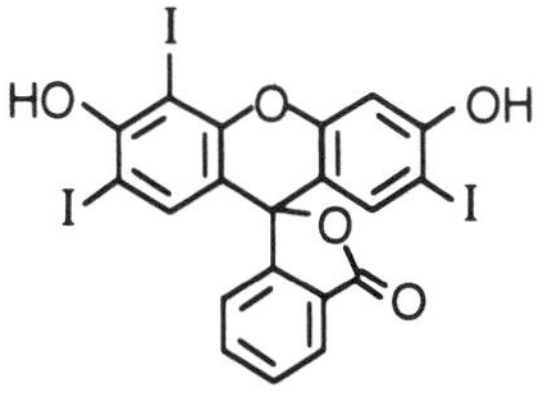   **23**

Note that the response before the first absorbance plateau (Fig. 12.20b, fractions 15–18) decreases in intensity until its level stabilizes (at fraction 25). The eluate corresponding to these fractions consisted of a suspension that contained 21 slightly contaminated with other impurities. The low yield obtained for the pure component 21 is due to the elution of the suspension in fractions 15–18. The elution of the suspension, as previously proposed (22) (see Section 12.4.4.5), was due to the fact that the quantity of retainer acid (TFA) present in the sample solution was insufficient to convert all the sodium salt of the dye into the acid form. By optimizing the quantity of retainer acid added to the sample solution (22) in future separations, the recovery of pure 21 may be improved.

12.4.5.2. Separation of 4',5'-Diiodo- and 2',4',5'-Triiodofluorescein from the Color Additive D&C Orange No. 10

D&C Orange No. 10 (Colour Index No. 45425:1) is a U.S.-certified color additive used in drugs and cosmetics. It is identified as a mixture containing principally three hydroxyxanthene dyes: 60–95% 4',5'-diiodofluorescein, 24;

Figure 12.20. Separation of a 3-g portion of the color additive FD&C Red No. 3 by pH-zone-refining CCC. (a) Reversed-phase HPLC chromatogram of the sample. (b) pH-Zone-refining CCC of the separation and the RP-HPLC chromatograms of the separated components. *Solvent system*: Et_2O/acetonitrile/0.01 M NH_4OAc (4:1:5 by volume). Lower phase adjusted to pH 7.53 with ammonium hydroxide. The TFA (400 μL) was added to the upper, stationary phase (500 mL). *Sample* is 3 g of FD&C Red No. 3 in a 40-mL solvent system (20 mL of the lower phase and 20 mL of the unacidified upper phase). The structures in Fig. 12.20b show the acid forms of the isolated compounds.

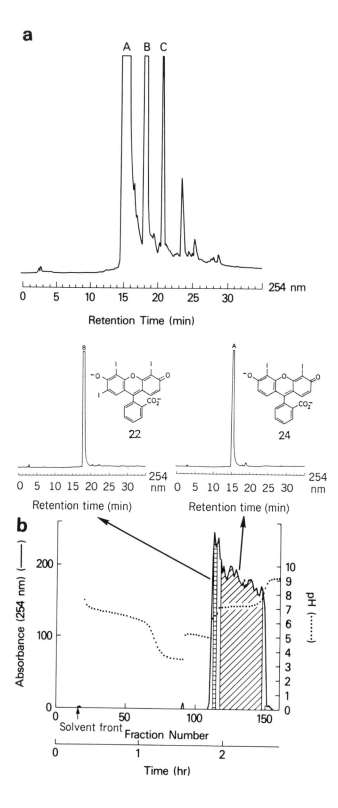

not more than 35% 2′,4′,5′-triiodofluorescein, **22**; and not more than 10% 2′,4′,5′,7′-tetraiodofluorescein, **21** (8). The pH-zone-refining CCC of 350 mg of a commercial sample of D&C Orange No. 10 is shown in Fig. 12.21b.

The two-phase solvent system used for this separation consisted of Et_2O/acetonitrile/0.01 M NH_4OAc (4:1:5 by volume). The pH of the lower phase was adjusted to 9.16 by addition of ammonium hydroxide. The sample mixture was prepared by partitioning 350 mg of D&C Orange No. 10 between 5 mL of each of the upper and lower phases. The retaining acid, 300 μL of TFA, was added to the sample mixture. The resulting suspension was injected into the column without filtering. The elution with the basic aqueous mobile phase resulted in a broad rectangularly shaped absorbance peak. Monitoring of the pH of the eluted fractions revealed a short and a long pH plateau. The pH plateaus corresponded to the separated compounds (fractions 113–114 and 117–147, respectively, Fig. 12.21b). The compounds were isolated in the lactone form (19) and identified by CIMS and ^1H NMR. The separation resulted in the isolation of 2′,4′,5′-triiodofluorescein, **22** (20 mg, contaminated with trace amounts of mono- and diiodofluorescein), and 4′,5′-diiodofluorescein, **24** (244 mg, peak A in Fig. 12.21a).

12.4.5.3. Separation of 2′,4′,5′,7′-Tetraiodo-4,5,6,7-Tetrachlorofluorescein from Commercial Rose Bengal

Commercial Rose Bengal (RB) is a dye that contains principally the disodium salt of 2′,4′,5′,7′-tetraiodo-4,5,6,7-tetrachlorofluorescein, **14**. It is manufactured by the iodination of TCF (see Scheme 3). Among the many applications of RB are its use as a biological stain (31, 32, 43) and its use as a photosensitizing agent (44–46). The pH-zone-refining CCC of the separation of 1.5 g of commercial RB is shown in Fig. 12.22b (47).

The separation was performed with the HSCCC system described in Section 12.4.4.3 (34, 35). The two-phase solvent system used consisted of diethyl ether/acetonitrile/water (4:1:5 by volume). The two phases were equilibrated, separated, and degassed by sonication. To the upper phase (500 mL) was added 400 μL (5 mmol) of TFA, and the pH of the upper phase was thus brought to approximately 2.45. The acidified upper phase was used as the stationary phase. To the lower phase (885 mL) was added 1.3 mL (1.19 g) of

◄

Figure 12.21. Separation of a 350-mg portion of the color additive D&C Orange No. 10 by pH-zone-refining CCC. (a) Reversed-phase HPLC chromatogram of the sample. (b) pH-Zone-refining CCC of the separation and the RP-HPLC chromatograms of the separated components. *Solvent system* consists of Et_2O/acetonitrile/0.01 M NH_4OAc (4:1:5 by volume). Lower phase is adjusted to pH 9.16 with ammonium hydroxide. *Sample* is 350 mg of D&C Orange No. 10 in 10 mL of solvent system (5 mL of each of the upper and lower phases). The TFA (300 μL) was added to the sample mixture.

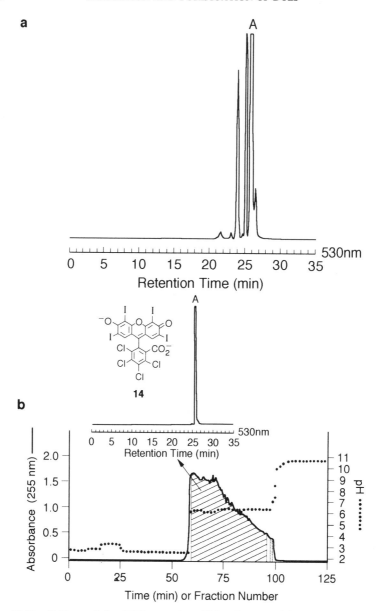

Figure 12.22. pH-Zone-refining CCC separation of 1.5 g of commercial Rose Beugal. (a) HPLC chromatogram of the sample. (b) pH-zone-refining CCC of the separation and the HPLC chromatogram of the separated main component. For experimental details, see the text.

ammonium hydroxide ($NH_3 > 25\%$, $d = 0.91$), and the pH of the lower phase was brought to approximately 10.65. The basified lower phase was used as the mobile phase. The sample solution was prepared by partitioning 1.5 g of commercial RB into a solvent system that consisted of 70 mL of (unacidified) upper phase and 30 mL of (unbasified) lower phase. To the sample solution was added 300 μL (4 mmol) of TFA. For this separation, the detector cell was set at 0.1 mm and the column was rotated at 1000 rpm. Under a flow-rate of 3 mL/min, the separation was completed in less than 2 h. The solvent front emerged at fraction 14 and the retention of the stationary phase, measured after the separation, was 60% of the total column content. One component eluted as a broad absorbance peak that corresponded to a pH plateau (Fig. 12.22b). The fractions (60–96) that corresponded to this plateau contained a single component ($>99\%$ purity, as shown by the associated HPLC chromatogram in Fig. 12.22b), which was isolated in the lactone form (42) (1.14 g) and was identified by CIMS, ^{1}H NMR, and ^{13}C NMR (47) as 2′,4′,5′,7′-tetraiodo-4,5,6,7-tetrachlorofluorescein (the lactone form of **14**).

12.5. CONCLUSIONS

It has been shown that HSCCC, in its conventional form and in its recently developed modified form known as pH-zone-refining CCC, is an effective method for the purification of hydroxyxanthene dyes. Conventional HSCCC is useful for purifying milligram quantities of dyes with yields comparable to those reported from use of other semipreparative-scale methods (48). pH-Zone-refining CCC offers a significant improvement over conventional HSCCC, as it results in excellent separations of the main components of acidic dyes at the multigram level. Indeed, no preparative-scale separation of multigram quantities of these dyes has been previously reported in the literature. With respect to solving the problem posed at the outset of this chapter, that is, isolating dye contaminants, pH-zone-refining CCC can readily be implemented. The resulting compounds can then be used as reference materials for methods development and in FDA batch certification. Furthermore, purified forms of hydroxyxanthene dyes that are used as biological stains can allow the comparison of samples stained in different laboratories (11, 49, 50) and, ultimately, facilitate standardization of the biological stains themselves (11, 12, 51).

In view of the recency of pH-zone-refining CCC's development, its potential has not yet been fully realized. The technique holds promise of widespread applicability, particularly if it is further scaled-up through modifications of the instrumentation used. There is need for further experimentation on various technical issues, such as determining the ideal quantity of retaining acid necessary for the separation of particular mixtures (52). Research efforts on these technical questions will be facilitated by application of the recently developed mathematical and equilibrium model (26, 53). Overall, pH-zone-refining CCC is a new and very effective preparative purification technique, and further exploration of its many challenging opportunities is clearly warranted.

ACKNOWLEDGMENTS

I am deeply grateful to Dr. Yoichiro Ito for his assistance and support during the preparation of this chapter. Most of the reported separations by pH-zone-refining CCC were conducted in Dr. Ito's laboratory with his guidance and direct participation. I would also like to express appreciation to Dr. Sandra J. Bell, Dr. D. Adele Dennis, and Dr. John E. Bailey, Jr., all of the Office of Cosmetics and Colors, FDA, Washington, DC, for their support of my involvement in CCC projects. Additional thanks are due to Dr. Sandra J. Bell for her review of this manuscript and to Ms. Alice L. Marcotte for her editorial efforts. I would also like to acknowledge the late Professor Dr. Julius I. Kleeberg, Haifa, Israel, for introducing me to the study of colors.

REFERENCES

1. H. M. Fales, L. K. Pannell, E. A. Sokoloski, and P. Carmeci, *Anal. Chem.*, **57**, 376 (1985).
2. H. S. Freeman and C. S. Williard, *Dyes Pigm.*, **7**, 407 (1986).
3. H. S. Freeman, Z. Hao, S. A. McIntosh, and K. P. Mills, *J. Liq. Chromatogr.*, **11**, 251 (1988).
4. W. D. Conway, *Countercurrent Chromatography. Apparatus, Theory and Applications*, VCH Publishers, New York, 1990.
5. J. M. Ruth and N. B. Mandava, in *Countercurrent Chromatography, Theory and Practice*, N. B. Mandava and Y. Ito, Eds., Chapter 12, Marcel-Dekker, New York, 1988.
6. P. N. Marshall, S. A. Bentley, and S. M. Lewis, *Stain Technol.*, **50**, 107 (1975).
7. D. Fompeydie and P. Levillain, *Talanta*, **31**, 1125 (1984).
8. *Code of Federal Regulations*, Title 21, Part 74, U.S. Government Printing Office, Washington, DC, 1995.
9. R. W. Horobin, *Histochem. J.*, **1**, 231 (1969).
10. P. N. Marshall and S. M. Lewis, *Stain Technol.*, **49**, 351 (1974).
11. E. K. W. Schulte, *Histochemistry*, **95**, 319 (1991).
12. H. O. Lyon, A. P. De Leenheer, R. W. Horobin, W. E. Lambert, E. K. W. Schulte, B. Van Liedekerke, and D. H. Wittekind, *Histochem. J.*, **26**, 533 (1994); and references cited therein.

13. Y. Ito, K. Shinomiya, H. M. Fales, A. Weisz, and A. L. Scher, "*PITTCON'93*," *The 44th Pittsburgh Conference and Exposition on Analytical Chemistry and Applied Spectroscopy*, March 8–12, 1993, Atlanta, GA; Abstract 54P in the Book of Abstracts.

14. A. Weisz, A. L. Scher, K. Shinomiya, H. M. Fales, and Y. Ito, *J. Am. Chem. Soc.*, **116**, 704 (1994).

15. Y. Ito and A. Weisz, U.S. Patent No. 5 332 504; July 26, 1994.

16. Y. Ito, J. Sandlin, and W. G. Bowers, *J. Chromatogr.*, **244**, 247 (1982).

17. Y. Ito, *J. Chromatogr.*, **301**, 387 (1984).

18. H. Oka, Y. Ikai, N. Kawamura, J. Hayakawa, M. Yamada, K.-I. Harada, H. Murata, M. Suzuki, H. Nakazawa, S. Suzuki, T. Sakita, M. Fujita, Y. Maeda, and Y. Ito, *J. Chromatogr.*, **538**, 149 (1991).

19. A. Weisz, A. L. Scher, D. Andrzejewski, Y. Shibusawa, and Y. Ito, *J. Chromatogr.*, **607**, 47 (1992).

20. Y. Ito, H. Oka, and J. L. Slemp, *J. Chromatogr.*, **475**, 219 (1989).

21. A. Weisz, A. L. Langowski, M. B. Meyers, M. A. Thieken, and Y. Ito, *J. Chromatogr.*, **538**, 157 (1991).

22. A. Weisz, D. Andrzejewski, and Y. Ito, *J. Chromatogr. A*, **678**, 77 (1994).

23. C. Graichen and J. C. Molitor, *J. Assoc. Off. Anal. Chem.*, **42**, 149 (1959).

24. Y. Ito, Y. Shibusawa, H. M. Fales, and H. J. Cahnmann, *J. Chromatogr.*, **625**, 177 (1992).

25. Y. Ma and Y. Ito, *J. Chromatogr. A*, **678**, 233 (1994).

26. A. L. Scher and Y. Ito, in Modern *Countercurrent Chromatography. ACS Symposium Series No. 593*, W. D. Conway and R. J. Petroski, Eds., Chapter 15, American Chemical Society, Washington, DC, 1995, pp. 184–202.

27. C. Denekamp, A. Mandelbaum, A. Weisz, and Y. Ito, *J. Chromatogr. A*, **685**, 253 (1994).

28. R. D. Stolow, *J. Am. Chem. Soc.*, **81**, 5806 (1959).

29. G. B. Barlin and D. D. Perrin, in *Techniques of Chemistry*, 2nd ed., Vol. 4, *Elucidation of Organic Structures by Physical and Chemical Methods*, K. W. Bentley and G. W. Kirby, Eds., Wiley-Interscience, New York, 1972, pp. 661–664.

30. Y. Ito, in *Journal of Chromatography Library*, Vol. 51A, *Chromatography*, 5th ed., Part A, E. Heftmann, Ed., Elsevier, New York, 1992, pp. A90–A91.

31. G. Clark, *Staining Procedures*, 4th ed., Williams and Wilkins, Baltimore, 1981.

32. F. J. Green, *The Sigma-Aldrich Handbook of Stains, Dyes and Indicators*, Aldrich Chemical Co., Inc., Milwaukee, WI, 1990, p. 305.

33. A. Weisz, D. Andrzejewski, R. J. Highet, and Y. Ito, unpublished results.

34. A. Weisz, A. L. Scher, D. Andrzejewski, and Y. Ito, *208th ACS National Meeting*, August 21–25, 1994, Washington, DC; Abstract ANYL 77 in the Book of Abstracts.

35. A. Weisz, A. L. Scher, and Y. Ito, *J. Chromatogr. A*, submitted.

36. A. Weisz, K. Shinomiya, and Y. Ito, "*PITTCON'93*," *The 44th Pittsburgh Conference and Exposition on Analytical Chemistry and Applied Spectroscopy*, March 8–12, 1993, Atlanta, GA; Abstract 865 in the Book of Abstracts.

37. A. Weisz, D. Andrzejewski, K. Shinomiya, and Y. Ito, in *Modern Countercurrent Chromatography. ACS Symposium Series No. 593*, W. D. Conway and R. J. Petroski, Eds., Chapter 16, American Chemical Society, Washington, DC, 1995, pp. 203–217.

38. W. R. Orndorff and W. A. Adamson, *J. Am. Chem. Soc.*, **40**, 1235 (1918).

39. M. Kamikura, *Shokuhin Eiseigaku Zasshi*, **9**, 348 (1968); *Chem. Abstr.*, **70**, 79128z (1969).

40. A. Weisz, D. Andrzejewski, R. J. Highet, and Y. Ito, unpublished results.

41. P. N. Marshall, *Histochem. J.*, **9**, 487 (1976); and references cited therein.

42. A. Weisz, D. Andrzejewski, R. J. Highet, and Y. Ito, *J. Chromatogr. A*, **658**, 505 (1994).

43. H. J. Conn, *Biological Staining*, 9th ed., R. D. Lillie, Ed., Williams and Wilkins, Baltimore, 1977.

44. J. J. M. Lamberts and D. C. Neckers, *Tetrahedron*, **41**, 2183 (1985).

45. D. C. Neckers, *J. Photochem. Photobiol., A: Chem.*, **47**, 1 (1989).

46. D. C. Neckers and O. M. Valdes-Aguilera, in *Advances in Photochemistry*, Vol. 18, D. Volman, G. S. Hammond, and D. C. Neckers, Eds., Wiley, New York, 1993.

47. A. Weisz, D. Andrzejewski, R. J. Highet, and Y. Ito, unpublished results.

48. E. Gandin, J. Piette, and Y. Lion, *J. Chromatogr.*, **249**, 393 (1982).

49. P. N. Marshall and S. M. Lewis, *Stain Technol.*, **49**, 235 (1974).

50. P. N. Marshall, S. A. Bentley, and S. M. Lewis, *J. Clin. Pathol.*, **28**, 920 (1975).

51. P. N. Marshall, S. A. Bentley, and S. M. Lewis, *Scand. J. Haematol.*, **20**, 206 (1978).

52. A. Weisz, A. Idina, A. Mandelbaum, and Y. Ito, "*PITTCON'95*," *The 46th Pittsburgh Conference and Exposition on Analytical Chemistry and Applied Spectroscopy*, March 5–10, 1995, New Orleans, LA; Abstract 1068 in the Book of Abstracts.

53. Y. Ito, K. Shinomiya, H. M. Fales, A. Weisz, and A. L. Scher, in *Modern Countercurrent Chromatography. ACS Symposium Series No. 593*, W. D. Conway and R. J. Petroski, Eds., Chapter 14, American Chemical Society, Washington, DC, 1995, pp. 156–183.

CHAPTER

13

SEPARATION OF PROTEINS BY HIGH-SPEED COUNTERCURRENT CHROMATOGRAPHY

YOICHI SHIBUSAWA

Tokyo College of Pharmacy, Hachioji, Tokyo, Japan

13.1. INTRODUCTION

Aqueous–aqueous polymer phase systems for the partitioning of biological macromolecules were first established by Albertsson in the 1950s (1). Partitions of macromolecules, such as proteins (2–6), nucleic acids (7–10), viruses (11, 12), bacteria (13, 14), cell organelles (15), and blood cells (16–22) have been carried out by the use of various aqueous polymer phase systems. Among many types of aqueous–aqueous polymer phase systems available, dextran–polyethylene glycol (PEG) and PEG-phosphate systems have been most commonly used for partition of biological samples.

While these polymer phase systems provide an ideal environment for biopolymers and live cells, high viscosity and low interfacial tension between the two phases tend to delay the phase separation and the operation of the conventional countercurrent distribution apparatus becomes tedious and requires a long partition time (23, 24). In the 1960s, Albertsson introduced the thin-layer countercurrent distribution apparatus that can conveniently carry out multistage partitioning with polymer phase systems in a unit gravitational field. However, in this thin-layer countercurrent distribution apparatus the partition process is carried out by three operational steps of mixing, settling, and transferring the mobile phase, hence it requires relatively long separation times.

Countercurrent chromatography (CCC) established by Ito and co-workers is essentially a form of liquid–liquid partition chromatography in which the stationary phase is retained in the column with the aid of gravity or a centrifugal force field (25–28). This method has been termed after two classic partition techniques, that is, the countercurrent distribution and liquid chromatography. The flow-through centrifuge systems (29, 30) have been developed

High-Speed Countercurrent Chromatography, Edited by Yoichiro Ito and Walter D. Conway.
Chemical Analysis Series, Vol. 132.
ISBN 0-471-63749-1   © 1996 John Wiley & Sons, Inc.

for performing CCC and provide various advantages for continuous elution through a rotating column. In the early 1980s a great advance in the CCC technology was made by the discovery of a new hydrodynamic phenomenon in a rotating coiled tube (31), which provided a basis for developing a highly efficient CCC system called high-speed CCC (HSCCC) (32). Recently, some CCC models, such as Type J horizontal coil planet centrifuge (CPC) and Type X CPC have been modified for performing CCC with highly viscous polar solvent systems (33, 34). A few reports have described the CCC separations of stable proteins using PEG–potassium phosphate aqueous two-phase systems by the Type J horizontal CPCs (35, 36). The cross-axis CPC with the column holders shifted laterally along the holder shaft enables retention of the stationary phase of polar solvent systems, such as PEG 1000–potassium phosphate with the aid of the laterally acting strong centrifugal field. The apparatus can retain the upper phase of polar solvent systems over 50% of the total column capacity even under a high flow-rate of 2.0 mL/min. The separation of cytochrome c, myoglobin, ovalbumin, and bovine hemoglobin was demonstrated by the XLL cross-axis CPC with a polymer phase system composed of PEG 1000 and dibasic potassium phosphate (37). The synthetic mixture of human plasma lipoproteins (high and low density lipoproteins) were separated from each other (38) by using PEG–potassium phosphate solvent systems and XLL cross-axis CPC. The recombinant enzymes, such as purine nucleoside phosphorylase, and uridine phosphorylase, were purified from *Escherichia coli* homogenate by the XLL cross-axis CPC in a one-step operation even without the preliminary clean-up by ultracentrifugation (39). The XLLL cross-axis CPC, which yields a stronger lateral centrifugal force than the XLL cross-axis CPC, has been developed to retain the stationary phase of extremely viscous polymer phase systems composed of PEG–dextran. The separation of basic histones from other histones and of serum proteins, globulins, and human serum albumin, were accomplished in these solvent systems by optimizing the partition coefficient of the proteins at different pH values (40).

Profilin, which regulates the actin polymerization, was purified from *Acanthamoeba* extracts with 4.4% (w/w) PEG 8000–7.0% (w/w) dextran T500 polymer phase system using the L cross-axis CPC (41). The improved retention of the stationary phase by this type of cross-axis CPC permits the use of aqueous–aqueous polymer phase systems for CCC fractionation of various stable and labile proteins. The details of the protein separations along with the design of the apparatus and applied solvent systems will be described in this chapter.

13.2. APPARATUS

Countercurrent chromatography separations of proteins were performed with the horizontal flow-through CPC (horizontal CPC) and the cross-axis CPCs.

These two CCC centrifuges share a common feature in that the system permits continuous elution of the mobile phase through the rotating column without the use of the conventional rotary seal device, which is a potential source of leakage and contamination (29, 30).

Figure 13.1. Horizontal CPC equipped with a set of three composite coil assemblies connected in series.

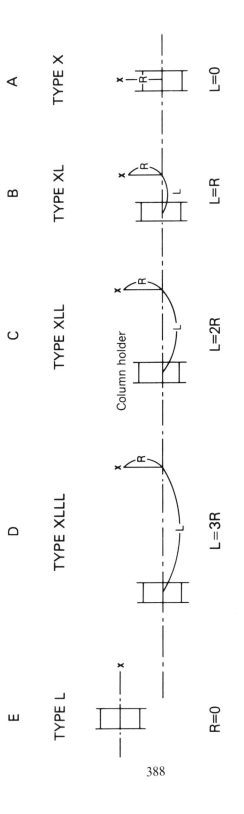

Figure 13.2. Orientation of the column holder in five different types of the cross-axis CPCs.

x: axis of revolution; ———— :axis of rotation; **L**: lateral shift; **R**: revolution radius

388

Figure 13.1 shows a photograph of the horizontal CPC. The apparatus holds a set of three column holders symmetrically arranged on the rotary frame. Each holder carries a composite coil assembly, which consists of a series of eight coil units and each coil unit is made up of a dual layer coil prepared from 1.6-mm i.d. polytetrafluoroethylene (PTFE) tubing with about 75 helical turns and 9-mL capacity. Eight coil units are arranged around each holder in parallel with and at a distance of 3 cm from the holder axis. Three coil assemblies on the rotary frame are serially connected with flow-tubes (0.85 mm i.d. PTFE) to make up a total capacity of about 220 mL.

The cross-axis CPCs used in the protein separations are modified versions of the high-speed CCC centrifuge. The cross-axis CPC has a unique feature among the CPC systems in that the system provides reliable retention of the stationary phase for viscous polymer phase systems. Figure 13.2 shows five different types of the cross-axis CPCs. A series of studies has shown that the stationary-phase retention is enhanced by laterally shifting the position of the coil holder along the holder shaft, apparently due to the asymmetry of the laterally acting force field between the upper and the lower halves of the rotating coil. The degree of the lateral shift of the column holder may be conveniently expressed by L/R, where L is the distance from the center of the holder axis to the coil holder and R is the distance from the centrifuge axis to the

Figure 13.3. The XLLL cross-axis CPC equipped with a pair of multilayer coils connected in series.

Table 13.1. Type of the Apparatus and Dimensions of the Columns Used in the Protein Separations

Apparatus	Coil Holder			Columns				
	$(L/R)^a$	Diameter (cm)	Width (cm)	ID (mm)	Length (m)	No. of Layers	Capacity (mL)	β Values[b]
Type XLL	(2.0)	3.8	5.1	2.6	47	8	250	0.25–0.60
		7.6	5.1	2.6	53	6	280	0.50–1.00
Type XLLL	(3.5)	3.8	5.1	2.6	57	9	300	0.50–1.30
		3.35	5.1	2.6	83	12	440	0.45–1.50
Type L	(∞)	3.6	5.1	2.6	25	5	130	0.16–0.27[c]
Horizontal CPC		0.6	12.0	1.6	109	2	220	0.3

[a] Distance from the center of the holder shaft to the coil holder = L; R = the distance from the centrifuge axis to the holder shaft.

[b] $\beta = r/R$

[c] $\beta = r/L$, where r is the distance from the holder axis to the coil.

holder axis. Among those, the XLL, XLLL, and L cross-axis CPCs have been successfully used for protein separation with various polymer phase systems.

Figure 13.3 shows a photograph of the XLLL cross-axis CPC ($L/R = 3.5$) equipped with a pair of multilayer coil separation columns. Each multilayer coil was prepared from 2.6-mm i.d. PTFE tubing by winding it onto a holder forming multiple layers of left-handed coils.

Table 13.1 lists various CPC models used for protein separation along with the dimensions of the columns and column holders, β values of the multilayer coils, and so on.

13.3. POLYMER PHASE SYSTEMS FOR PROTEIN SEPARATION

Countercurrent chromatography utilizes a pair of immiscible solvent phases preequilibrated in a separatory funnel: One phase is used as the stationary phase and the other as the mobile phase. Table 13.2 shows the compositions of four different PEG 1000–potassium phosphate systems. The polymer phase systems were prepared by dissolving 150 or 192 g of PEG 1000 and 150 g of potassium phosphate in water. The ratio of monobasic to dibasic potassium phosphates determines the pH of the solvent system and affects the partition coefficients of the proteins. These solutions form two layers, the upper layer is rich in PEG and the lower layer is rich in potassium phosphate.

The PEG 8000–dextran T500 systems are similarly prepared by dissolving 44 g of PEG 8000 and 70 g of dextran T500 in 886 g of 10-mM potassium phosphate buffer solution. The pH of the system is also adjusted by choosing the proper ratio between the monobasic and dibasic potassium phosphate. This two-phase system consists of the PEG-rich upper phase and dextran-rich lower phase.

13.4. COUNTERCURRENT CHROMATOGRAPHY FRACTIONATION OF PROTEINS

13.4.1. PEG–Potassium Phosphate Systems

Using PEG–potassium phosphate polymer phase systems, the synthetic mixture of four stable proteins was separated by two types of CPCs: the Type J horizontal CPC and XLL cross-axis CPC to evaluate their capability. The method has also been applied to separation of low and high density lipoproteins (LDL and HDL), purification of recombinant enzymes such as uridine phosphorylase (UrdPase) and purine nucleoside phosphorylase (PNP) directly from crude *E. coli* homogenate. Table 13.3 summarizes the experimental conditions including the protein samples, the apparatus, and solvent systems used for separation.

Table 13.2. Preparation of Polymer Two-Phase Solvent Systems

	Concentration (% w/w)				Weight (g)			
Solvents	PEG 1000	K_2HPO_4	KH_2PO_4		PEG 1000	K_2HPO_4	KH_2PO_4	H_2O
1	12.5	12.5	0		150	150	0	900
2	12.5	9.375	3.125		150	112.5	37.5	900
3	12.5	8.33	4.17		150	100	50	900
4	16.0	6.25	6.25		192	75	75	858

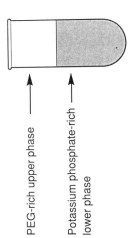

PEG-rich upper phase →

Potassium phosphate-rich
lower phase →

Table 13.3. Protein Separation by PEG–Potassium Phosphate Systems

Proteins	Apparatus	Solvent Systems	Mobile Phase[a]	Reference
Cytochrome c Lysozyme	Type J Horizontal	5% PEG 8000, 5% PEG 1000, 10% K_2HPO_4	LP	35
Cytochrome c Lysozyme	Type J Horizontal	12.5% PEG 1000, 12.5% K_2HPO_4	LP	36
Cytochrome c Myoglobin Ovalbumin Hemoglobin	Type J Horizontal	12.5% PEG 1000, 12.5% K_2HPO_4	LP $(UP)^b$	37
Cytochrome c Myoglobin Ovalbumin Hemoglobin	Type XLL X axis	12.5% PEG 1000, 12.5% K_2HPO_4	LP $(UP)^b$	37
Lipoproteins (HDL and LDL)	Type XLL X axis	16.0% PEG 1000, 12.5% K_2HPO_4	LP	38
Recombinant Enzymes	Type XLL X axis	16.0% PEG 1000, 6.25% K_2HPO_4, 6.25% KH_2PO_4	UP	39

[a]Lower phase = LP; upper phase = UP.
[b]Upper phase mobile in a reversed elution mode after the first three peaks were eluted = (UP).

393

13.4.1.1. Selection of PEG–Potassium Phosphate Systems Based on the Partition Coefficients of the Proteins

Countercurrent chromatography is a liquid–liquid partition method where the separation is based on the difference in partition coefficient of solutes. For achieving efficient separations of proteins, it is essential to optimize the partition coefficient (K) of each component by selecting the proper composition of the polymer phase system. Since the PEG–phosphate polymer phase system partitions small molecules unilaterally in either upper or lower phase, purification of proteins can be efficiently performed by selecting their K values close to unity. Below, a suitable solvent composition for the separation of cytochrome c, myoglobin, ovalbumin, and bovine hemoglobin is determined by a set of partition data.

In Fig. 13.4, the partition coefficient values of four proteins are plotted on a logarithmic scale against the ratios of monobasic-to-dibasic potassium

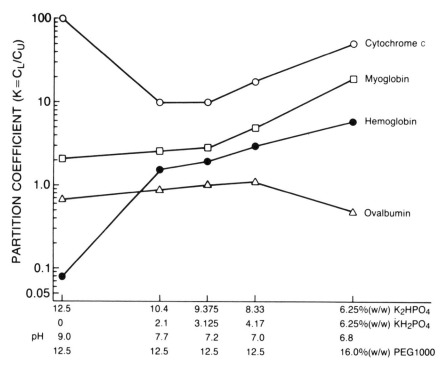

Figure 13.4. Partition coefficients (K) of four protein samples in various polymer phase systems composed of PEG 1000 and potassium phosphate. The partition coefficient is the solute concentration in the lower phase divided by that in the upper phase.

phosphates in the PEG 1000 polymer phase system (37). In this figure, partition coefficient values are expressed as C_L/C_U, that is, protein concentration in the lower phase divided by that in the upper phase, for the potassium phosphate-rich lower phase was more often used as the mobile phase. The partition coefficients of these four proteins generally increase as the relative amount of monobasic potassium phosphate increases. This result is apparently due to the pH shift toward the isoelectric points of the applied proteins. When the relative concentration of the monobasic-to-dibasic potassium phosphates exceeds a 1:1 ratio, however, the solvent mixture forms a single phase unless the polymer concentration is further increased. In this figure, an evenly scattered ideal distribution of the four partition coefficient is observed in the solvent system composed of 12.5% (w/w) dibasic potassium phosphate and 12.5% (w/w) PEG 1000 at pH 9.0.

13.4.1.2. *Stationary-Phase Retention*

The effects of the revolution speed and flow-rate of the mobile phase on the retention of the PEG-rich upper phase were investigated using both the horizontal and cross-axis CPCs (37). The results are summarized in Figs. 13.5 and 13.6.

In the horizontal CPC, the effects of revolution speed on the retention of the stationary phase were studied using the lower phase as the mobile phase at a flow-rate of 0.51 mL/min (Fig. 13.5A). The optimum condition for retention is found at 800 rpm. An increased revolution speed at 1000 rpm resulted in reduction of the retention volume probably due to emulsification caused by

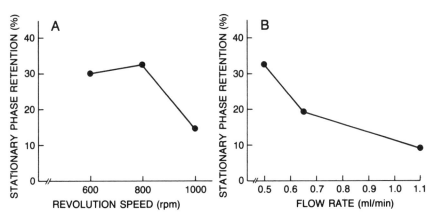

Figure 13.5. Effects of revolution speed (A) and flow-rate (B) on the stationary-phase retention in the horizontal CPC. (A) 0.5 mL/min flow-rate and (B) 800 rpm revolution speed.

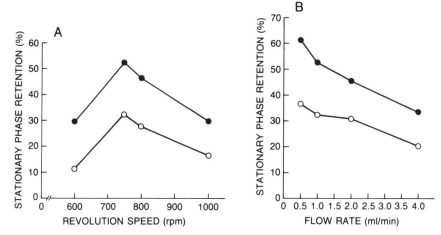

Figure 13.6. Effects of revolution speed (A) and flow-rate (B) on the stationary-phase retention in the XLL cross-axis CPC. —○—: large column ($\beta = 0.50$–1.00); —●—: small column ($\beta = 0.25$–0.60); (A) 1 mL/min flow-rate and (B) 750 rpm revolution speed.

excessive mixing of the two phases. On the other hand, a reduced revolution speed down to 600 rpm gave unsatisfactory retention apparently due to lack of the strength in the centrifugal force field to sustain the stationary phase in the column. The effect of the flow-rate was then investigated at the optimum revolution speed of 800 rpm (Fig. 13.5B). The retention of the stationary phase rapidly decreases with increased flow-rate. As the flow-rate is increased from 0.5 to 1.1 mL/min, the stationary-phase retention becomes about one third of the value observed at 0.5 mL/min. From these data, the operating conditions for the horizontal CPC may be optimized at 800 rpm and a 0.5 mL/min flow-rate for the present PEG–potassium phosphate polymer phase system.

Similar experiments were performed to optimize the operating conditions for the XLL cross-axis CPC. The results are shown in Fig. 13.6, where open circles indicate the retention data obtained from the larger column ($\beta = 0.5$–1.0) and solid circles, those obtained from the smaller column ($\beta = 0.25$–0.5). The effects of the revolution speed on the stationary-phase retention was studied at a flow-rate of 1.0 mL/min (Fig. 13.6A). The retention sharply rises with the increased revolution speed from 600 to 750 rpm to reach the maximum level. Further increase of the revolution speed up to 1000 rpm results in a decline of the retention in a linear fashion. In Fig. 13.6B, the effects of the flow-rate of the mobile phase on the retention of the stationary phase was investigated at a revolution speed of 750 rpm. In both columns, the retention decreases with an increased flow-rate from 0.5 to 4.0 mL/min. The retention level in the smaller column is much greater than that in the larger column. Even at a high flow-rate of 2.0 mL/min, the smaller column holds the stationary phase near 45% of the total column capacity. This finding promises an efficient peak resolution of proteins in a short elution time.

13.4.1.3.   Separation of Four Stable Proteins

To study the effects of the revolution speed and flow-rate on the protein separation, a series of experimental runs was performed with a two-phase solvent system composed of 12.5% (w/w) PEG 1000 and 12.5% (w/w) dibasic potassium phosphate using the lower phase as the mobile phase (37). From the obtained chromatogram, peak resolution (R_s) between peaks 1, 2, and 3 was computed from the conventional formula

$$R_s = 2(R_2 - R_1)/(W_1 + W_2) \qquad (13.1)$$

where R_1 and R_2 are the retention times or volumes of the two adjacent peaks, and W_1 and W_2 are the peak widths of the corresponding peaks expressed in the same unit as R_1 and R_2. The results are summarized in Tables 13.4 and 13.5, where R_s values between the first (cytochrome c) and second (myoglobin) peaks and those between the second and third (ovalbumin) peaks are listed along with the percentage retention of the stationary phase.

In the horizontal CPC (Table 13.4), a fixed flow-rate of 0.5 mL/min produced the highest peak resolution at 800 rpm as shown in the top three rows. The shift of the revolution speed in either direction results in a lower R_s value. This loss of peak resolution may be secondary to the reduced stationary-phase volume retained in the column, because the percentage retention values appear to bear a somewhat significant correlation with the R_s values. At the

Table 13.4. Peak Resolution of Three Proteins Obtained by Horizontal CPC[a]

| Revolution Speed (rpm) | Flow Rate (mL/min) | Peak Resolution[b] | | Stationary Phase Retention (%) |
		cy/myo	myo/ov	
600	0.5	1.63	0.43	29.8
800	0.5	2.22	0.43	32.6
1000	0.5	1.71	0.29	14.1
800	0.5	2.22	0.43	32.6
800	0.7	1.43	0.60	19.3
800	1.1	0.73	—	9.0

[a]Column capacity: 220 mL; $\beta = 0.3$.
[b]Cytochrome c = cy; myoglobin = myo; ovalbumin = ov.

Table 13.5. Peak Resolution of Three Proteins Obtained by XLL Cross-Axis CPC

Column Capacity (mL) (β)	Revolution (rpm)	Flow Rate (mL/min)	Peak Resolution[a] cy/myo	myo/ov	Stationary Phase Retention (%)
250	600	1.0	1.59	0.56	29.1
(0.25–0.60)	750	1.0	2.10	0.85	52.1
	800	1.0	1.88	0.51	46.4
	1000	1.0	1.75	0.45	29.6
	750	0.5	2.26	0.82	61.4
	750	1.0	2.10	0.85	52.1
	750	2.0	2.28	0.86	45.4
	750	4.0	1.01	—	33.5
280	600	1.0	2.46	0.81	11.3
(0.50–1.00)	750	1.0	2.29	0.90	32.3
	800	1.0	2.53	0.68	27.5
	1000	1.0	2.11	0.53	16.6
	750	0.5	3.35	0.99	36.6
	750	1.0	2.29	0.90	32.3
	750	2.0	1.73	—	30.9
	750	4.0	0.18	—	20.2

[a]Cytochrome c = cy; myoglobin = myo; ovalbumin = ov.

optimum revolution speed of 800 rpm, the flow-rate was varied from 0.5 to 1.1 mL/min (bottom three rows in Table 13.4). With one exception, the peak resolution decreases with an increased flow-rate and, again, R_s values between the first and second peaks show close correlation with the percentage retention of the stationary phase.

These experimental results strongly suggest that, in the horizontal CPC, the retention of the stationary phase may be a major factor that determines the peak resolution of proteins within the applied experimental conditions.

Figure 13.7 shows the chromatogram of the four proteins obtained by the horizontal CPC under the optimum experimental conditions. The separation was performed at 800 rpm at a flow-rate of 0.65 mL/min using the lower phase as the mobile phase. The four components were eluted in the order of their partition coefficient values within 15 h. After the elution of cytochrome c ($K = 103.7$), myoglobin ($K = 2.08$), and ovalbumin ($K = 0.63$), the PEG-rich upper phase was pumped into the column in the reversed direction to facilitate rapid elution of the hemoglobin ($K = 0.08$) still remaining in the column. From the obtained chromatogram, the separation efficiency may be computed and expressed in terms of theoretical plate (TP) number using the conventional

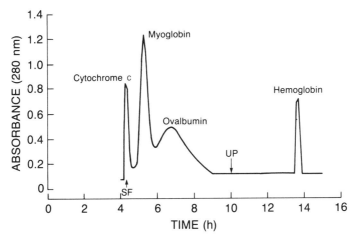

Figure 13.7. Chromatogram of proteins obtained by the horizontal CPC. Experimental conditions: Column is an eccentric dual layer coil assemblies × 3, 1.6-mm i.d. PTFE, 220-mL capacity; Sample consists of 10–200 mg of each protein in 4-mL solvent; solvent system consists of 12.5% (w/w) PEG 1000–12.5% (w/w) K_2HPO_4 in distilled water; mobile phase is the lower phase; flow-rate = 0.65 mL/min; revolution = 800 rpm; SF = solvent front; UP = upper phase eluted in the reversed direction.

chromatographic equation,

$$N = (4R/W)^2 \tag{13.2}$$

where N denotes the theoretical plate number, R is the retention time or volume of the peak maximum, and W is the peak width expressed in the same unit as R. With the use of Eq. (13.2), the separation efficiency obtained from the third peak (ovalbumin) was 50 TP, while the R_s between the second and the third peaks was 0.60.

The peak resolution of proteins in the two different columns on the XLL cross-axis CPC have also been studied under various experimental conditions with the same standard sample mixture and polymer phase system. The results are summarized in Table 13.5. In the smaller coil ($\beta = 0.25$–0.60), the effects of the revolution speed on the peak resolution were examined at a 1.0-mL/min flow-rate (see top 4 rows). The maximum peak resolution was obtained at 750 rpm. The lower (600 rpm) or higher (1000 rpm) revolution speed results in considerable decrease of peak resolution probably due to the lower stationary-phase retention as observed in the horizontal CPC. At a 750 rpm revolution speed (5th–8th rows in Table 13.5), an increase of the flow-rate up to 2 mL/min does not affect the peak resolution, despite a considerable decrease in stationary phase retention. However, further increase of the flow-rate to 4.0 mL/min results in a detrimental loss in peak resolution probably due to a sharp decline in the phase retention level.

The effects of revolution speed on the peak resolution in the XLL cross-axis CPC were similarly investigated in the larger column ($\beta = 0.50$–1.00) at a flow-rate of 1.0 mL/min (9th–12th rows in Table 13.5). The results show that the R_s values between the first and second peaks (cy/myo) are rather insensitive to the revolution speed, while the best resolution between the second and third peaks (myo/ov) is found at 750 rpm associated with the highest retention level of the stationary phase as observed in the small column. At a constant revolution speed of 750 rpm (bottom 4 rows), the lowest flow-rate of 0.5 mL/min yields the best peak resolution, while the highest flow-rate at 4.0 mL/min results in a serious loss of peak resolution apparently due to the lowered stationary phase retention.

Figure 13.8 shows a chromatogram of the four proteins obtained by the XLL cross-axis CPC using the smaller column ($\beta = 0.25$–0.60). The separation was performed at 750 rpm and at a high flow-rate of 2 mL/min. All compo-

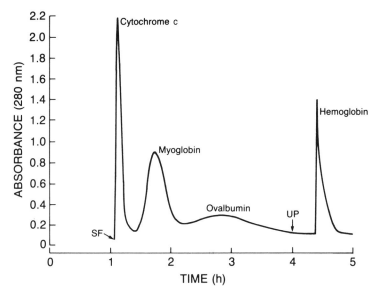

Figure 13.8. Chromatogram of proteins obtained by the XLL cross-axis CPC. Experimental conditions: Column is a 2.6-mm i.d. PTFE multilayer coils × 2, $\beta = 0.25$–0.60, 250-mL capacity; Sample consists of four proteins each protein 10–200 mg in 4 mL of solvent; solvent system is 12.5% (w/w) PEG 1000–12.5% (w/w) K_2HPO_4 in distilled water; mobile phase is the lower phase; flow-rate = 2 mL/min; revolution = 750 rpm; SF = solvent front; UP = upper phase eluted in the reversed direction.

nents were resolved in less than 5 h. The hemoglobin with a low partition coefficient was collected from the column by applying the reversed elution as described earlier. The partition efficiencies computed from the chromatogram range from 550 TP for the first peak to 35 TP for the third peak, while $R_s = 0.86$ between the second and third peaks.

13.4.1.4. Applications

The performance of the XLL cross-axis CPC was evaluated in separation of HDL and LDL from human plasma using a polymer phase system composed of 16.0% (w/w) PEG 1000 and 12.5% (w/w) dibasic potassium phosphate in distilled water using the lower phase as the mobile phase (38). The separations were performed in a pair of multilayer coils coaxially mounted around the column holders with 3.8-cm hub diameter. Human HDL and LDL isolated by ultracentrifugation were used for preparation of the sample solution. A lipoprotein suspension, a mixture of 3 mL of HDL (44.2 mg/mL of protein) and 2 mL of LDL (11.4 mg/mL of protein), was loaded on the column. The apparatus was rotated at 750 rpm while the mobile lower phase was pumped into the column at a flow-rate of 0.5 mL/min in the proper elution mode (33).

CCC SEPARATION OF LIPOPROTEINS

A

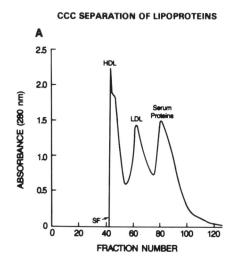

AGAROSE GEL PROFILE OF CCC FRACTIONS

B

Lane 1—Corresponds to the CCC fractions # 43-50
Lane 2—Corresponds to the CCC fractions # 60-70
Lane 3—Corresponds to the CCC fractions # 80-90

Figure 13.9. Separation of lipoproteins by the XLL cross-axis CPC (A) and 0.6% (w/v) agarose gel electrophoretic profile of the CCC fractions. Experimental conditions: Column is a 2.6-mm i.d. PTFE multilayer coils × 2, $\beta = 0.25$–0.60, 250-mL capacity; sample is a mixture of 3 mL of HDL (44.2 mg/mL of protein), 2 mL of LDL (11.4 mg/mL of protein); solvent system consists of 16.0% (w/w) PEG 1000–12.5% (w/w) K_2HPO_4 (pH 9.2); mobile phase is the lower phase; flow-rate = 0.5 mL/min; revolution = 750 rpm.

As shown in Fig. 13.9A, HDL and LDL were eluted from the column in the decreasing order of their partition coefficient values, $K = C_L/C_U$, and also partially separated from other plasma proteins. The K values of HDL and LDL are 3.8 and 1.8, respectively. These values are substantially different from other plasma proteins such as albumin ($K = 0.12$), α-globulin ($K = 0.05$), and γ-globulin ($K = 0.02$). The separation was completed within 12 h and the volume of the upper stationary phase retained in the column was 45% of the total column capacity (250 mL). Figure 13.9B shows the 0.6% agarose gel electrophoretic patterns of each peak. The lipid moiety of the lipoproteins was stained by oil red 7B. The fractions 43–50 and 60–70, corresponding to center cuts of the first and second peaks in the chromatogram, contained HDL and LDL, which migrated to the respective positions as indicated. The third peak in the chromatogram may represent other plasma proteins, because the partition coefficient value computed from the chromatogram is much lower than those of lipoproteins and the lipid staining of the fractions was negative.

The capability of the XLL cross-axis CPC was further examined in the purification of some recombinant enzymes from crude *Escherichia coli* homogenates. The polymer phase system used was 16.0% (w/w) PEG 1000–6.25% (w/w) monobasic and 6.25% (w/w) dibasic potassium phosphate at pH 6.8. The potassium phosphate-rich lower phase was used as the stationary phase. About 1.0 mL of crude purine nucleoside phosphorylase (PNP) in 10 mL of the above solvent system was loaded in the multilayer coil and eluted with the PEG-rich upper phase at a flow-rate of 0.5 mL/min in the $P_{II}HI$ elution mode (see Tables 1.2 and 13.8).

Figure 13.10 shows the chromatogram of a crude PNP extract obtained by the XLL cross-axis CPC. A 3-mL volume was collected in each fraction. The solvent front emerged at the 46th fraction and purified PNP was harvested from fractions 65–80. The lower diagram in Fig. 13.10 shows the 12.0% sodium dodecylsulfate (SDS) gel electrophoresis patterns of CCC fractions obtained from the crude PNP extract. Lane 1 is from the molecular weight markers, lane 2 is from the original sample, lane 3 is from fraction 46 (solvent front), and lane 4 is from fraction 71. The gel electrophoresis clearly demonstrates that PNP in the crude *E. coli* homogenate was highly purified by HSCCC via a single pass through the column.

Purification of recombinant uridine phosphorylase (UrdPase) from *E. coli* was similarly performed as shown in Fig. 13.11 (39). The polymer phase system was the same as that used for the purification of recombinant PNP above. About 2.0 mL of a crude UrdPase extract in 4 mL of the solvent (1 mL of upper phase and 3 mL of lower phase containing 16.0% PEG 1000 and 12.5% potassium phosphate) was loaded on the column and eluted with the PEG-rich upper phase at 0.5 mL/min. In Fig. 13.11, the amount of proteins (solid line) in the fractions and their uridine phosphorylase activity (dotted line) are

CCC PURIFICATION OF CRUDE EXTRACT OF PNP

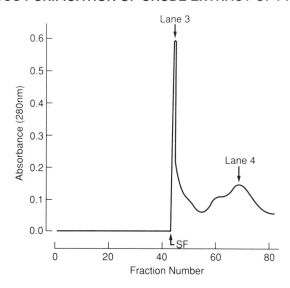

SDS GEL PROFILE OF THE CCC FRACTIONS OF PNP

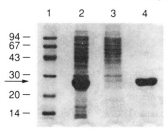

Lane 1 - Contains the molecular weights markers
Lane 2 - Represents the crude extract
Lane 3 - Corresponds to the HSCCC fraction #46
Lane 4 - Corresponds to the HSCCC fraction #71

Figure 13.10. Countercurrent chromatographic purification of purine nucleoside phosphorylase (PNP) from crude extract of *E. coli* (A) and SDS gel electrophoresis profile of CCC fractions (B). Experimental conditions: Column is a 2.6 mm i.d. PTFE multilayer coils × 2, $\beta = 0.25$–0.60, 250-mL capacity; sample consists of crude PNP in 10 mL of solvent; solvent system is 16.0% (w/w) PEG 1000–6.25% (w/w) KH_2PO_4 + 6.25% (w/w) K_2HPO_4 (pH 6.8); mobile phase is the upper phase; flow-rate = 0.5 mL/min; revolution = 750 rpm.

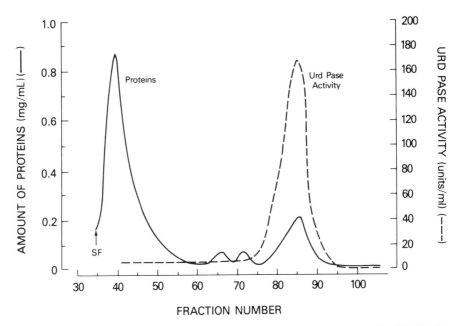

Figure 13.11. Countercurrent chromatographic purification of uridine phosphorylase (UrdPase) from crude homogenate of *E. coli*. Experimental conditions: Column is a 2.6-mm i.d. PTFE multilayer coils × 2, $\beta = 0.25$–0.60, 250-ml capacity; sample consists of crude UrdPase 2 mL in 4 mL of solvent; solvent system is 16.0% (w/w) PEG 1000–6.25% (w/w) KH_2PO_4 + 6.25% (w/w) K_2HPO_4 (pH 6.8); mobile phase is the lower phase; flow-rate = 0.5 mL/min; revolution = 750 rpm.

plotted against the fraction number. The chromatogram shows four protein peaks. The majority of protein mass was eluted immediately after the solvent front in fractions 36–55, whereas the enzyme activity of the UrdPase coincides with the fourth protein peak corresponding to fractions 77–95. The results indicate that the recombinant UrdPase can be highly purified from the crude *E. coli* extract in a one-step operation by the XLL cross-axis CPC.

13.4.2. PEG–Dextran Polymer Phase System

The polymer solution containing PEG and dextran can form two layers without any other additives. Consequently, the system provides high solubility for various macromolecules, which would be largely salted out in the PEG–phosphate systems due to a high salt concentration. The PEG–dextran system also permits adjustment of the osmotic pressure, pH, and ionic composition to meet the physiological requirements for vulnerable mammalian cells to preserve their biological functions.

On the other hand, the PEG–dextran system has a serious drawback in its application to CCC. A high concentration of dextran in the system increases the viscosity and a close polarity between the two polymers results in extremely low interfacial tension between the two phases. These physical properties of the solvent system produce a problem in retention of the stationary phase in the XLL cross-axis CPC. In addition, high mass transfer resistance between the two phases lowers the partition efficiency considerably compared with the PEG–phosphate solvent system previously described.

For the efficient use of the PEG–dextran polymer phase system, the XLLL and L cross-axis CPCs have been designed to improve the retention of the stationary phase. In these centrifuges, the position of the column holders are further shifted along the rotary shaft to enhance the laterally acting centrifugal field (see Fig. 13.2). The capability of these instruments has been evaluated in separations of various proteins such as histones, serum proteins, and profilin–actin complex.

Table 13.6 summarizes the CCC separation of several proteins with the PEG–dextran polymer phase systems using these two cross-axis CPCs.

13.4.2.1. Partition Coefficient Values of Proteins

In contrast to the PEG–phosphate systems, which partition small molecules unilaterally into one of the phases, the PEG–dextran polymer phase systems tend to distribute the majority of small molecules rather evenly into both phases. In this circumstance, the efficient purification of the proteins can be achieved by adjusting the partition coefficient of the desired protein so that it is unilaterally distributed in either upper or lower phase. Usually, this can be done by manipulating the pH of the solvent system.

Table 13.7 shows the partition coefficient values ($K = C_U/C_L$) of proteins including six different types of histones, α- and γ-globulins, and human serum albumin in the 4.4% (w/w) PEG 8000–7.0% (w/w) dextran T500 polymer

Table 13.6. Protein Separation by PEG-Dextran Systems[a]

Proteins	Apparatus	Mobile Phase[b]	pH	Reference
Histones	Type-XLLL CPC	LP	5.7	40
α-Globulin, HSA	Type-XLLL CPC	UP	9.2	40
α-, γ-Globulins	Type-XLLL CPC	UP	6.8	40
Profilin-actin	Type-L CPC	UP	6.8	41

[a]Solvent system: 4.4% (w/w) PEG 8000–7.0% (w/w) dextran T500-10 mM potassium phosphate buffer.
[b]Lower phase = LP; upper phase = UP.

Table 13.7. Partition Coefficients ($K = C_U/C_L$) of the Several Proteins in the PEG 8000–Dextran T500 Polymer Phase System at Various pH Values[a]

Proteins		Partition Coefficient (C_U/C_L)					
	pH	4.7	5.7	6.5	6.8	7.7	9.2
Histone 2A		0.79	0.69	0.22	0.19	0.13	0.84
Histone 3S		0.29	0.17^b	0.23	0.35	0.44^b	0.55
Histone 5S		0.85^b	0.78	0.07^b	0.14	0.08^b	0.05^b
Histone 6S		0.47^b	0.20^b	0.04^b	0.06	0.05^b	0.43
Histone 7S		1.02	1.02	0.13^b	0.64	0.15	0.28
Histone 8S		0.12	0.03	—	0.01	1.00^b	0.75^b
Human serum albumin		0.36	0.19	0.31	0.87	1.25	1.20
α-Globulin		1.27^b	1.11^b	0.93	7.14	12.50	8.33
γ-Globulin		0.35	0.14^b	0.33^b	1.09	1.00^b	0.80

[a] Solvent system: 4.4% (w/w) PEG 8000, 7.0% (w/w) dextran T500 in 10 mM potassium phosphate buffer.
[b] Slightly soluble; — insoluble.

phase system at various pH values ranging from 4.7 to 9.2. The K values of histones are always smaller than 1.0, except for histone 7S at pH 4.7–5.7 and histone 8S at pH 7.7. Human serum albumin is fairly evenly distributed in the two phases between pH 6.8 and 9.2. The K values of α-globulin are close to unity at acidic pH from 4.7 to 6.5, but sharply increase at basic pH above 6.8. The K values of γ-globulin are similar to those of serum albumin, remaining low at acidic pH and increasing to unity at neutral pH 6.8. These results suggest that the K values of proteins, especially of α-globulin, are greatly altered at the isoelectric points.

In Table 13.7, there are large differences in K values between histones 8S and 2A at pH 5.7, which is considered to be the optimum pH for the separation of basic histone (8S, arginine-rich subgroup f3) from another type of histone (histone 2A, calf thymus histone). Among the serum proteins (Table 13.7, bottom), α-globulin shows very high K values consistently at pH above 6.8 suggesting that it may be well separated from human serum albumin and β-globulin in this pH range.

13.4.2.2. Retention of the Stationary Phase

As mentioned earlier, the PEG–dextran polymer phase system has high viscosity and extremely low interfacial tension that lead to emulsification under violent mixing unless a strong lateral force field is applied across the diameter of the tube to accelerate phase separation. Both the XLLL and L cross-axis CPCs have been developed mainly for the use of the PEG–dextran polymer phase system.

A series of experiments was performed to measure the retention of the stationary phase for the 4.4% (w/w) PEG 8000–7.0 (w/w) dextran T500 polymer phase systems in the pair of small multilayer coils mounted on the XLLL cross-axis CPC. The separations were carried out at a 1.0-mL/min flow-rate in various elution modes using both upper PEG-rich and lower dextran-rich phases as the mobile phase. The results are summarized in Table 13.8 along with those obtained from the XLL cross-axis CPC for comparison.

In Table 13.8, P_I and P_{II} indicate the direction of the planetary motion, H and T are the head–tail elution mode, and O and I are the inward–outward elution mode along the holder axis. The parameter P_I indicates the counterclockwise revolution and P_{II} is the clockwise revolution. In the inward elution mode (I), the mobile phase is eluted against the lateral force field and in the outward elution mode (O), this flow direction is reversed. These three factors yield a total of four combinations for the left-handed coils as indicated in the table. Among these elution modes, the inward–outward elution mode plays the most important role in stationary-phase retention. To obtain a satisfactory retention of the stationary phase, the lower phase should be eluted outwardly along the direction of the lateral ateral force field and the upper phase in the opposite direction.

Table 13.8. Retention of Stationary Phase by Two X-Asix Coil Planet Centrifuges[a]

X-Axis CPC	Elution Mode[b]	Revolution (rpm)	Stationary Phase[c]	% Retention
Type-XLLL	P_IHO	900	UP	24.6
	P_{II}TO	900	UP	24.6
	P_{II}HI	900	LP	41.1
	P_ITI	900	LP	45.1
Type-XLL	P_IHO	750	UP	21.1
	P_{II}TO	750	UP	19.6
	P_{II}HI	750	LP	28.4
	P_ITI	750	LP	28.9

[a] Solvent system: 4.4% (w/w) PEG 8000, 7.0% (w/w) dextran T500 in a 10 mM potassium phosphate buffer (pH 6.8); flow-rate: 1.0 mL/min.
[b] Combination of three parameters: P = direction of revolution (P_I = counterclockwise; P_{II} = clockwise); H = head-to-tail elution; T = tail-to-head elution; O = outward elution; I = inward elution.
[c] Upper phase = UP; lower phase = LP.

In both types of cross-axis CPC, the best results are obtained by eluting the lower phase inward (P_IHO and $P_{II}TO$) or the upper phase outward ($P_{II}HI$ and P_ITI), while the retention of the dextran-rich lower phase substantially exceeds that of the PEG-rich upper phase in all cases. Note that different from other types of the cross-axis CPC, such as the Type XL and Type X, the direction of the planetary motion and the head–tail elution mode produce little effect in retention of the stationary phase in the XLLL cross-axis CPC. When the percentage retention from the two types of the apparatus is compared the XLLL cross-axis CPC yields 20% (in upper phase retention) to 45% (in lower phase retention) greater retention over the XLL cross-axis CPC in a given elution mode. A high retention value of 45.1% was obtained from the XLLL cross-axis CPC by eluting the upper PEG-rich phase in the P_ITI mode. The overall results indicate that the XLLL cross-axis CPC is more suitable for separations of proteins with the PEG–dextran polymer phase system than the XLL cross-axis CPC.

13.4.2.3.   CCC Separation of Proteins

In order to demonstrate the capability of the XLLL cross-axis CPC in protein separation, various mixtures of selected proteins were subjected to CCC fractionation in the dextran–PEG polymer phase system composed of 4.4% (w/w) PEG 8000–7.0% (w/w) dextran T500 at the optimum pH determined by the preliminary studies (Table 13.8).

Figure 13.12 shows a chromatogram of the histone mixture obtained from the large column (440-mL capacity, $\beta = 0.45$–1.50) mounted on the XLLL cross-axis CPC. The separation was performed at a flow-rate of 1.0 mL/min using the lower dextran-rich phase as the mobile phase in the P_IHO elution mode. The pH of the polymer phase was adjusted at 5.7 by a 10-mM potassium phosphate buffer. Histone 8S with a small K value ($C_U/C_L = 0.03$) was eluted immediately after the solvent front, being separated from the histone 2A with a higher K value ($C_U/C_L = 0.69$). The CCC run was completed within 9 h.

A semipreparative-scale separation of serum proteins was performed with the large multilayer coil (440-mL capacity). A mixture of 65.8 mg of α-globulin and 114.1 mg of human serum albumin was eluted with the PEG-rich upper phase (pH 9.2) in the P_ITI elution mode (Fig. 13.13). The revolution speed and the flow-rate were same as those applied in the separation of histones. The α-globulin with a high K value (8.33) was eluted near the solvent front, whereas human serum albumin with a lower K value (1.20) emerged at a retention volume close to the total column capacity. A 50 mg amount of proteins was well separated within 9 h.

In order to shorten the separation time of α- and γ-globulins, the small multilayer coil (300-mL capacity) was used and a mixture of 94.5 mg of

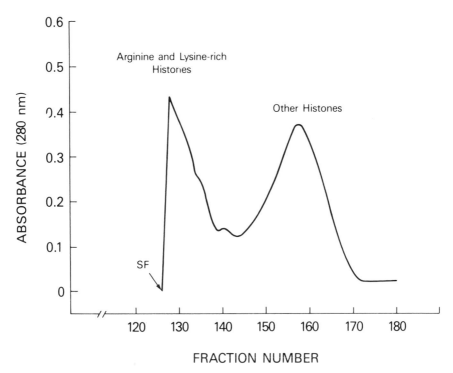

Figure 13.12. Chromatogram of histones obtained by the XLLL cross-axis CPC. Experimental conditions: Column is a 2.6-mm i.d. PTFE multilayer coils × 2, $\beta = 0.45$–1.50, 440-mL capacity; sample consists of 10.8 mg of arginine-rich, 12.5 mg of lysine-rich, and 80 mg calf thymus histones in 4 mL of solvent; solvent system is 4.4% (w/w) PEG 8000–7.0% (w/w) dextran T500 in a 10 mM potassium phosphate buffer at pH 5.7; mobile phase is the lower phase; flow-rate = 1 mL/min; revolution = 900 rpm ($P_I HO$); SF = solvent front.

α-globulin and 107.3 mg of γ-globulin was eluted with a PEG-rich upper phase at pH 6.8. The separation was completed within 6.5 h as shown in Fig. 13.14. The retention of the stationary phase was 40% of the total column capacity.

The overall results of these studies indicate that the XLLL cross-axis CPC (L/R = 3.5) has substantially improved the retention of the viscous PEG–dextran polymer phase systems over the XLL cross-axis CPC (L/R = 2). This improved retention is apparently caused by the increased lateral shift of the column position from L/R = 2–3.5. These results suggest that the retention of the stationary phase may be further increased in the L cross-axis CPC (Fig. 13.2) with the column holder position at L/R → ∞ or R = 0 in which the column holder axis crosses the central axis of the centrifuge. In fact, the L cross-axis CPC recently constructed has produced substantially higher

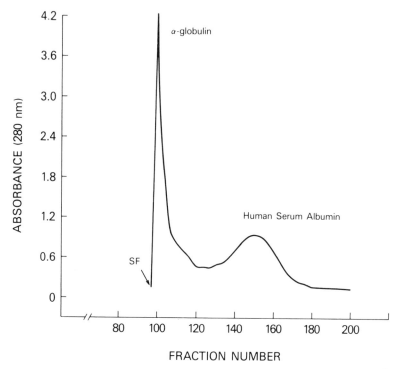

Figure 13.13. CCC separation of human serum albumin and α-globulin. Experimental conditions: Column is a 2.6-mm i.d. PTFE multilayer coils × 2, $\beta = 0.45–1.50$, 440-mL capacity; sample consists of 65.8 mg of α-globulin and 114.1 mg of HSA in a 4 mL of solvent; solvent system is 4.4% (w/w) PEG 8000–7.0% (w/w) dextran T500 in a 10 mM potassium phosphate buffer at pH 9.2; mobile phase is the upper phase; flow-rate = 1 mL/min; revolution = 900 rpm (P_ITI); SF = solvent front.

stationary-phase retention for the viscous PEG–dextran polymer phase systems compared with the XLLL cross-axis CPC.

13.4.2.4. Application of PEG–Dextran Polymer Phase System

Using the L cross-axis CPC equipped with a pair of multilayer coils (130-mL capacity), profilin–actin complex from the *Acanthamoeba* extract was purified by a polymer phase system composed of 4.4% (w/w) PEG 8000–7.0% (w/w) dextran T500 at pH 6.8.

Figure 13.15 shows the chromatogram of a crude *Acanthamoeba* extract obtained by the L cross-axis CPC. The sample solution was prepared by adding proper amounts of PEG 8000 and dextran T500 to 12.5 g of the

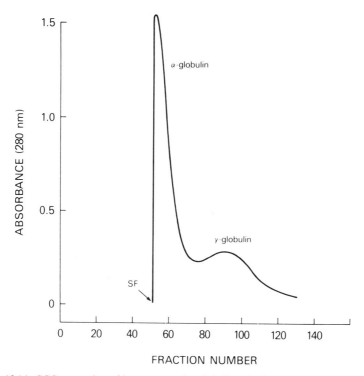

Figure 13.14. CCC separation of human α- and γ-globulins obtained by the XLLL cross-axis CPC. Experimental conditions: Column is a 2.6-mm i.d. PTFE multilayer coils × 2, $\beta = 0.50$–1.30, 300-mL capacity; sample consists of 94.5 mg of α- and 107.3 mg of γ-globulins in 4 mL of solvent; solvent system is 4.4% (w/w) PEG 8000–7.0% (w/w) dextran T500 in a 10 mM potassium phosphate buffer at pH 6.8; mobile phase is the upper phase; flow-rate = 1.0 mL/min; revolution = 900 rpm (P_1TI); SF = solvent front.

Acanthamoeba crude extract to bring the polymer composition similar to that of the solvent phases used for separation. The separation was carried out by pumping the PEG-rich upper phase into the head of the column at 0.5 mL/min under a high-revolution speed of 1000 rpm. The results are shown in Fig. 13.15 (top). The solvent front emerged at the 14th fraction (3 mL per fraction) whereas profilin–actin complex was eluted in fractions 20 and 28 and was well separated from the impurities, most of which were eluted later with a retention volume close to the total column capacity (around fraction 33). Some impurities were also found near the solvent front (fraction 15). Identification of profilin–actin complex was made by 12.0% SDS gel electrophoresis, as illustrated in Fig. 13.15 (bottom). The retention of the lower stationary phase was 69.0% of the total column capacity.

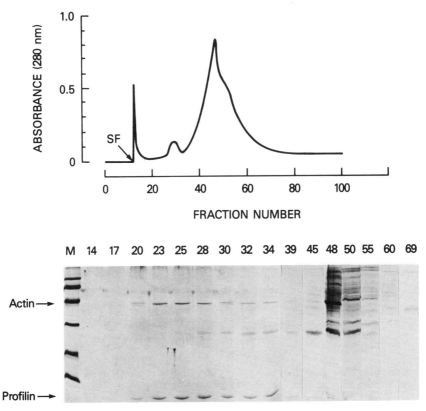

Figure 13.15. Purification of profilin–actin complex from *Acanthamoeba* soluble extract. Experimental conditions: Column is a 2.6-mm i.d. PTFE multilayer coils × 2, $\beta = 0.16$–0.27; 130-mL capacity; sample consists of 12.5 g of *Acanthamoeba* soluble extract; solvent system is 4.4% (w/w) PEG 8000–7.0% (w/w) dextran T500 in a 10 mM potassium phosphate buffer at pH 6.8; mobile phase is the lower phase; flow-rate = 0.5 mL/min; revolution = 900 rpm ($P_{II}HO$); SF = solvent front.

13.5. CONCLUSION

The capability and the efficiency of the cross-axis CPCs in performing CCC have been demonstrated in the separation and purification of several protein samples. The unique features of the cross-axis CPCs provide excellent retention of the stationary phase of viscous, low interfacial tension polar solvent systems, such as aqueous–aqueous polymer phase systems. Purification of the recombinant proteins from the *E. coli* crude extract has been successfully performed by the XLL cross-axis CPC with a PEG–potassium phosphate polymer phase system. The profilin–actin complex, which has a low solubility in the above polymer phase system has been purified with PEG–dextran polymer phase system by the L cross-axis CPC. The combined use of the aqueous–aqueous polymer phase system and a suitable cross-axis CPC enables efficient separation and purification of a variety of proteins with minimum denaturation and loss of the enzymatic activity. The method may be further extended to affinity partitioning by the use of various ligand-bound polymers.

REFERENCES

1. P. Å. Albertsson, Partition of Cell Particles and Macromolecules, 3rd ed., Wiley, New York, 1986.

2. P. Å. Albertsson, *Nature (London)*, **182**, 709 (1958).

3. P. Å. Albertsson and E. J. Nyns, *Nature (London)*, **184**, 1465 (1959).

4. P. Å. Albertsson and E. J. Nyns, *Akr. Kemi*, **17**, 197 (1961).

5. P. Å. Albertsson, S. Sasakawa, and H. Walter, *Nature (London)*, **228**, 1329 (1970).

6. P. Å. Albertsson, *Adv. Protein Chem.*, **24**, 309 (1970).

7. P. Å. Albertsson, *Biochim. Biophys. Acta*, **103**, 1 (1965).

8. G. Frick and T. Lif, *Arch. Biochem. Biophys. Suppl.*, **1**, 271 (1962).

9. T. Lif, G. Frick, and P. Å. Albertsson, *J. Mol. Biol.*, **3**, 727 (1961).

10. P. Å. Albertsson, *Eur. J. Biochem.*, **3**, 25 (1967).

11. S. Bengtsson, L. Phillipson, and P. Å. Albertsson, *Biochim. Biophys. Res. Commun.*, **9**, 318 (1962).

12. E. C. J. Norrby and P. Å. Albertsson, *Nature (London)*, **188**, 1047 (1960).

13. P. Å. Albertsson, *Anal. Biochem.*, **11**, 121 (1965).

14. P. Å. Albertsson and G. D. Baird, *Exp. Cell Res.*, **28**, 296 (1962).

15. I. Ericson, *Biochim. Biophys. Acta*, **356**, 100 (1974).

16. P. Å. Albertsson and G. D. Baird, *Exp. Cell Res.*, **28**, 296 (1962).

17. H. Walter and F. W. Selby, *Biochim. Biophys. Acta*, **112**, 146 (1966).

18. H. Walter and F. W. Selby, *Biochim. Biophys. Acta*, **148**, 517 (1967).

19. H. Walter, E. J. Krob, and R. Garza, *Biochem. Biophys. Acta*, **165**, 507 (1968).

20. H. Walter, E. J. Krob, R. Garza, and G. S. Ascher, *Exp. Cell Res.*, **55**, 57 (1969).

21. H. Walter, E. J. Krob, and G. S. Ascher, *Exp. Cell Res.*, **55**, 279 (1969).

22. H. Walter, F. W. Selby, and R. Garza, *Biochim. Biophys. Acta*, **136**, 148 (1967).

23. L. C. Craig, in P. Alexander and R. J. Block, (Eds.), *A Laboratory Manual of Analytical Methods of Protein Chemistry*, Vol. I, Pergamon Press, Oxford, 1950, p. 121.

24. L. C. Craig and D. Craig, in Weissberger, (Ed.), *Technique of Organic Chemistry*, Vol III, Part 1, 2nd ed., Interscience, New York, 1956.

25. Y. Ito and R. L. Bowman, *Science*, **167**, 281 (1970).

26. Y. Ito and R. L. Bowman, *J. Chromatogr. Sci.*, **8**, 315 (1970).

27. T. Tanimura, J. J. Pisano, Y. Ito, and R. L. Bowman, *Science*, **169**, 54 (1970).

28. Y. Ito and R. L. Bowman, *Anal. Chem.*, **43**, 69A (1971).

29. Y. Ito and R. L. Bowman, *Science*, **173**, 420 (1972).

30. R. E. Hurst and Y. Ito, *Clin. Chem.*, **18**, 814 (1972).

31. Y. Ito, *J. Chromatogr.*, **207**, 161 (1981).

32. Y. Ito, J. Sandlin, and W. G. Bowers, *J. Chromatogr.*, **244**, 247 (1982).

33. Y. Ito, E. Kitazume, and J. L. Slemp, *J. Chromatogr.*, **538**, 81 (1991).

34. Y. Ito, E. Kitazume, and M. Bhatnagar, *J. Chromatogr.*, **538**, 59 (1991).

35. Y. Ito and T.-Y. Zhang, *J. Chromatogr.*, **437**, 121 (1988).

36. Y. Ito and H. Oka, *J. Chromatogr.*, **457**, 393 (1988).

37. Y. Shibusawa and Y. Ito, *J. Chromatogr.*, **550**, 695 (1991).

38. Y. Shibusawa, Y. Ito, K. Ikewaki, D. J. Rader, and B. Brewer, *J. Chromatogr.*, **596**, 118 (1992).

39. Y.-W. Lee, Y. Shibusawa, F. T. Chen, J. Meyers, J. M. Schooler, and Y. Ito, *J. Liq. Chromatogr.*, **15**, (15, 16), 2831 (1992).

40. Y. Shibusawa and Y. Ito, *J. Liq. Chromatogr.*, **15** (15, 16) 2787 (1992).

41. S. Holliday, Y. Shibusawa and Y. Ito, unpublished data.

CHAPTER

14

SEPARATION OF RARE EARTH AND CERTAIN INORGANIC ELEMENTS BY HIGH-SPEED COUNTERCURRENT CHROMATOGRAPHY

EIICHI KITAZUME

Faculty of Humanities and Social Sciences, Iwate University, Ueda, Morioka, Iwate, 020 Japan

14.1. INTRODUCTION

Countercurrent chromatography (CCC) is a useful method of separating many organic materials, such as biologically active substances, natural and synthetic peptides, and various plant hormones. Recently, high-speed countercurrent chromatography (HSCCC) has been developed by Ito (1). It is the most advanced form of CCC and has been widely applied to the separation of various organic materials (1–3). Its advantages over other CCC methods include high partition efficiency, speedy separation, and versatility in performing true CCC and foam CCC. Like other CCC methods, it is also free from problems based on solid support, such as adsorption or irreversible binding and contamination of the sample. In addition, highly efficient chromatographic separation has been achieved using a set of three multilayer coils (4–6).

In spite of these advantages, there have been no applications to the separation of inorganic elements until quite recently. However, many of the features of HSCCC have convinced us that this method can be successfully used in separation of inorganic elements and inorganic analytical chemistry.

This chapter describes such application of HSCCC to the separation and analysis of rare earth elements and other inorganic materials.

High-Speed Countercurrent Chromatography, Edited by Yoichiro Ito and Walter D. Conway.
Chemical Analysis Series, Vol. 132.
ISBN 0-471-63749-1 © 1996 John Wiley & Sons, Inc.

14.2.   SEPARATION OF RARE EARTH ELEMENTS BY HIGH-SPEED COUNTERCURRENT CHROMATOGRAPHY

Besides being widely used in electronic and glass industries, rare earth elements, such as lanthanum, cerium, neodymium, samarium, gadolinium, dysprosium, erbium, and ytterbium, have recently been found to have important biological effects including the ability to stabilize and enhance interferon activity (7). However, lanthanoid elements are particularly difficult to separate due to their formation of trivalent cations with almost equal diameters (8). The solvent extraction of the trivalent lanthanoids by reagents of different types has been studied extensively since the 1950s, and this technique is now widely accepted in the commercial processing of rare earth compounds (9). However, it takes too much time to separate and multistage extraction processes are required.

In the past, chromatographic separations of lanthanoid elements were performed by using high-performance liquid chromatography (HPLC) (10, 11), ion chromatography (12), centrifugal droplet countercurrent chromatography (CDCCC) (13–16). However, sample size was extremely limited in both HPLC and ion chromatography, whereas separation was inefficient in CDCCC.

High-speed CCC has a great potential for the chromatographic separation of chemically similar materials and recently, a new HSCCC centrifuge equipped with a set of three multilayer coils has achieved highly efficient chromatographic separations (4–6).

In this section, some results and the capability based on our experience with the improved HSCCC centrifuge are described for the separation of rare earth elements (17–19). Under optimum conditions, excellent separation of rare earth elements was achieved in an isocratic elution, while versatility of the method was demonstrated in gradient elution using an exponential gradient of hydrochloric acid concentration, to resolve all 14 rare earth elements in one operation.

14.2.1.   Description of the Instrument

The design of the CCC apparatus employed in this study has been described by Ito in Chapter 1 and therefore only a brief statement is given here. The apparatus symmetrically holds a set of three identical columns on the rotary frame at a distance of 7.6 cm from the central axis of the centrifuge. Each column holder is equipped with two planetary gears, one of which engages an identical stationary sun gear mounted around the central stationary pipe of the centrifuge. This gear arrangement produces a desired planetary motion of each column holder, that is, one rotation about its own axis per one revolution around the central axis of the centrifuge in the same direction. The other gear on the column holder is engaged to an identical gear on a rotary tube support mounted between the column holders. This gear arrangement produces counterrotation of the tube support to prevent twisting of the flow-tubes on the rotary frame.

All column holders can be removed from the rotary frame by loosening a pair of screws on each bearing block, thus facilitating the mounting of the coiled column on the holder. Each multilayer coil was prepared from a single piece of approximately 100 m long, 1.07-mm i.d. polytetrafluoroethylene (PTFE) tubing (Zeus Industrial Products, Orangeburg, SC) by winding it directly onto the holder hub (7.6 cm diameter), making 13 layers of the coil between a pair of flanges spaced 5 cm apart. The β value ranged from 0.5 at the internal terminal to 0.75 at the external terminal. [Beta (β) is an important parameter used to determine the hydrodynamic distribution of the two solvent phases in the rotating coil and $\beta = r/R$, where r is the distance from the column holder axis to the centrifuge.] Each multilayer coil consists of about 400 helical turns with an approximately 90-mL capacity. To prevent dislocation of the multilayer coil on the column holder, the innermost layer of the coil was glued onto the holder hub with an RTV silicone rubber adhesive sealant (General Electric Company, Waterford, NY), while the whole column and the peripheral portion of the flanges were wrapped with a heat shrinkable polyvinylchloride (PVC) tube.

Each terminal of the multilayer coil was connected to a flow tube, 0.55-mm i.d. and 0.45-mm wall thickness. The use of this thick-wall small-bore tubing is required at the flexing portion of the flow-tube to withstand the back pressure created from the long narrow-bore semianalytical columns. Serial connection between the three multilayer coils was made in such a way that the internal terminal of the first column is connected to the external terminal of the second column, and the internal terminal of the second column is attached to the external terminal of the third column. In this way, all three columns are subjected to the same head–tail elution mode. As previously described, each interconnection flow-tube runs across the rotary frame along the rotary tube support where it is secured with nylon ties. Each flexing portion of the flow-tubes is lubricated with grease and protected with a sheath of Tygon tubing to prevent direct contact with metal parts. Both inlet and outlet flow tubes are secured on the centrifuge wall with a silicone–rubber-padded clamp.

The HSCCC can be operated at the revolutional speed of 900 rpm. A Shimadzu LC-6A pump was used to pump the mobile phase, a Rainin 1011/B-100-S pump to add a postcolumn reagent, Shimadzu SPD-6AV spectrophotometer to monitor the absorbance of effluent, and an LKB fraction collector (Ultrorac) to fractionate the effluent. Another Shimadzu LC-6A pump was introduced between HSCCC and the spectrophotometer to pump and stabilize the effluent for the precise monitoring. The total system was shown in Fig. 14.1.

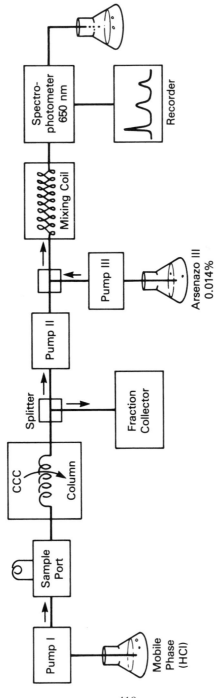

Figure 14.1. Flow diagram of instrumentation assembly for separation of rare earth elements by HSCCC.

A Buchler Instruments KCI-23A2A1 gradient maker was used for gradient elution of 14 lanthanoid elements.

14.2.2. Preparation of Two-Phase Solvent and Sample Solutions

Di-(2-ethylhexyl)phosphoric acid (DEHPA) was washed several times with 1 M hydrochloric acid followed by washing twice with deionized water and then dissolved in n-heptane (stationary phase). One normal hydrochloric acid was diluted with deionized water (mobile phase).

Sample solutions were prepared by dissolving various lanthanoid chlorides in 0.02 M hydrochloric acid.

14.2.3. Separation Procedure

Each separation was initiated by filling the entire column with the stationary nonaqueous phase followed by injection of a 100-μL sample solution containing 25 μg of each component through the sampling port. Then, the mobile phase was eluted through the column at a flow-rate of 5 mL/min in the proper elution mode while the apparatus was rotated at 900 rpm. Continuous detection of the rare earth elements was effected by means of a postcolumn reaction with arsenazo III (11) and the elution curve was obtained by monitoring the effluent at 650 nm using a Shimadzu SPD-6AV spectrophotometer equipped with an analytical flow-cell. The effluent was divided into two streams with a tee adapter and a low-dead-volume Shimadzu LC-6A pump (pump II). Pump II was used to deliver a portion of the effluent at a flow-rate of 1.4 mL/min to the spectrophotometer (see Fig. 14.1). At the outlet of this pump, the arsenazo III–ethanol solution (0.014%, w/v) was continuously added to the effluent stream at a flow-rate of 2.7 mL/min with a Rainin metering pump (pump III). The resulting effluent was first passed through a narrow mixing coil (PTFE tube, 1 m × 0.55 mm i.d.) immersed in a water bath heated to about 40°C, and then led through an analytical flow-cell (1-cm light path) in a Shimadzu SPD-6AV spectrophotometer set to monitor the absorbance at 650 nm. The other effluent stream through the tee adapter was either collected or discarded (see Fig. 14.1).

14.2.4. Measurement of Partition Coefficients K_T and K_E

The distribution ratio K_T of each rare earth element was obtained using a simple test tube method (19).

$$K_T = (A_T - A_L)/(A_L - A_0) \qquad (14.1)$$

where A_T is the total absorbance of sample solution in lower phase before equilibration with upper phase, A_L is the absorbance of the sample solution in the lower phase after equilibration, A_0 is the absorbance of equilibrated lower phase only, $A_T - A_L$ is the absorbance of the upper phase after equilibration, and $A_L - A_0$ is the net absorbance of the lower phase.

The partition coefficient, K_E, was also calculated from the elution curve using the following equation:

$$K_E = (R - R_{sf})/(V_c - R_{sf}) \qquad (14.2)$$

where R is the retention volume of a peak maximum, R_{sf} is the retention volume of solvent front, and V_c is the total column capacity.

The optimum range of the partition coefficient values (K_T) for neodymium, praseodymium, cerium, and lanthanum was obtained with a solvent pair composed of 0.02 M DEHPA in n-heptane (stationary phase) and 0.02 M HCl (mobile phase). The K_T values of these elements each plotted against the sample concentration from 0.005 to 0.025 mM in the mobile phase are presented in Fig. 14.2. The left chart (A) was obtained from the ligand (DEHPA) treated with 1 M hydrochloric acid before being dissolved in n-heptane and the right chart (B) was obtained from the ligand without the hydrochloric acid treatment. Lanthanum, praseodymium, and neodymium were separated with the use of the above solvent systems at a flow-rate of

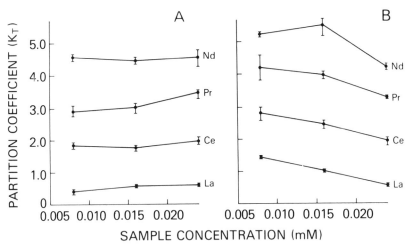

Figure 14.2. Effects of hydrochloric acid treatment of ligand on the partition coefficients of light rare earth elements. (A) After washing with 1 M hydrochloric acid and (B) without pretreatment.

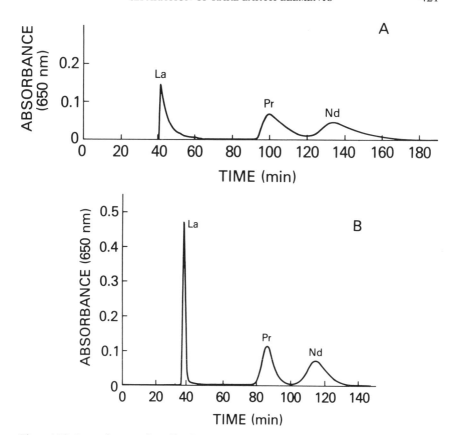

Figure 14.3. Isocratic separation of lanthanum, praseodymium, and neodymium with nontreated (A) and hydrochloric acid treatment (B) ligands by HSCCC. The experimental conditions were as follows: apparatus consists of a HSCCC centrifuge with 7.6-cm revolution radius; column consists of three multilayer coils connected in series, 300 m × 1.07 mm i.d., 270-mL capacity; stationary phase is 0.02 M DEHPA in n-heptane; mobile phase is 0.02 M hydrochloric acid; sample is LaCl$_3$, PrCl$_3$, and NdCl$_3$ each 0.001 M in 100 μL of 0.02 M hydrochloric acid; revolution = 900 rpm; flow-rate = 5 mL/min; pressure = 300 psi.

5 mL/min at a revolution speed of 900 rpm. A chromatogram obtained from the solvent without the hydrochloric acid pretreatment is given in Fig. 14.3A. The three lanthanoid elements were well resolved in 2.5 h but each peak displays marked skewness as expected since K_T increases with decreased sample concentration (Fig. 14.2B). This effect was largely eliminated by using the hydrochloric acid treated ligand, as illustrated in Fig. 14.3B. Figure 14.4 shows the chromatograms of lanthanum, cerium, praseodymium, and neo-

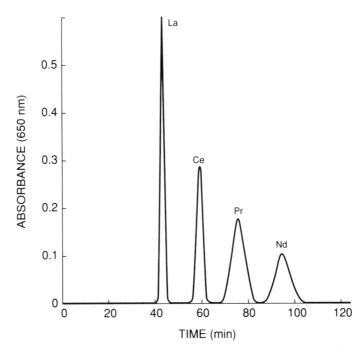

Figure 14.4. Isocratic separation of lanthanum, cerium, praseodymium, and neodymium with hydrochloric acid treatment. The experimental conditions were as follows: apparatus consists of a HSCCC centrifuge with 7.6-cm revolution radius; column consists of three multilayer coils connected in series, 300 m × 1.07 mm i.d., 270-mL capacity; stationary phase is 0.02 M DEHPA in n-heptane; mobile phase is 0.02 M hydrochloric acid; sample is $LaCl_3$, $PrCl_3$, and $NdCl_3$ each 0.001 M in 100 μL of 0.02 M hydrochloric acid; revolution = 900 rpm; flow-rate = 5 mL/min; pressure = 300 psi.

dymium obtained with the hydrochloric acid pretreatment (18, 19). The maximum column pressure during these separations was about 300 psi.

From these chromatograms, the partition coefficients (K_E) were computed for each peak using Eq. (14.2) and compared with the K_T values obtained from the test tube measurements (Fig. 14.2). As shown in Table 14.1, the two partition coefficient values, K_E and K_T, are generally in close agreement in both groups. With both methods the nontreated groups show substantially higher K values than the hydrochloric acid treated group, especially with lanthanum and praseodymium (19).

14.2.5. Partition Efficiencies

From the chromatograms obtained, the partition efficiency of the separation was calculated and expressed in terms of theoretical plates (TP) according to

Table 14.1. Partition Coefficient Values[a]

Element	K_E (Average)		K_T	
	No Washing[b]	Washing[c]	No Washing[b]	Washing[c]
La	0.5	0.3	0.48–1.40	0.33–0.59
Pr	3.2	2.8	3.25–4.23	2.75–3.51
Nd	4.7	4.5	4.15–5.72	4.25–4.78

[a] Upper phase: 0.02 M DEHPA in n-heptane; lower phase: 0.02 M hydrochloric acid.
[b] DEHPA was dissolved in n-heptane without pretreatment.
[c] DEHPA was washed several times with 1 M hydrochloric acid before dissolving in n-heptane.

the following equation:

$$N = (4R/W)^2 \qquad (14.3)$$

where N is the partition efficiency in TP, R is the retention time or volume of the peak maximum, and W is the peak width expressed in the same unit as R.

The partition efficiency can also be expressed in terms of peak resolution according to the following formula:

$$R_s = 2(R_1 - R_2)/(W_1 + W_2) \qquad (14.4)$$

where R_s is the resolution of two adjacent peaks expressed in the units of 4σ in a Gaussian distribution, R_1 and R_2 are the retention time or volumes of two adjacent peaks ($R_1 > R_2$), and W_1 and W_2 are the widths (4σ) of the corresponding peaks. When $R_s = 1.5$, base line separation (99.7% pure) is indicated.

Peak resolution was studied using a standard solution of praseodymium and neodymium. Figure 14.5 shows the effect of the ratio of the hydrochloric acid concentration in the mobile phase to the [DEHPA] concentration in the stationary phase on the peak resolution at the various concentrations of [DEHPA]. Each curve indicates the concentration of [DEHPA] in the stationary phase as labeled. All curves show the maximum R_s value at the ratio around 1.0, namely, at nearly the same concentration between [H$^+$] and [DEHPA], regardless of the applied [DEHPA] concentration. These results suggest that the ratio of acidity to the concentration of [DEHPA] provides a convenient parameter for optimizing solvent compositions. At a given concentration of [DEHPA], the time required for separation increased as the value of [H$^+$]/[DEHPA] decreased. The use of 0.005 M [DEHPA] produced skewed peaks below 1.2 for [H$^+$]/[DEHPA]. The best peak resolution between praseodymium and neodymium was attained between 0.01 and 0.02 M [DEHPA].

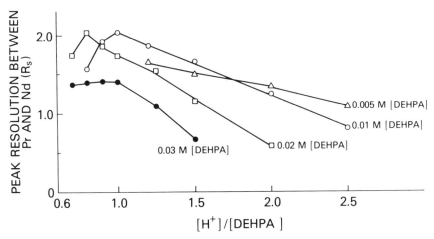

Figure 14.5. Effect of the ratio of [H$^+$] in the mobile phase to [DEHPA] in the stationary phase on the peak resolution between praseodymium and neodymium.

The partition efficiencies of the separation were determined from the chromatogram (Fig. 14.3B) using Eqs. (14.3) and (14.4): These range from 5900 TP for the first peak (lanthanum) to 520 TP for the third peak (neodymium), while the peak resolution between the first and the second peak is 6.94 and that between the second and the third is 1.74. Another set of heavy lanthanoid elements (thulium, ytterbium, and lutetium), which are most difficult to resolve (13–16), was also separated with the present method using a modified solvent composition of 0.003 M DEHPA in n-heptane as the stationary phase and 0.1 M hydrochloric acid as the mobile phase. The results showed greatly reduced partition efficiencies ranging from 170 TP for the first peak (thulium) to 60 TP for the third peak (lutetium).

In Table 14.2 (19), these results are compared with those obtained by other CCC methods where N is the partition efficiency in TP, R_s is the peak resolution between the indicated two peaks, and CDCCC stands for centrifugal droplet CCC (or centrifugal partition chromatography). In the N and R_s groups, the second column (CDCCC^d) (13) and the third column (CDCCC^e) (14), respectively, indicate the efficiencies obtained using the same solvent pairs used in the HSCCC method. The fourth column (CDCCC^f) in both the N and R_s groups (15) and the fifth column (CDCCC^g) in the R_s group (16) indicate the improved efficiencies obtained with a modified stationary phase containing 2-ethylhexyl-phosphonic acid mono-2-ethylhexyl ester (EHPA) at a higher temperature (55°C), and Cyanex 272 [bis(2,4,4-trimethylpentyl)phosphinic acid], respectively. As seen from Table 14.2, HSCCC yields much higher

Table 14.2. Comparison in Partition Efficiencies between HSCCC and Previous Methods

Element	N^a				Rs^b				
	HSCCC[c]	CDCCC[d]	CDCCC[e]	CDCCC[f]	HSCCC[c]	CDCCC[d]	CDCCC[e]	CDCCC[f]	CDCCC[g]
La	5900	34		169.0	6.94	1.4		2.70	2.7
Pr	770	54		85.8	1.74	0.44		0.77	1.2
Nd	520	41		86.2					
Tm	170		8^h	19.5	1.24				
Yb	74		$21^h,4$				0.87^h		0.7
Lu	61		5		0.79		0.22		

[a]Theoretical plate numbers = N.

[b]Peak resolution = R_s.

[c] A solution of 0.02 M DEHPA in n-heptane (stationary phase) and 0.02 M hydrochloric acid (mobile phase) were used for the separation of La, Pr, and Nd. A solution of 0.003 M DEHPA in n-heptane and 0.1 M hydrochloric acid were used for the separation of Tm, Yb, and Lu.

[d,e]The values were obtained from centrifugal droplet CCC with the same solvent system (13, 14).

[f]The values were obtained from centrifugal droplet CCC with 2-ethylhexyl-phosphonic acid mono-2-ethylhexyl ester (EHPA) at 55°C (15).

[g]The values were obtained from centrifugal droplet CCC with 0.1 M Cyanex 272 [bis(2,4,4-trimethylpentyl)phosphinic acid] (16).

[h]Ethylene glycol (20%) was added to the mobile phase at 45°C (14).

efficiencies than the other CCC methods. The separation time is also much shorter in HSCCC (2 h) than in the other methods (3–30 h).

14.2.6. Gradient Elution of Fourteen Rare Earth Elements

A one step separation of all 14 lanthanoids (except for promethium) was performed by applying an exponential gradient of hydrochloric acid concentration in the mobile phase (Fig. 14.6) (17, 19). The main problem in gradient elution is that the optimum range of the ligand concentration in the stationary phase is substantially different between the lighter and heavier groups of the rare earth elements. Because the separation of the heavy lanthanoid elements including thulium, ytterbium, and lutetium is more difficult, the ligand concentration in the experiment was selected at 0.003 M for best resolution. Consequently, the resolution of the light lanthanoid elements, such as lanthanum, cerium, praseodymium, and neodymium, became less efficient compared with that observed in the isocratic separation shown in Fig. 14.3B. Figure 14.6 shows the chromatogram of all 14 lanthanoid elements resolved in less than 5 h.

As indicated in Table 14.2, the HSCCC method radically improved the results obtained by the other CCC methods. The higher performance of the HSCCC over the CDCCC may be explained on the basis of the hydrodynamics inherent to each system. In the hydrostatic equilibrium system such

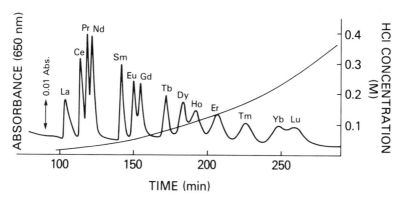

Figure 14.6. Gradient separation of 14 lanthanoids obtained by HSCCC. The experimental conditions were as follows: apparatus consists of a HSCCC centrifuge with 7.6-cm revolution radius; column consists of three multilayer coils connected in series, 300 m × 1.07 mm i.d., 270-mL capacity; stationary phase is 0.003 M DEHPA in n-heptane; mobile phase is the exponential gradient of hydrochloric acid concentration from 0 to 0.3 M as indicated in the chromatogram; sample is 14 lanthanoid chlorides each 0.001 M in 100 μL of water; revolution = 900 rpm; flow-rate = 5 mL/min; pressure = 300 psi.

as CDCCC, mixing of the two solvent phases relies entirely on the flow of the mobile phase (20). Due to the high interfacial tension between the two solvent phases used in this system, lack of a mixing force tends to form large droplets of the mobile phase, limiting the partition efficiency. On the other hand, HSCCC is a typical hydrodynamic equilibrium system where two solvent phases are vigorously mixed by the effect of planetary motion of the coil (1). Small droplets of the mobile phase produced by the vigorous mixing reduce the mass-transfer resistance between the two phases resulting in a high partition efficiency. Compared with HPLC and ion chromatography, HSCCC will provide a higher capacity for preparative-scale separations.

There have been no reports about theoretical or chemical kinetic investigation of HSCCC for the separation of inorganic elements. However, more recently, the influence of the kinetic properties on the separation of several elements including rare earth element was investigated by Fedotov et al. (41, 42). On the other hand, a mathematical model for countercurrent extraction of rare earth elements by a hydrochloric acid–2-ethylhexylphosphonate–kerosine system was derived by Huang (21). An increasing theoretical study of HSCCC might be expected in the future.

14.3. ANALYTICAL CAPABILITY OF HIGH-SPEED COUNTERCURRENT CHROMATOGRAPHY FOR RARE EARTH ELEMENTS

Regarding the analysis of rare earth elements, gravimetric analysis and volumetric analysis cannot be applied unless each element is separated completely (18). However, as described in Section 14.2, lanthanoid elements are quite difficult to separate from each other. Although spectrophotometry has high sensitivity (molar absorptivity is sometimes more than 10^4), there is less selectivity. Thus, their absorption spectra are apt to overlap each other.

Numerous works have been reported regarding the determination of rare earth elements with spectroscopy, such as X-ray fluorescence (XRF) spectroscopy (22–24), atomic absorption spectroscopy (AAS) (25, 26), and atomic emission spectroscopy (AES) (27, 28). Of these analytical methods, inductively coupled plasma atomic emission spectroscopy (ICP-AES) is one of the most popular methods for the determination of rare earth elements. However, there may be problems with spectral interference because of the many spectral lines. To ensure sufficient precision and accuracy of analytical results, especially for the determination of trace elements, separation of matrix elements that interfere with the determination must be undertaken to minimize the analytical error. High-speed CCC has great possibility as a separation technique that preconcentrates trace elements before determination using instrumental multielement analysis, such as ICP-AES (determination of rare earth elements in some macrocomponents by ICP-AES after preconcentration is described in Section 14.4.8).

However, HSCCC has a great possibilities, not only as a separation method, but also as an analytical method, especially when combined with a continuous detection method like spectrophotometry or spectroscopy. If a sampling loop is introduced between the pump for the mobile phase and the HSCCC column, continuous and reproducible measurement is possible, as in HPLC. It was from this point of view that the analytical capability of HSCCC was studied (18). This section describes the capabilities of this method and some analytical results based on our experience using an improved HSCCC centrifuge for the analysis of rare earth elements.

14.3.1.   Reproducibility of Chromatogram

The arrangement of the apparatus for the experiment is similar to that described in Section 14.2 (Fig. 14.1). For precise and accurate analysis, a stable

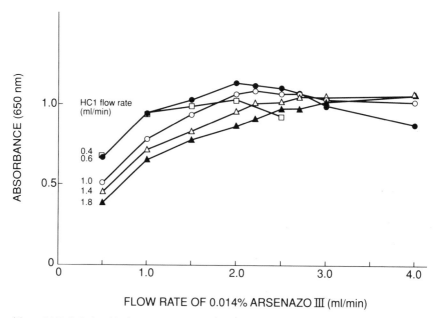

Figure 14.7. Relationship between pump speed and absorbance. Sample is 0.001 M LaCl$_3$ in 100 μL of 0.02 M HCl.

Table 14.3. Reproducibility of Peak Height of Lanthanum, Cerium, and Neodimium

Element	Absorbance[a]	Mean	RSD[b] (%)
La	0.503, 0.491, 0.506, 0.496, 0.514	0.502	1.6
Pr	0.177, 0.180, 0.178, 0.164, 0.182	0.176	3.6
Nd	0.118, 0.122, 0.118, 0.112, 0.119	0.118	2.8

[a]Measurement was performed by successive sample charge without changing column content.
[b]Relative standard deviation.

detection system is important. Figure 14.7 shows the relationship between the flow-rate of 0.014% Arsenazo III (Pump III in Fig. 14.1), and the absorbance at 650 nm for various flow-rates of hydrochloric acid eluent after the splitter (Pump II in Fig. 14.1). It was found that the values of absorbance were nearly constant beyond 2.5 mL/min of Arsenazo III, for an effluent flow of over

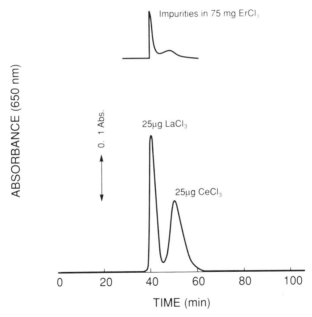

Figure 14.8. Chromatograms of impurities in a large quantity of $ErCl_3$ (top) and lanthanum chloride, cerium chloride standard mixture (bottom). The experimental conditions were as follows: apparatus consists of a HSCCC centrifuge with 7.6-cm revolution radius; column consists of three multilayer coils connected in series, 300 m × 1.07 mm i.d., 270-mL capacity; stationary phase is 0.02 M DEHPA in n-heptane; mobile phase is 0.02 M hydrochloric acid; sample is 75 mg of $ErCl_3$ in 100 μL of 0.02 M HCl and a standard mixture of $LaCl_3$ and $CeCl_3$ (each 25 μg in 100 μL of 0.02 M HCl); revolution = 900 rpm; flow-rate = 5 mL/min; pressure = 300 psi.

Table 14.4. Quantitative Determination of Cerium in Erbium Chloride

Found[a]	Mean	RSD (%)	ICP–AES
34.9, 35.6, 33.3	34.6	2.8	30.0

[a]Results are given as ppm; $ErCl_3 \cdot 6H_2O$ was dissolved in 0.2 M HCl. A 100-μL aliquot containing 75 mg of $ErCl_3 \cdot 6H_2O$ was injected into the HSCCC column.

1.0 mL/min. Therefore in this work measurements were made at a flow-rate of 2.7 mL/min for Pump III and 1.4 mL/min for Pump II.

The reproducibility of this method was studied using standard solutions of lanthanum, praseodymium, and neodymium. After equilibration between the mobile and stationary phases was established, a 100 μL sample was successively injected five times and the peak height (absorbance) for each injection was measured. As shown in Table 14.3, the absorbance values are quite reproducible, with small relative standard deviations for each element, indicating that HSCCC may be used for the quantitative analysis of microgram-to-milligram amounts of rare earth elements.

14.3.2.   Quantitative Determination of Cerium Impurities in Erbium Chloride

Use of HSCCC in trace analysis of impurities present in rare earth elements was also studied using erbium chloride as an example. Figure 14.8 shows a chromatogram of the standard mixture of lanthanum chloride and cerium chloride (bottom), and that of impurities in a large quantity of the erbium chloride sample (top). The first peak in the sample chromatogram could not be used for the determination of lanthanum because the peak height varies with the concentration of DEHPA. However, as the second peak was stable, cerium could be determined with good reproducibility. As shown in Table 14.4, the results are in good agreement with those obtained by inductively coupled plasma–atomic emission spectrometry (ICP–AES) (29).

14.4.   SEPARATION OF OTHER INORGANIC ELEMENTS BY HIGH-SPEED COUNTERCURRENT CHROMATOGRAPHY

High-speed CCC has only recently been applied to the separation of inorganic elements, including rare earth elements. Separation of zirconium and hafnium was achieved by Pavlenko et al. (30). Separation of ortho- and pyrophosphate ions (31, 32), cesium and strontium (31), and zinc and cadmium (33) were studied by Maryutina et al. (32). However, there have been no complete isocratic separations except that of cesium and strontium. On the other hand, we achieved the mutual separations of nickel, cobalt, magnesium and copper; copper, cadmium and manganese; and iron(II) and iron(III); using isocratic elution (34, 35).

Centrifugal droplet CCC, which has a separation mechanism based on liquid–liquid partition, the same as for HSCCC, was applied for separation of rare-earth elements, as described in Section 14.2 (13–16). However, its partition efficiency is inferior to that of HSCCC. Regarding other elements, palladium(II) has been separated from platinum(II), rhodium(III), and iridium(III) by centrifugal droplet CCC (36). The chromatograms of palladium(II)–trioctylphosphine oxide and 3-picoline were compared to study the influence of chemical kinetic factors on separation efficiency (37).

Recently there have also been good results using HSCCC in the separation of several inorganic elements other than rare earth elements.

14.4.1. Separation of Ortho- and Pyrophosphate Ions

This investigation was made on a system consisting of a planetary centrifuge with a vertical column drum developed in the Institute of Analytical Instrumentation of the USSR Academy of Science (Leningrad), a peristaltic pump PPI-05 (Poland), and a fraction collector FG-60 (Czechoslovakia). This HSCCC instrument had the following design parameters: revolution radius $R = 140$ mm, rotation radius $r = 50$ mm. The column was made of a Teflon F4-MB tubing with an i.d. of 1.5 mm and a wall thickness of 0.75 mm. The total inner capacity of the column was 20 mL. The rotation and revolution speed of the column was equal to 350 rpm. (The same system was used in Sections 14.4.2–14.4.4 and 14.4.8.)

Dinonyltin dichloride in methylisobutylketone permitted the concentration of ortho- and pyrophosphate and their separation by elution with hydrochloric acid solution (Fig. 14.9) (31, 32). Phosphate was determined in eluates using a flow-injection analyzer (FIA Star 5020-003, Sweden) combined with spectrophotometry based on the formation of molybdophosphoric heteropoly acid in the presence of a reducing agent. This method for concentrating phosphate ions also made it possible to separate them from large amounts of other metal ions. It was possible to separate phosphate ions from in excess of 10000-fold Cu(II), 5000-fold Fe(III), Ni(II), Cr(III), 500-fold Si(IV), V(V), and 200-fold Cr(VI). Only As(V) was extracted together with P(V).

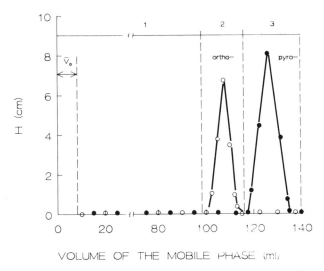

Figure 14.9. Concentration and separation of ortho- and pyrophosphate ions. The experimental conditions were as follows: apparatus consists of a HSCCC centrifuge with 14.0-cm revolution radius; column consists of one monolayer coils 11.3 m × 1.5 mm i.d., 20-mL capacity; stationary phase is 5% dinonyltin dichloride in methylisobutylketone; Kr is 50% (retention factor equal to the ratio of the stationary-phase volume to the total capacity of the column); mobile phase is (1) 1 M HNO$_3$, (2) 0.5 M NaCl in 0.5 M HCl, (3) 3 M HCl; sample is 0.5 μg of phosphate; revolution = 350 rpm; flow-rate = 1.5 mL/min.

14.4.2. Separation of Cesium and Strontium

High-speed CCC was used for the separation of cesium and strontium radionuclides (31, 33). Cobalt dicarbolide solution in nitrobenzene was used as the stationary phase. Separation of cesium and strontium radionuclides was carried out in weakly acidic solutions. Their elution into the mobile phase, containing barium nitrate, polyethylene glycol, and nitric acid was studied (Fig. 14.10). Introduction of polyethylene glycol (PEG-300) into the mobile phase considerably increased the partition coefficient for strontium (38). Figure 14.11 shows a chromatogram for isocratic separation of cesium and strontium.

14.4.3. Separation of Zirconium and Hafnium

Separation of zirconium and hafnium (30, 33) was achieved using 1-phenyl-3-methyl-4-benzoylpyrazolone-5 (PMBP) in methylisobutylketone as the stationary phase (Figure 14.12).

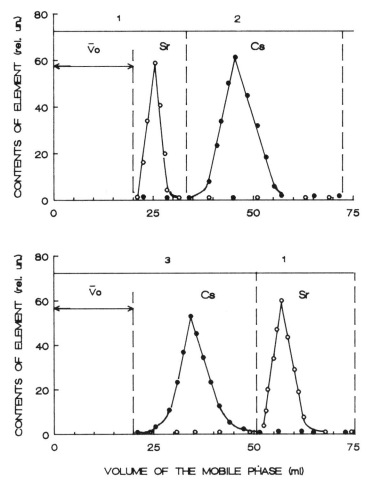

Figure 14.10. Separation of cesium and strontium. The experimental conditions were as follows: apparatus consists of a HSCCC centrifuge with 14.0-cm revolution radius; column consists of one monolayer coils 11.3 m × 1.5 mm i.d., 20-mL capacity; stationary phase is 0.01 M cobalt dicarbolide in nitrobenzene; Kr is 12%; mobile phase is (1) 1×10^{-3} M $Ba(NO_3)_2$ in 0.1 M HNO_3, (2) 5×10^{-4} M $Ba(NO_3)_2 + 8 \times 10^{-3}$ M polyethylene glycol-300 in 0.5 M HNO_3, (3) 8×10^{-3} M polyethylene glycol-300 in 0.1 M HNO_3; sample is 100 μg of Sr, 200 μg of Cs; revolution = 350 rpm; flow-rate = 2.0 mL/min.

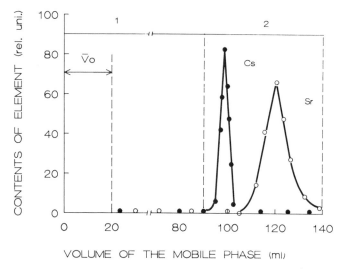

Figure 14.11. Concentration and separation of cesium and strontium. The experimental conditions were as follows: apparatus consists of a HSCCC centrifuge with 14.0-cm revolution radius; column consists of one monolayer coils 11.3 m × 1.5 mm i.d., 20-mL capacity; stationary phase is 0.01 M cobalt dicarbolide in nitrobenzene; Kr is 12%; mobile phases is (1) 5×10^{-4} M HNO_3, (2) 1×10^{-3} M $Ba(NO_3)_2 + 8 \times 10^{-3}$ M polyethylene glycol-300 in 0.1 M HNO_3; sample: each 100 μg; revolution = 350 rpm; flow-rate = 2.0 mL/min.

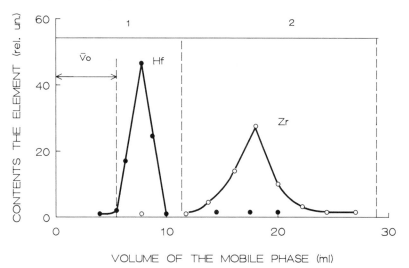

Figure 14.12. Separation of zirconium and hafnium. The experimental conditions were as follows: apparatus consists of a HSCCC centrifuge with 14.0-cm revolution radius; column consists of one monolayer coils 11.3 m × 1.5 mm i.d., 20-mL capacity; stationary phase is 1×10^{-3} M PMBP in methylisobutylketone; Kr is 75%; mobile phase is (1) 3 M HNO_3, (2) 0.1 M Na_2SO_4 in 3 M HNO_3; sample: each 50 μg; revolution = 350 rpm; flow- rate = 2.0 L/min.

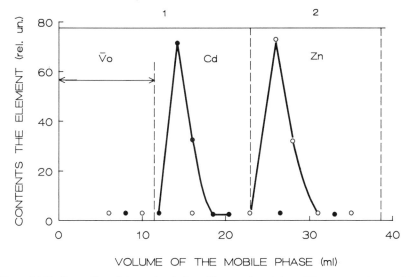

Figure 14.13. Separation of zinc and cadmium. The experimental conditions: apparatus consists of a HSCCC centrifuge with 14.0-cm revolution radius; column consists of one monolayer coil, 11.3 m × 1.5 mm i.d., 20-mL capacity; stationary phase is 30% TBP in heptane; Kr is 55%; mobile phase is (1) 3 M KBr in 0.25 M H_2SO_4, (2) 0.25 M H_2SO_4; sample is 15 μg of Cd and 7.5 μg of Zn; revolution = 350 rpm; flow-rate = 2.0 mL/min.

14.4.4.   Separation of Cadmium and Zinc

Trace amounts of cadmium and zinc were separated (33) using tributyl phosphate (TBP) in heptane as the stationary phase (Fig. 14.13).

14.4.5.   Separation of Nickel, Cobalt, Magnesium, and Copper

Investigation was conducted using nearly the same instrumentation as in Section 14.2, except for the HSCCC instrument and the detection system for each element (34, 35). A prototype of Shimadzu CCC (HSCCC-1A) holding a column on the rotary frame at a distance of 10.0 cm from the central axis of the centrifuge was used. The multilayer coil was prepared from a single piece of approximately 150 m long, 1.60-mm i.d. PTFE tubing by winding it directly onto the holder hub (10 cm diameter). The β values ranged from 0.5 at the internal terminal to 0.60 at the external terminal. The coil capacity was approximately 300 mL. The total system was shown in Fig. 14.14.

A direct current plasma atomic emission spectrometer (DCP, Spectra-Metrics Model SpectraSpan IIIB system with fixed-wavelength channels) was used for observation of the elution profile. For profile measurement of a single

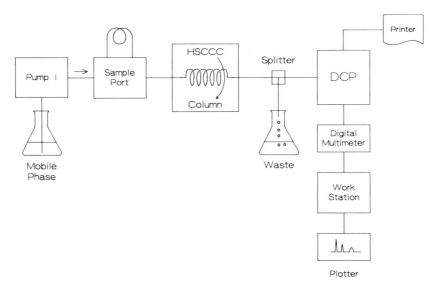

Figure 14.14. Flow-diagram of instrumentation assembly for separation of inorganic elements by HSCCC.

element, an analog recorder signal from the DCP was converted into a digital signal using a Hewlett Packard 3478A digital multimeter. The digital data was stored in a Hewlett Packard Vectra-D Work Station through an HP-IB Bus, and the elution profile was plotted using a Hewlett Packard 7440A graphic plotter and Advanced Graphics Software SlideWrite Plus. For simultaneous multielement measurement, the emission signal for each channel was integrated for 10 s at intervals of 20 s, and the integrated data were printed out. The data were entered manually to the work station and the elution profile was

Table 14.5. Observed Wave-lengths Used for Elemental Emission

Element	Line (nm)
Cd	226.5
Co	345.3
Cu	324.7
Fe	259.9
Mg	280.2
Mn	257.6
Ni	341.4

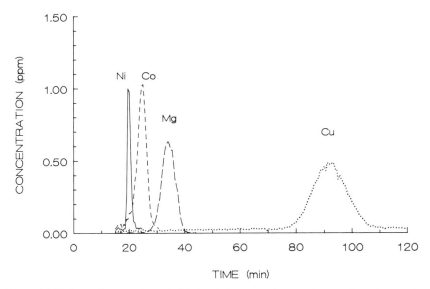

Figure 14.15. Isocratic separation of nickel, cobalt, magnesium, and copper by HSCCC. The experimental conditions were as follows: apparatus consists of a HSCCC centrifuge with 10.0-cm revolution radius; column consists of one multilayer coil, 150 m × 1.6 mm i.d., 300-mL capacity; stationary phase is 0.2 M DEHPA in heptane; mobile phase is 7 mM citric acid; sample is 10 μg of Ni, each 20 μg of Co and Mg, 40 μg of Cu; revolution is 800 rpm; flow-rate = 5.0 mL/min; pressure = 140 psi; observed wavelength = 341.4 nm (Ni), 345.3 nm (Co), 280.2 nm (Mg), 324.7 nm (Cu).

plotted. (The same installation was used in Sections 14.4.6 and 14.4.7.) Warelengths measured for this work are presented in Table 14.5.

An n-heptane solution of DEHPA was used as the stationary phase. Metal ions eluted by the aqueous mobile phase (diluted citric acid) were detected by their emission intensity with DCP. Figure 14.15 shows a typical chromatogram of $Ni(NO_3)_2$, $Co(NO_3)_2$, $Mg(NO_3)_2$, and $CuSO_4$ obtained by eluting with 7 mM citric acid at a flow-rate of 5 mL/min. The partition efficiencies ranged from 1600 (Ni) to 200 (Cu) in theoretical plates.

14.4.6. Separation of Copper, Cadmium, and Manganese

Cadmium and manganese were separated from copper using 50 mM citric acid as an eluent (34, 35). Figure 14.16 shows a chromatogram of $CuSO_4$, $Cd(NO_3)_2$, and $M_n(NO_3)_2$. The partition efficiencies ranged from 630 (Cu) to 270 (Mn) in theoretical plates.

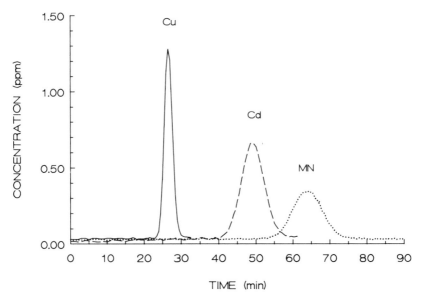

Figure 14.16. Isocratic separation of copper, cadmium, and manganese by HSCCC. The experimental conditions were as follows: apparatus consists of a HSCCC centrifuge with 10.0-cm revolution radius; column consists of one multilayer coil, 150 m × 1.6 mm i.d. 300-mL capacity; stationary phase is 0.2 M DEHPA in heptane; mobile phase is 50 mM citric acid; sample: each 14.1 μg; revolution is 800 rpm; flow-rate = 5.0 mL/min; pressure = 140 psi; observed wavelength is 324.7 nm (Cu), 226.5 nm (Cd), 257.6 nm (Mn).

14.4.7. Separation of Iron(II) and Iron(III)

Separation of iron(II) and iron(III) (34, 35) was achieved at lower DEHPA (stationary phase) concentration and higher acid (mobile phase) concentration because iron(III) was strongly retained in the stationary phase (Fig. 14.17). The partition efficiency of the separation was 4200 for the first peak ($FeCl_2$) and 410 for the second peak ($FeCl_3$) in theoretical plates, while the resolution between the two peaks was 2.06. The greater difference in the plate count for iron(II) and iron(III) probably indicates slow kinetics involved in the distribution of iron(III).

14.4.8. Preconcentration of Trace Elements from Certain Rock Macrocomponents

Direct application of simultaneous multielement determination techniques for trace elements is frequently difficult when substances that interfere with the

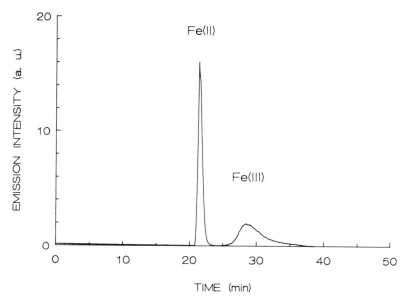

Figure 14.17. Isocratic separation of iron(II) and iron(III) by HSCCC. The experimental conditions were as follows: apparatus consists of a HSCCC centrifuge with 10.0-cm revolution radius; column consists of one multilayer coil, 150 m × 1.6 mm i.d., 300-mL capacity; stationary phase is 0.002 M DEHPA in heptane; mobile phase is 140 mM citric acid; sample: each 0.5 mg; revolution = 800 rpm; flow-rate = 4.5 mL/min; pressure = 130 psi; observed wavelength = 259.9 nm.

determination exist in the sample. Quantitative group separation of the rare earth elements from the constituents of rocks, which interfere with the determination of trace elements by ICP-AES, was achieved (Fig. 14.18) (40). A 0.5 M solution of DEHPA in decane was used as the stationary phase. Alkaline, alkaline earth elements, and iron(II) were separated from the total amount of rare earth elements at the first stage of the eluent concentration with 0.1 M hydrochloric acid. Then the total amount of rare earth elements, including yttrium, was eluted from the stationary phase with 3 M hydrochloric acid. To elute other elements, including iron(III), from the stationary phase, 5 M hydrochloric acid was introduced into the column. Separation of the mixture was performed within 40 min at a pumping rate of the mobile phase equal to 2 mL/min.

Table 14.6 shows the results of the determination of rare earth elements in an SY-3 reference rock sample by ICP-AES after selective separation of the elements from all other constituents of the sample in its solution, obtained by rock fusion and dissolution. Direct analysis for the sample solution was impossible.

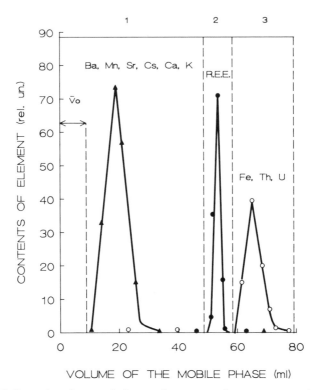

Figure 14.18. Separation of rare earth elements from some rock macrocomponents. The experimental conditions were as follows: apparatus consists of a HSCCC centrifuge with 14.0-cm revolution radius; column consists of one monolayer coil, 11.3 m × 1.5 mm i.d., 20-mL capacity; stationary phase is 0.5 M DEHPA in decane; Kr is 30%; mobile phase is (1) 0.1 M HCl, (2), 3 M HCl, (3) 5 M HCl; sample is Mn-0.3, Fe-6.4; Ca-8.2, Na-4.2, K-4.1 (%) and Ba-440, La-1350, Ce-2290, Nd-760, Pr-239, Sm-97, Eu-18.3, Gd-120, Tb-15, Dy-139, Ho-29, Er-88, Tm-12.5, Yb-71, Lu-8.6, Y-716, Cs-2.6, Cs-307, U-682, Th-940 (ppm) (39); revolution = 350 rpm; flow-rate = 2.0 mL/min.

As mentioned above, HSCCC has a great potential as a separation technique that preconcentrates trace elements before determination using instrumental multielement analysis, such as ICP-AES and inductively coupled plasma-mass spectrometry (ICP-MS) (34, 40). Enrichment of the desired trace elements prior to determination can not only overcome such problems as interferences, toxic, or radioactive samples, but also provide highly sensitive determination of trace elements.

Table 14.6. Determination of Rare Earth Elements in International SY-3 Standard Reference Sample by ICP-AES after Group Separation of the Elements by HSCCC ($n = 6^a$, $P = 0.95^b$)

Element	Rare Earth Contents in		
	SY-3 Sample, Mass % $\times 10^{-4}$	Found, Mass % $\times 10^{-4}$	RSD^c
La	1350	1337 ± 14	0.01
Ce	2290	2285 ± 42	0.02
Pr	232	229 ± 4	0.02
Nd	760	854 ± 37	0.04
Sm	134	138 ± 8	0.06
Eu	18	18 ± 1	0.04
Gd	120	118 ± 4	0.03
Tb	15	15 ± 1	0.04
Ho	29	28 ± 1	0.03
Er	88	144 ± 3	0.03

^{a}Times of replicates.
^{b}95% confidence level.
^{c}Relative standard deviation.

REFERENCES

1. Y. Ito, *CRC Crit. Rev. Anal. Chem.*, **17**, 65 (1986).

2. Y. Ito, in N. B. Mandava and Y. Ito, Ed., *Countercurrent Chromatography, Theory and Practice*, Marcel-Dekker, New York, 1988, Chapter 3, pp. 79–442.

3. Y. Ito, J. L. Sandlin, and W. G. Bowers, *J. Chromatogr.*, **244**, 247 (1982).

4. Y. Ito, H. Oka, and J. L. Slemp, *J. Chromatogr.*, **475**, 219 (1989).

5. Y. Ito, H. Oka, and Y.-W. Lee, *J. Chromatogr.*, **498**, 169 (1990).

6. Y. Ito, H. Oka, E. Kitazume, M. Bhatnagar, and Y.-W. Lee, *J. Liq. Chromatogr.*, **13**, 2329 (1990).

7. J. J. Sedmak and S. E. Grossberg, *J. Gen. Virol.*, **52**, 195 (1981).

8. D. F. Peppard, G. W. Maeson, J. L. Maier, and W. J. Driscoll, *J. Inorg. Nucl. Chem.*, **4**, 334 (1957).

9. G. V. Korpusov, *Proc. Int. Solvent Extr. Conf. ISEC 88*, Moscow, USSR, Vol. III, 1988, p. 120.

10. P. Duker, M. Vobecky, J. Holik, and J. Valasek, *J. Chromatogr.*, **435**, 259 (1988).

11. D. J. Barkley, M. Blanchette, R. M. Cassidy, and W. Elchuk, *Anal. Chem.*, **58**, 2222 (1986).

12. S. S. Heberling, J. M. Riviello, M. Shifen, and A. W. Ip, *Res. Dev.*, **74** (1987).

13. T. Araki, T. Okazawa, Y. Kubo, H. Ando, and H. Asai, *J. Liq. Chromatogr.*, **11**, 267 (1988).

14. T. Araki, H. Asai, H. Ando, N. Tanaka, K. Kimata, K. Hosoya, and H. Narita, *J. Liq. Chromatogr.*, **13**, 3673 (1990).

15. K. Akiba, S. Sawai, S. Nakamura, and W. Murayama, *J. Liq. Chromatogr.*, **11**, 2517 (1988).

16. S. Muralidharan, R. Cai, and H. Freiser, *J. Liq. Chromatogr.*, **13**, 3651 (1990).

17. E. Kitazume, M. Bhatnagar, and Y. Ito, Separation of rare earth elements by high-speed countercurrent chromatography, *the 1990 Pittsburgh Conference on Analytical Chemistry and Applied Spectroscopy* in New York, Abstract No. 298.

18. E. Kitazume, M. Bhatnagar, and Y. Ito, Mutual separation of rare earth elements by high-speed countercurrent chromatography, Presented at *Proc. Internat. Trace Anal. Symposium*, 103, 1990 (in Sendai, Japan).

19. E. Kitazume, M. Bhatnagar, and Y. Ito, *J. Chromatogr.*, **538**, 133 (1991).

20. A. Berthod and D. W. Armstrong, *J. Liq. Chromatogr.*, **11**, 547 (1988).

21. W. Huang, *Zhongguo Xitu Xuebao*, **8**, 213 (1990).

22. I. Roelandts, *Anal. Chem.*, **53**, 676 (1981).

23. A. T. Kosuba and C. R. Hines, *Anal. Chem.*, **43**, 1758 (1971).

24. G. N. Edy, *Anal. Chem.*, **44**, 2137 (1972).

25. K. Dittrich, E. John, and I. Rohde, *Anal. Chim. Acta*, **94**, 75 (1977).

26. B. V. L'vov and L. A. Pelieva, *Zh. Anal. Khim.*, **34**, 1744 (1979).

27. A. G. I. Dalvi, C. S. Deodhar, and B. D. Joshi, *Talanta*, **24**, 143 (1977).

28. J. A. C. Broekaent, F. Leis, and K. Laqua, *Spectrochim. Acta*, **34B**, 167 (1979).

29. Ruth Wantz, "Product Information," 1987, Aldrich Chemical Co.

30. I. P. Pavlenko, V. L. Bashlov, B. Ya, Spivakov, and Yu. A. Zolotov, *Zh. Anal. Khim.*, **44**, 827 (1989).

31. Yu. A. Zolotov, B. Ya. Spivakov, T. A. Maryutina, V. L. Bashlov, and I. V. Pavlenko, *Z. Anal. Chem.*, **335**, 938 (1989).

32. T. A. Maryutina, B. Ya. Spivakov, L. K. Shpigun, I. V. Pavlenko, and Yu. A. Zolotov, *Zh. Anal. Khim.*, **45**, 665 (1990).

33. B. Ya. Spivakov, T. A. Maryutina, V. M. Pukhovskaya, V. L. Bashlov, V. M. Pukhovskaya, and Yu. A. Zolotov, Countercurrent Chromatography: A status report and application in inorganic analysis, pressented at *the Proc. Internat. Trace Anal. Symposium*, 241 1990 (in Sendai, Japan).

34. E. Kitazume, N. Sato, Y. Saito, and Y. Ito, Separation of heavy metals by high-speed countercurrent chromatography, pressented at *the 1992 Pittsburgh Conference on Analytical Chemistry and Applied Spectroscopy* in New Orleans, Abstract No. 392.

35. E. Kitazume, N. Sato, Y. Saito, and Y. Ito, *Anal. Chem.*, **65**, 2225 (1993).

36. V. Surakitbanharn, S. Muralidharan, and H. Freiser, *Solv. Extr. Ion Exch.*, **9**, 45 (1991).

37. Y. Surakitbanharn, S. Muralidharan, and H. Friser, *Anal. Chem.*, **63**, 2642 (1991).

38. V. Koprda and V. Scasnar, *J. Radioanal. Chem.*, **77**, 71 (1983).

39. S. Abbey and F. S. Gladney, *Geostandards Newslett.*, **10**, 3 (1986).

40. V. M. Pukhovskaya, T. A. Maryutina, O. N. Grebneva, N. M. Kuz'min, and B. Ya. Spivakov, *Spectrochim. Acta*, **48B**, 1365 (1993).

41. P. S. Fedotov, T. A. Maryutina, A. A. Pichugin, and B. Ya. Spivakov, Russian Journal of Inorganic Chemistry, **38**, 1958 (1993). (Translated from Zhurnal Neoganichesko; Khimii, Vol. 38, No. 11, 1993, pp. 1878–1884.)

42. P. S. Fedotov, T. A. Maryutina, V. M. Pukhovskaya, and B. Ya. Spivakov, *J. Liq. Chromatogr.*, **17**, 3491 (1994).

INDEX

445